ADDITION, ELIMINATION AND SUBSTITUTION: MARKOVNIKOV, HOFMANN, ZAITSEV AND WALDEN

ADDITION, ELIMINATION AND SUBSTITUTION: MARKOVNIKOV, HOFMANN, ZAITSEV AND WALDEN

Discovery and Development

DAVID E. LEWIS

Professor Emeritus, Department of Chemistry and Biochemistry, University of Wisconsin-Eau Claire, Eau Claire, WI, United States

ELSEVIER

Elsevier
Radarweg 29, PO Box 211, 1000 AE Amsterdam, Netherlands
The Boulevard, Langford Lane, Kidlington, Oxford OX5 1GB, United Kingdom
50 Hampshire Street, 5th Floor, Cambridge, MA 02139, United States

Notices

Knowledge and best practice in this field are constantly changing. As new research and experience broaden our understanding, changes in research methods, professional practices, or medical treatment may become necessary.

Practitioners and researchers must always rely on their own experience and knowledge in evaluating and using any information, methods, compounds, or experiments described herein. In using such information or methods they should be mindful of their own safety and the safety of others, including parties for whom they have a professional responsibility.

To the fullest extent of the law, neither the Publisher nor the authors, contributors, or editors, assume any liability for any injury and/or damage to persons or property as a matter of products liability, negligence or otherwise, or from any use or operation of any methods, products, instructions, or ideas contained in the material herein.

ISBN: 978-0-12-821027-7

Publisher: Susan Dennis
Editorial Project Manager: Hilary Carr
Production Project Manager: Kiruthika Govindaraju
Cover Designer: Miles Hitchen

Typeset by TNQ Technologies

Contents

CHAPTER ONE

Introduction: Organic Chemistry in the Nineteenth Century

Nucleophilic substitution, base elimination, and electrophilic addition occupy a central place among the synthetic organic chemist's toolbox. They are among the first reactions taught in introductory courses, and all were first reported in the last half of the 19th century, placing them among the earliest organic reactions to be characterized in detail (Scheme 1.1). Organized chronologically, the first[1] came in 1851, when August Wilhelm (von) Hofmann (1818–92) reported the elimination of tertiary amines from quaternary ammonium salts by pyrolyzing the quaternary ammonium hydroxides. The second[2] came in 1869, when Vladimir Vasil'evich Markovnikov (1838–1904) reported his rule (Scheme 1.2) for the addition of unsymmetrical electrophilic reagents to 1-alkenes and isobutylene. The third[3] came in 1875, when Aleksandr Mikhailovich Zaitsev (1841–1910) reported his empirical rule for base elimination from alkyl halides. In 1896, Paul Walden (Pavel Ivanovich Val'den, 1863–1957) reported the last of these reactions,[4] the cycle that demonstrated inversion of configuration in substitution of α-chloro- and α-hydroxy acids that now bears his name. The chemists responsible for the discovery of these rules and reactions are shown in Fig. 1.1.

The first reports of the reactions that were the foundations for these regiochemical and stereochemical rules are shown in Scheme 1.1; where there was a graphic in the original, that graphic is given explicitly.

A 19th-century chronology of organic chemistry

The work of these four chemists encompasses a time of dramatic change that saw organic chemistry mature from its serendipitous beginnings with the synthesis of urea by Friedrich Wöhler (1800–82, Fig. 1.2)[5,6] and the much more purposeful synthesis of acetic acid from its elements[7] (Scheme 1.2) by Hermann Kolbe (1818–84, Fig. 1.2); the key reactions here were Regnault's pyrolysis[8] of carbon tetrachloride (**19**) to give

Addition, Elimination and Substitution: Markovnikov, Hofmann, Zaitsev and Walden
ISBN: 978-0-12-821027-7
https://doi.org/10.1016/B978-0-12-821027-7.00005-0

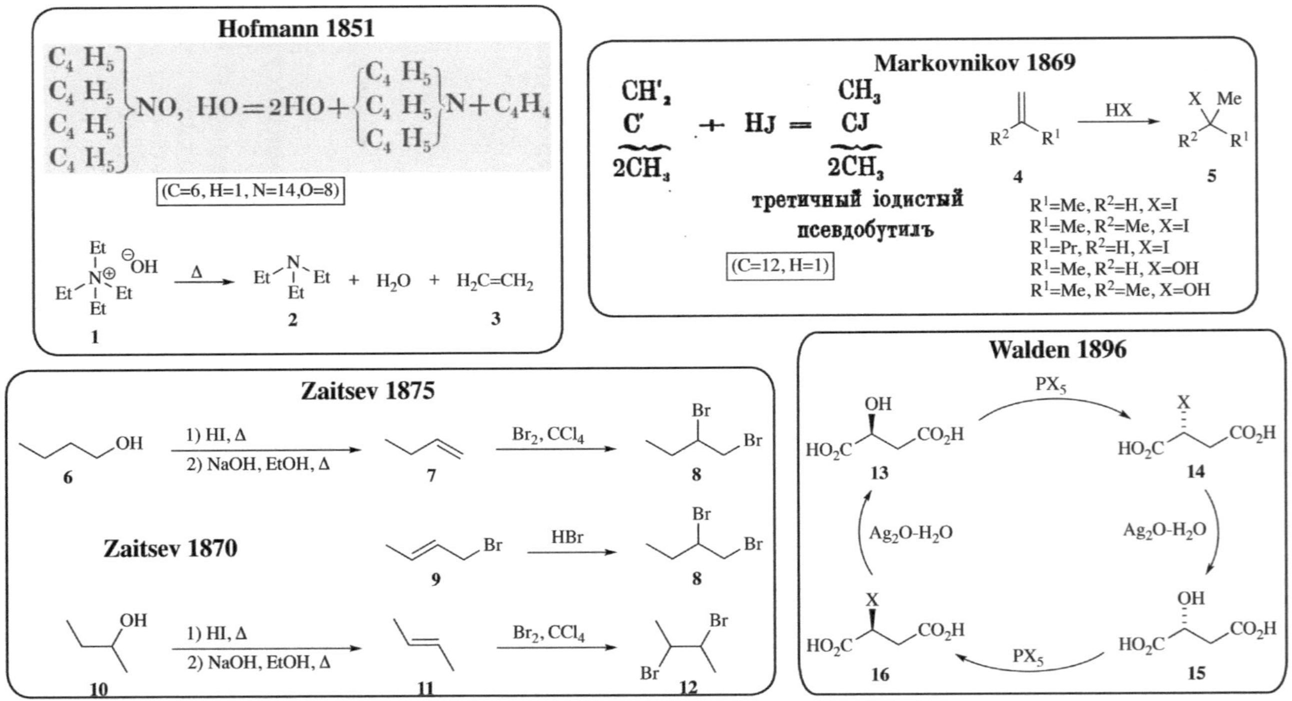

Scheme 1.1 Prototype examples of the Hofmann elimination, Markovnikov addition, Zaitsev elimination, and the Walden inversion.

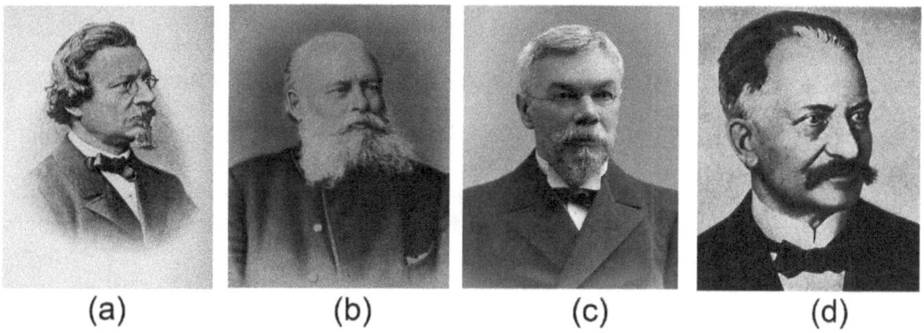

Figure 1.1 (A) A. W. (von) Hofmann (taken in 1902). (B) V. V. Markovnikov (late 19th century). (C) A. M. Zaitsev (late 19th century). (D) P. Walden (Val'den, taken in 1910). *(A) https://wellcomecollection.org/works/hy87xfwwlicensed under the Creative Commons Attribution 4.0 International license. (B) Public domain image downloaded from https://commons.wikimedia.org/wiki/File:VladimirMarkovnikov.jpg. (C) Public domain image downloaded from https://en.wikipedia.org/wiki/Alexander_Mikhaylovich_Zaytsev#/media/File:Alexander_Mikhaylovich_Zaytsev.jpg. (D) Public domain image downloaded from https://commons.wikimedia.org/wiki/File:Paul_Walden.jpg.*

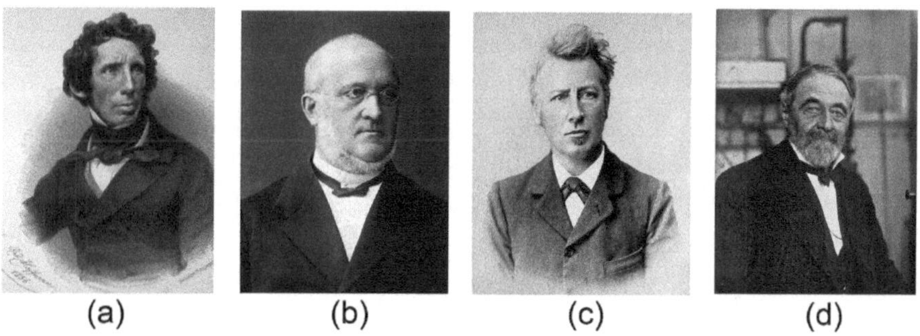

Figure 1.2 (A) Friedrich Wöhler 1856. (B) Hermann Kolbe. (C) Jacobus van't Hoff, 1904. (D) Joseph-Achille Le Bel. *(A) Wöhler Lithograph by Rudolf Hoffmann, 1856, after a photograph by Petri (Göttingen). From "Gallerie ausgezeichneter Naturforscher" published in Vienna. (B) https://commons.wikimedia.org/wiki/File:Hermann_Kolbe2.jpg. (C) Public domain image from the Library of Congress http://hdl.loc.gov/loc.pnp/cph.3b12511. (D) Joseph Achille Le Bel. Photograph. Credit: Wellcome Collection. Attribution 4.0 International (CC BY 4.0).*

tetrachloroethylene (**20**), and the chlorination of this alkene in the presence of water and sunlight to give trichloroacetic acid (**21**); through the development and popularization of the Structural Theory of Organic Chemistry;[9–15] and on to the end of the century, by which time the stereochemical theories of Jacobus Henricus van't Hoff (1852–1911, Fig. 1.2)[16,17] and Joseph-Achille Le Bel (1847–1930, Fig. 1.2)[18] had achieved universal acceptance. Fig. 1.2 shows a brief chronology of the development of organic chemistry in the 1st century of the discipline.

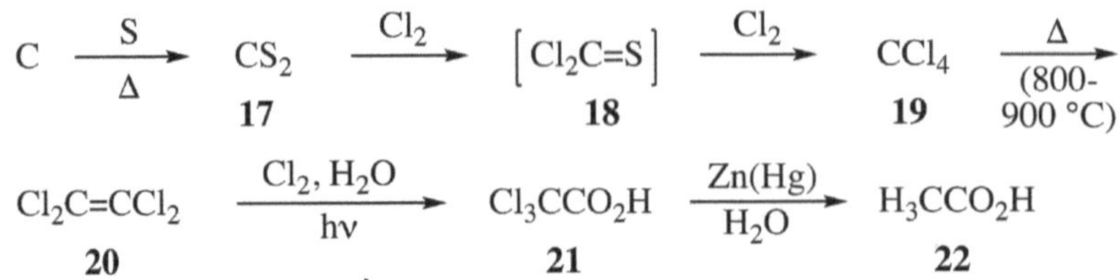

Scheme 1.2 Kolbe's synthesis of acetic acid from demonstrably inorganic precursors.

Atomic theory: a brief history

For the entire century, the atomic model of John Dalton (1766–1844, Fig. 1.3),[19–22] with its invisible, indivisible atoms, was still dominant form of atomic theory used by chemists, although as late as 1870, there was disagreement over whether atoms were physical objects—physical atoms—or just a manifestation of the chemical properties of the element—chemical atoms.[23,24] In fact, the idea that the atom could be divided was still more in the realm of science fiction. The electron, discovered by J. J. Thomson (1856–1940, Fig. 1.3)[25] in 1897, was the first hint that there might be particles smaller than atoms. The nuclear atom,[26] proposed in 1911 by Ernest Rutherford (1871–1937, Fig. 1.3), was still over a decade into the new century, and the first really scientific models of chemical bonding would not appear until its second decade.[27] The modern model of the atom and chemical bonding required another two decades for its development. The chronology of important developments in organic chemistry during the 19th century is given in Fig. 1.4.

(a) (b) (c)

Figure 1.3 (A) John Dalton. (B) J. J. Thomson. (C) Ernest Rutherford in 1908. *(A) http://hdl.loc.gov/loc.pnp/pga.12996. (B) http://hdl.loc.gov/loc.pnp/ggbain.35256. (C) http://hdl.loc.gov/loc.pnp/ggbain.03392.*

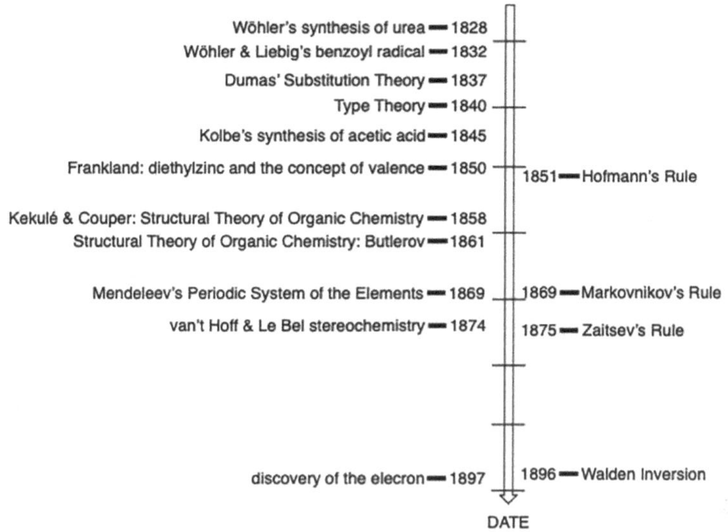

Figure 1.4 Chronology of landmark developments in organic chemistry in the 19th century.

Dalton's Law of Definite Proportions[22] allowed him to provide a set of atomic weights of the elements. Dalton's atomic weights were actually equivalent weights based on the gravimetric analysis of water (to which he gave the simplest possible formula, HO). Equivalent weights held sway until the Karlsruhe Conference,[28] where Stanislao Cannizzaro (1826–1910) publicized his paper[29,30] promoting the views of his countryman, Amadeo Avogadro (1776–1856).[31] By the end of the 1860s, the Cannizzaro atomic weights, which were used by Dmitrii Ivanovich Mendeleev (1834–1907, Fig. 1.5) in developing his Periodic System of the Elements,[32,33] had moved into the dominant position.

Theories of organic chemistry

Organic radicals: Theory of substitutions

One of the first advances in the new science was reported by Wöhler and Justus Liebig (1803–73, Fig. 1.6) in their pioneering 1832 paper[34] on the benzoyl radical where they showed that the group C_7H_5O remained unchanged through a series of transformations, thus demonstrating the existence of "complex radicals." In 1837, the same two chemists reported the hydrolysis of amygdalin to give oil of bitter almonds (benzaldehyde), prussic acid (hydrogen cyanide), and sugars.[35,36] Very shortly after this landmark

Figure 1.5 (A) Stanislao Cannizzaro. (B) Amadeo Avogadro. (C) Dmitrii Mendeleev. *(A) https://commons.wikimedia.org/wiki/File:Cannizzaro_Stanislao.jpg. (B) https://commons.wikimedia.org/wiki/File:Amadeo_Avogadro.png. (C) https://en.wikipedia.org/wiki/Dmitri_Mendeleev#/media/File:DIMendeleevCab.jpg.*

Figure 1.6 (A) Justus von Liebig. (B) Jean-Baptiste André Dumas. (C) Auguste Laurent. (D) Charles Gerhardt. (A) https://commons.wikimedia.org/wiki/File:Justus_von_Liebig_NIH.jpg. (B) https://commons.wikimedia.org/wiki/File:Jean_Baptiste_Andr%C3%A9_Dumas.jpg. (C) https://commons.wikimedia.org/wiki/File:Auguste_Laurent2.JPG. (D) https://commons.wikimedia.org/wiki/File:Gerhardt_Charles.jpg.

paper, the process of systematization of organic chemistry began; the state of organic chemistry in 1837 was described in a joint paper by French chemist, Jean Baptiste André Dumas (1800–84, Fig. 1.6) and Liebig in the *Comptes rendus*.[37] Two years later, in his report of the strong similarity between the physical and chemical properties of acetic acid and chlorinated acetic acids,[38–40] Dumas proposed his short-lived Substitution Theory. The work of two younger French chemists: Dumas' student, Auguste Laurent

(1807–53, Fig. 1.6),[41,42] and Charles Frédéric Gerhardt (1816–56, Fig. 1.6),[43–46] bolstered the standing of substitution theory. This was not without criticism. In 1838, Liebig wrote,[47] in a paper titled, "On Laurent's theory of organic compounds":

"We call cyanogen a radical (1) because it is a non-varying constituent in a series of compounds, (2) because in these latter it can be replaced by other simple substances, and (3) because in its compounds with a simple substance, the latter can be turned out and replaced by equivalents of other simple substances."

Clearly, this statement is in defense of the definitions that he and Wöhler had developed 6 years earlier.

As Substitution Theory continued to gain adherents at the expense of the radical theory, it became the target of a satirical paper[48] by Wöhler, writing under the pseudonym S. C. H. Windler (i.e., *Schwindler*, a swindler or mountebank).

For its first quarter century or so, organic chemistry was enriched by the discovery of organic reactions (what is now called *chemoselectivity*) and the elucidation of their limitations. The reactions of organic compounds with acids and bases, their oxidation with chromium and manganese compounds and nitric acid, their reduction with sodium metal or sodium amalgam in alcohol, and their reactions with halogens and their analogues to give both addition and substitution products were all the subjects of intense interest on the part of organic chemists.

Theory of types

The Theory of Substitutions was in direct conflict with the dualistic theory of Berzelius:[49,50] Berzelius' theory, based on inorganic chemistry, specified that the replacement of an electropositive element (e.g., H) with an electronegative element (e.g., Cl) should result in a profound change in physical and chemical properties (e.g., compare NaH with NaCl), whereas Substitution Theory specified that such a substitution would have minimal impact on the properties of the two compounds (e.g., compare CH_3CO_2H and CCl_3CO_2H). Within 3 years, the deficiencies in the Theory of Substitutions had become obvious, so the time was ripe for the next advance. The result was the Theory of Types, reported by Dumas in 1840.[51] This theory was augmented with the "ammonia Type," which arose from Hofmann's researches on the amines, published in English[52–58] and German,[59–66] beginning in 1849, and the "water Type," which arose from the research of Alexander William Williamson[67–71] (1824–1904, Fig. 1.7).

Valence and the structural theory of organic chemistry

Sir Edward Frankland (1825–99, Fig. 1.7) was responsible for the next major advance in chemistry. In 1850, he reported[72,73] the formation of a new class of organic compounds containing metals—the dialkylzincs—by heating a primary alkyl iodide with zinc metal in the presence of a small amount of sodium. He expanded his work to other metal alkyls (Scheme 1.3), and using his data, he developed the concept of valence.[74]

(a) (b)

Figure 1.7 (A) Alexander Williamson, ca. 1862. (B) Edward Frankland, ca. 1860. *(N.d.). Retrieved June 2020, from (A) https://commons.wikimedia.org/wiki/File:Williamson_Alexander.jpg. (B) https://commons.wikimedia.org/wiki/File:Edward_Frankland.jpg.*

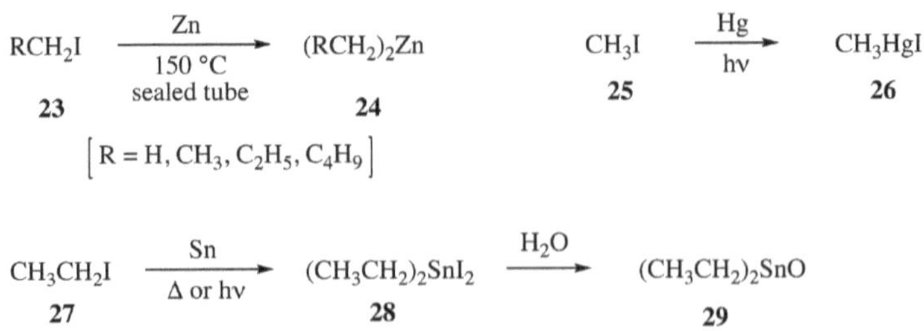

Scheme 1.3 Organometallic compounds prepared by Frankland.

(a) (b) (c) (d)

Figure 1.8 (A) August Kekulé, 1890s. (B) Archibald Scott Couper. (C) Adolphe Wurtz. (D) Aleksandr Butlerov. *(N.d.). (A) https://en.wikipedia.org/wiki/August_Kekulé#/media/File: Frkekulé.jpg. (B) https://commons.wikimedia.org/wiki/File:Couper_Archibald_Scott.jpg. (C) https://en.wikipedia.org/wiki/Charles_Adolphe_Wurtz#/media/File:Charles-Adolphe_Wurtz.jpg. (D) Frontispiece, Stat'i po mediumizmu. S vospominan'em ob A. M. Butlerove, N. R. Vagnera. Spb.: A. N. Aksakov, 1889; https://en.wikipedia.org/wiki/Alexander_Butlerov#/media/File:Butlerov,_A._M._1828-1886.jpg.*

At this point in time, all the theoretical concepts required for the development of the Structural Theory of Organic Chemistry were in place. Two versions of the same theory were published in 1858. The first was published by Friedrich August Kekulé (1829–96, Fig. 1.8) in the *Annalen*[10] on March 16, 1858. The second, published by Scotsman Archibald Scott Couper (1831–92, Fig. 1.8) in the *Comptes rendus,*[11] was read before the Académie des Sciences on June 14, 1858, 13 weeks later.

The delay in the reading of Couper's paper, which led to such rancor, may be traced to the fact that his mentor, Charles Adolphe Wurtz (1817–84, Fig. 1.8), was not at the time a member of the Académie, so he needed a sponsor to present the paper. Couper blamed Wurtz for his loss of priority, and things got so bad between the two men that the usually placid Wurtz summarily dismissed Coper from his laboratory. Shortly after his return to Britain, Couper had a nervous breakdown, effectively ending his career.

The value of Structural Theory was really revealed by the Russian, Aleksandr Mikhailovich Butlerov (1828–86, Fig. 1.8), who published his version of Structural Theory in 1861.[9] This was the theory that both Markovnikov and Zaitsev used in their work. The theory was based on the chemical atom, but within 5 years, the idea that atoms chemically close to each other might also be physically close to each other became more mainstream. It culminated in the idea that the physical location within molecules could actually be specified, leading to the stereochemical theories of van't Hoff[16,17] and Le

Bel,[18] in which a molecular-level explanation of the concept of chirality was provided. This change from chemical atoms to physical atoms provided the basis for the work of Walden[4] and the first study of the stereochemistry of organic reactions.

References

1. Hofmann, A. W. Beiträge Zur Kenntniss Der Flüchtigen Basen. Uebergang Der Flüchtigen Basen in Eine Reihe Nichtflüchtiger Alkaloïde. *Ann. Chem. Pharm* **1851,** *78,* 253–286.
2. Morkovnikov, V. K Voprosu O Vniamnom Vlyanii Atomov V Khimicheskikh Soedineniyakh [On the question of the mutual influence of atoms in chemical compounds]. *Zh. Russ. Khim. O-va.* **1869,** *1,* 242–247.
3. Saytzeff, A. Zur Kenntniss der Reihenfolge der Analgerung und Ausscheidung der Jodwasserstoffelemente in organischen Verbindungen. *Justus Liebigs Ann. Chem.* **1875,** *179* (3), 296–301. https://doi.org/10.1002/jlac.18751790304.
4. Walden, P. Ueber die gegenseitige Umwandlung optischer Antipoden. *Ber. Dtsch. Chem. Ges.* **1896,** *29* (1), 133–138. https://doi.org/10.1002/cber.18960290127.
5. Wöhler, F. Ueber Künstliche Bildung Des Harnstoffs. *Ann. Phys. Chem.* **1828,** 253–256. https://doi.org/10.1002/andp.18280870206.
6. Wöhler, F. Sur La Formation Artificielle de l'urée. *Ann. Chem. Phys.* **1828,** *37,* 330–334.
7. Kolbe, H. Beitrage Zur Kentniss Der Gepaarten Verbindungen. *Ann. Chem. Pharm.* **1845,** *54,* 145–188. https://doi.org/10.1002/jlac.18450540202.
8. Regnault, V. De l'action Du Chlore Sur Les Éthers Hydrochloriques de l'alcohol et l'esprit de Bois, et de Plusieurs Points de La Théorie Des Éthers. *Ann. Chem. Phys.* **1839,** *71,* 353–430.
9. Butlerow, A. Einiges Über Die Chemische Structur Der Körper. *Z. Chem. Pharm* **1861,** *4,* 549–560.
10. Kekulé, A. Ueber Die Constitution Und Die Metamorphosen Der Chemischen Verbindungen Und Über Die Chemische Natur Des Kohlenwasserstoffs (Vol. 1858, pp. 109-131. In *Classics of Science;* Benfey, O. T., Ed.; Dover: New York, 1963.
11. Couper, A.-S. Sur Une Nouvelle Théorie Chimique. *C. R. Hebd. Séances Acad. Sci.* **1858,** *46,* 1157–1160.
12. Kekulé, A. Ueber Die s.G. Gepaarten Verbindungen Und Die Theorie Der Mehratomigen Radicale. *Ann. Chem. Pharm* **1857,** *104,* 129–150. https://doi.org/10.1002/jlac.18571040202.
13. Couper, A. S. On a new chemical theory. *Phil. Trans. Roy. Soc.* **1858,** 104–116.
14. Couper, A. S. Sur Une Nouvelle Théorie Chimique. *Ann. Chem. Phys.* **1858,** *53* (3), 469–489.
15. Couper, A. S. Ueber eine neue chemische Theorie. *Justus Liebigs Ann. Chem.* **1859,** *110* (1), 46–51. https://doi.org/10.1002/jlac.18591100105.
16. van't Hoff, J. H. *Voorstel Tot Uitbreiding Der Tegenwoordig in Die Scheikunde Gebruikte Struktuur-Formels in Die Ruimte; Benevens Een Daarmeé Samenhangende Opmerkung Omtrent Het Verband Tusschen Optisch Actief Vermogen En Chemische Constitutie Van Organische Verbindingen [Proposal for the Extension of Current Chemical Structural Formulas into Space, Together with Related Observation on the Connection between Optically Active Power and the Chemical Constitution of Organic Compounds];* Greven: Utrecht, 1874.
17. van't Hoff, J. H. Sur Les Formules de Structure Dans l'espace. *Arch. Néerland. Sci. Exactes Natur.* **1874,** *9,* 445–454.

18. Le Bel, J. A. Sur Les Relations Qui Existent Entre Las Formules Atomiques Des Corps Organiques et Le Pouvoir Rotaire de Leurs Dissolutions. *Bull. Soc. Chim. Paris* **1874,** *22,* 337–347.
19. Dalton, J. On the tendency of elastic fluids to diffusion through each other. *Mem. Manch. Lit. Philos. Soc.* **1805,** *1,* 259–270.
20. Dalton, J. On the absorption of gases by water and other liquids. *Mem. Manch. Lit. Philos. Soc.* **1805,** *1,* 271–287.
21. Dalton, J. A. *New System of Chemical Philosophy;* Bickerstaff: London, 1805.
22. Dalton, J. Experimental enquiry into the proportion of the several gases or elastic fluids, constituting the atmosphere. *Mem. Manch. Lit. Philos. Soc.* **1805,** *1,* 244–258.
23. Rocke, A. J. *Atomism in the Nineteenth Century: From Dalton to Cannizzaro;* Ohio State University Press: Columbus, OH, 1984.
24. Schütt, H.-W. Chemical atomism and chemical classifications. In *Cambridge History of Science, Vol. 5. The Modern Physical and Mathematical Science;* Nye, M. J., Ed.; Cambridge University Press: Cambridge, 2002; pp 237–254.
25. Thomson, J. J. Cathode rays. *Philos. Mag.* **1897,** *44,* 293–316.
26. Rutherford, E. The scattering of α and β particles by matter and the structure of the atom. *Philos. Mag. A* **1911,** *21,* 669–688.
27. Lewis, G. N. The atom and the molecule. *J. Am. Chem. Soc.* **1916,** *38* (4), 762–785. https://doi.org/10.1021/ja02261a002.
28. Ihde, A. J. The Karlsruhe congress: a centennial retrospect. *J. Chem. Educ.* **1961,** *38* (2), 83–86. https://doi.org/10.1021/ed038p83.
29. Cannizzarro, S.; ; Sunto Di Un Corso Di Filosofia Chimica.; Fatto Nella, R. Universita' Di Genova. *Il Nuovo Cimento* **1858,** *7,* 321–368.
30. Cannizzarro, S. Sketch of a course of chemical philosophy. In *Alembic Club Reprints, No. 18;* E. & S. Livingstone: Edinburgh, 1910.
31. Avogadro, A. Essai d'une Manière de Déterminer Les Masses Des Molécules Élémentaires Des Corps et Les Proportions Selon Lesquelles Elles Entrent Dans Ces Combinaisons. *J. Phys. Chim. Hist. Nat.* **1811,** *73,* 58–76.
32. Mendeleev, D. Sootnoshenie Svoistv s Atomnym Vesom Elementov [The Correlation of the properties of the elements with their atomic weights]. *Zh. Russ. Khim. O-va.* **1869,** *1,* 60–77.
33. Mendelejeff, D. Ueber Die Beziehungen Der Eigenshaften Zu Den Atomgewichten Der Elemente. *Z. Chem.* **1869,** *5,* 405–406.
34. Wöhler, F.; Liebig, J. Untersuchungen Über Das Radikal Der Benzoesäure. *Ann. Chem. Pharm* **1832,** *3,* 249–287. https://doi.org/10.1002/jlac.18320030302.
35. Woehler, F.; Liebig, J. Sur La Formation de l'huile d'amandes Amères. *Ann. Chem. Phys.* **1837,** 185–209.
36. Liebig, J.; Wöhler, F. Ueber Die Bildung Des Bittermandelöls. *Ann. Chem. Pharm* **1837,** 1–24.
37. Dumas, (J.; Liebig, (J. Note Sur l'état Actuel de La Chimie Organique. *C. R. Hebd. Séances Acad. Sci.* **1837,** *5,* 567–572.
38. Dumas, J. Ueber die Chloressigsäure, die Constitution einiger organischer Körper und über die Substitutionstheorie. *Ann. Pharm.* **1839,** *32* (1), 101–119. https://doi.org/10.1002/jlac.18390320109.
39. Dumas, J. Mémoire Sur La Constitution de Quelques Corps Organiques et Sur La Théorie Des Substitutions. *C. R. Hebd. Séances Acad. Sci.* **1839,** *8,* 609–622.
40. Dumas, J. Recherches Sur Les Combinaisons de l'hydrogène et Du Carbone. *Ann. Chem. Phys.* **1832,** 182–197.

41. Gerhardt, C. Sur La Constitution Des Sels Organiques à Acides Complexes, et Leurs Rapports Avec Les Sels Ammoniacaux. *Ann. Chem. Phys.* **1839,** *72,* 184–214.
42. Gerhardt, C. Sur Une Nouvelle Classe Des Composés Organiques. *C. R. Hebd. Séances Acad. Sci* **1839,** 1031–1038.
43. Laurent, A. Théorie Des Combinaisons Organiques. *Ann. Chem. Phys.* **1836,** 125–146.
44. Laurent, A. Essai Sur l'action Du Chlore Sur La Liqueur Des Hollandais et Sur Quelques Éthers. *Ann. Chem. Phys.* **1836,** 377–389.
45. Laurent, A. Suite Des Recherches Diverses de Chimie Organique. *Ann. Chem. Phys.* **1837,** 314–335.
46. Laurent, A. *Thèse de Docteur; Paris* **1837,** *Vol. 88; p 102; p. 11.*
47. Liebig, J. Ueber Laurent's Theorie der organischen Verbindungen. *Ann. Pharm.* **1838,** *25* (1), 1–31. https://doi.org/10.1002/jlac.18380250102.
48. Windler, S. C. H. Ueber Das Substitutionsgesetz Und Die Theorie Der Typen. *Ann. Chem. Pharm.* **1840,** 308–310.
49. Berzelius, J. J. *Essai Sur La Théorie Des Proportions Chimiques et Sur l'influence Chimique de l'éléctricité.; Paris,* 1819.
50. Berzelius, J. J. *Versuch Über Die Theorie Der Chemischen Proportionen Und Über Die Chemische Wirkungen Der Elektrizität;* Arnold: Dresden, 1820.
51. Dumas, J. Mémoire Sur La Loi Des Substitutions et La Théorie Des Types. *C. R. Hebd. Séances Acad. Sci.* **1840,** *10,* 149–178.
52. Hofmann, A. W. Researches on the volatile organic bases. *Quart. J. Chem. Soc.* **1849,** *1,* 269–281. https://doi.org/10.1039/QJ8490100159.
53. Hofmann, A. W. Researches on the volatile organic bases. *Quart. J. Chem. Soc.* **1850,** *2,* 36–51.
54. Hofmann, A. W. Researches on the Volatile Bases. *Quart. J. Chem. Soc.* **1849,** *1,* 159–173.
55. Hofmann, A. W. Researches on the volatile organic bases. *Quart. J. Chem. Soc.* **1849,** *1,* 285–317. https://doi.org/10.1039/QJ8490100285.
56. Hofmann, A. W. Researches on the volatile organic bases. *Quart. J. Chem. Soc.* **1850,** 300–335.
57. Hofmann, A. W. Researches on the volatile organic bases. *Quart. J. Chem. Soc.* **1851,** 231–240.
58. Hofmann, A. W. Researches regarding the molecular constitution of the volatile organic bases. *Phil. Trans. Roy. Soc. Lond.* **1850,** 93–131.
59. Hofmann, A. W. Beiträge Zur Kenntniss Der Flüchtigen Basen. IX. Notiz Über Das Verhalten Des Anilins Und Der Alkohol-Basen Gegen Saltpetrige Säure. *Ann. Chem. Pharm.* **1850,** *75,* 356–368.
60. Hofmann, A. W. Beiträge Zur Kenntniss Der Flüchtigen Basen. III. Ueber Die Einwirkung Des Jods Auf Anilin Chlorcyans, Bromcyans Und Jodcyans Auf Anilin. *Ann. Chem. Pharm.* **1848,** *69,* 129–173.
61. Hofmann, A. W. Beiträge Zur Kenntniss Der Flüchtigen Basen. IV. Anilide. *Ann. Chem. Pharm.* **1849,** *70,* 129–149.
62. Hofmann, A. W. Ueber Zwei Neuer Reihe Der Flüchtigen Basen. *Ann. Chem. Pharm.* **1850,** *73,* 91–92.
63. Hofmann, A. W. Beiträge Zur Kenntniss Der Flüchtigen Basen V. Ueber Das Verhalten Des Cyananilins Gegen Säuren Und Basen. *Ann. Chem. Pharm.* **1873,** *50,* 180–189.
64. Hofmann, A. W. Beiträge Zur Kenntniss Der Flüchtigen Basen. VI. Metamorphosen Des Dicyanomelanilins. Bildung Des Der Cyansäure Entsprecheenden Gliedes Der Anilinreihe. *Ann. Chem. Pharm.* **1850,** *74,* 1–33.
65. Hofmann, A. W. Beiträge Zur Kenntniss Der Flüchtigen Basen. VII. Einwirkung Der Wasserfreien Phosphorsäure Auf Verschiedene Anilinsalz Unf Anilide. *Ann. Chem. Pharm.* **1850,** *74,* 33–40.

66. Hofmann, A. W. Beiträge Zur Kenntniss Der Flüchtigen Basen. VIII. Constitution Dieser Körper, Verhalten Des Anilins Und Ammoniaks Zu Den Bromüren Des Methyls, Aethyls Und Amyls. *Ann. Chem. Pharm.* **1850,** *74,* 117–177.
67. Williamson, A. W. On the constitution of salts. *Chem. Gaz* **1851,** *9,* 334–339.
68. Williamson, A. W. Theory of etherification. *Philos. Mag.* **1850,** *37,* 350–356.
69. Williamson, A. W. Suggestion for the dynamics of chemistry derived from the theory of etherification. *Chem. Gaz* **1851,** *9,* 294–298.
70. Williamson, A. W. XXII.—On etherification. *Quart. J. Chem. Soc.* **1852,** *4,* 229–239.
71. Williamson, A. W. On the constitution of salts. *Quart. J. Chem. Soc.* **1852,** *4,* 350–355.
72. Frankland, E. On the isolation of organic radicals. *Quart. J. Chem. Soc.* **1850,** *2,* 263–296.
73. Frankland, E. On a new series of bodies containing metals and phosphorus. *Quart. J. Chem. Soc.* **1850,** *2,* 297–299.
74. Frankland, E. On a new series of organic bodies containing metals. *Phil. Trans. Roy. Soc. Lond.* **1852,** *142,* 417–444.

CHAPTER TWO

Vladimir Vasil'evich Markovnikov and his rule for addition

Vladimir Vasil'evich Markovnikov (1837–1904)

Vladimir Vasil'evich Markovnikov (1837–1904, Fig. 2.1) was one of the few Russian organic chemists to attain international distinction during his lifetime. He was a brilliant theoretician who was also something of a hothead (especially toward those in authority over him), as well as both a genius and a gadfly.

The details of Markovnikov's birth are disputed, with several claims for the date and place appearing to be reasonable. His headstone (Fig. 2.2) records his birth year as 1838, in agreement with Kablukov's memorial,[1,2] which reports that he was born in the village of Chernorech'e, near Nizhnii Novgorod, on December 10(22), 1838. Markovnikov's student, Nikolai Matveevich Kizhner (1867–1935), cites the same date and place,[3] as does the Encyclopedia Britannica and articles online that draw on this source. The Biographical Dictionary of the Professors and Teachers of Kazan Imperial University, the article based on autobiographical data[4] gives the same information as Kablukov. In the Brockhaus & Ephron Encyclopedic Dictionary, his date of birth is given as December 25, 1838, but the birth year was changed to 1839 in a subsequent edition of the encyclopedia.[5]

In the biography by Platé, Bykov and Eventova,[6] on the other hand, his date and place of birth are given as December 13(25) 1837, in the village of Knyaginino, in the Nizhnii-Novgorod region; this is almost a year earlier than the Kablukov dates. The same information is given by Manolov in his two-volume treatise, Great Chemists.[7] The Platé biography cites a comment that Markovnikov made on December 16(28) 1892 that it was his birthday and that he was 55 years old; this would make his birth year 1837. An article published in the newsletter of a public company in Dzerzhinskii (Note 1) makes a strong argument in favor of the Kablukov version, based on sources from Kazan (although there may be an ulterior motive here). Very recently, evidence has come to light in support of the Platé date (Note 2). Thus, the date on his headstone notwithstanding, we will

Addition, Elimination and Substitution: Markovnikov, Hofmann, Zaitsev and Walden
ISBN: 978-0-12-821027-7
https://doi.org/10.1016/B978-0-12-821027-7.00004-9

Figure 2.1 Vladimir Vasil'evich Markovnikov. *Downloaded June 2020, from https://commons.wikimedia.org/wiki/File:VladimirMarkovnikov.jpg.*

use the Platé date, making Markovnikov a year older than in most current biographies. Markovnikov's family (originally spelled, "Morkovnikov") had noble origins. His grandfather, Vasilii Ivanovich, was a Senior Accountant in the treasury of the Board of Trustees of St. Petersburg and was descended from the children of a subaltern officer. His father, Vasilii Vasil'evich was recruited into the 71st Belevskii Jaeger Regiment at the age of 18 and had risen to the rank of Lieutenant by the time his son was born. At that time, the 71st Belevski Jaeger Regiment was stationed near Chernorech'e (Note 1). Shortly after the birth of his son, Vasilii Vasil'evich (now a Staff Captain) retired from his military post and moved the family to the estate at Ivanovskoe, a village in the Knyaginin district of the Nizhnii Novgorod region, which his wife had brought to him as part of her dowry. Markovnikov spent his early childhood in the village, and here he was first taught to read by the village priest. He quickly became an avid reader, reading not only novels, but also papers on military tactics on his father's desk. Little is known about his mother, Lyubov Nikolaevna, and the fact

Figure 2.2 Markovnikov's grave at the Bagan'kovskii Cemetery in Moscow. *Image from 2013 by Sergei Semyonov downloaded July 29, 2019, from МогилахимикаВладимираМарковниковаhttps://commons.wikimedia.org/wiki/File:Могилахимика Владимира Марковникова.JPG licensed under the Creative Commons Attribution 3.0 Unported license and under the terms of the GNU Free Documentation License, Version 1.2 or any later version published by the Free Software Foundation.*

that Markovnikov did not recall much of her suggests that her relationship with him may have ended when he was very young, perhaps by her early death. In 1844, Vasilii Vasil'evich was elected to a bureaucratic position by the nobility, and he therefore moved the family to Knyaginino.

The education of the chemist

In 1846, Markovnikov left home for the Alexander II Institute for Nobles (Fig. 2.3). By this time, he could read Russian, German, and French. He spent the next decade at the Nobles Institute. In 1856, he graduated from the Nobles Institute and entered the Juridical Faculty of the Imperial Kazan University (Fig. 2.4) as a student in economics. At this time, possibly in response to her humiliating defeat in the Crimea, Russia was instituting the Cameral System of education in an effort to generate government

Figure 2.3 The Alexander II Nobles Institute from a watercolor by Andrei Osipovich Karelin (1837–1906), 1870. *From (1) НижегородскийдворянскийинститутФайлАлександровскийдворянскиiiинститутНижнийНовгородhttps://ru.wikipedia.org/wiki/Нижегородский_дворянский_институт#/media/Файл:Александровский_дворянскии_институт._Нижний_Новгород.GIF (accessed Jul 2019).*

Figure 2.4 The imperial Kazan University in 1832. *Public domain image; Retrieved from https://commons.wikimedia.org/wiki/File:Kazan_University,_1832.jpg.*

workers and officials with the skills needed to manage a centralized economy in an industrialized nation. Cameralism[8] basically advocated a strong public administration to oversee a centralized, industrial economy. The goals of cameralism were to maximize the efficiency in the ways the state could acquire and use wealth. As Russia moved into the 19th century, its industrialization (although slow) led to the belief that the ideas of cameralism would provide the framework to allow the government to ensure that Russia would have a sufficient number of technologically literate bureaucrats to move the nation forward. Obviously, cameral students at Kazan were required to study subjects such as Russian history, political economy and statistics, public law and finance. However, they were also required to study scientific subjects (technology, mechanics, chemistry, and other subjects) for 2 years as part of their graduation requirements. This was where Markovnikov's career took a dramatic turn. Although he graduated in 1860 with a degree in economics, his required chemistry course, taught by Butlerov, moved him to become a chemist. When Markovnikov entered Kazan University, Modest Yakovlevich Kittary (1825–80, Fig. 2.5) was the Professor of Technology. Kittary's lectures inspired the young Markovnikov, who decided to study technology and had resolved to take his required technology courses early. In 1857, however, Kittary received the call to the Chair of Technology at Moscow University, and he left Kazan.[9] Kittary's departure for Moscow led to Markovnikov choosing to delay taking his science courses until his third year.

(a) (b)

Figure 2.5 (A) Modest Yakovlevich Kittary and (B) Aleksandr Mikhailovich Butlerov. *(N.d.). (A) Downloaded July 2019 from http://zz-project.ru/personalii/1777-kittary-modest-yakovlevich (B) Downloaded July 2019 from https://commons.wikimedia.org/wiki/File:Butlerov,_A._M._1828-1886.jpg.*

In 1859, Aleksandr Mikhailovich Butlerov (1828–86, Fig. 2.5), the Professor of Chemistry, returned from his komandirovka (paid study leave) in western Europe, where he had witnessed the birth of the new Structural Theory of Organic Chemistry. In Butlerov, Markovnikov quickly found a kindred spirit so, in the 1859–60 academic year, he took organic chemistry from Butlerov. We are fortunate in having Markovnikov's notes from this course, which he published with Butlerov's consent, as a lithographed copy.[10] This set of notes represents the last lecture notes we have from Butlerov's course before his conversion to Structural Theory. They have never been transcribed into modern printed Russian nor translated into English. In his reminiscence of Butlerov,[11] Markovnikov recalled, "I had the pleasure of making his personal acquaintance in his home in the first year of my enrollment in the laboratory, i.e., the third year of my studies."[12] Markovnikov had had the good fortune to meet Butlerov at the time when the latter's scientific research career was just taking off. Within a year of Markovnikov beginning his kandidat research with him, Butlerov had had the epiphany that changed him from a Type Theorist to a committed Structuralist, and he had proposed what became the dominant form of Structural Theory.[13,14] Butlerov continued to work on the version of the theory that would be the basis of the first textbook of organic chemistry based entirely on it, his, *Introduction to the Complete Study of Organic Chemistry*,[15] (Fig. 2.6) during the first 4 years of Markovnikov's Fig. 2.6 Magistr Khimii research.[16–18]

After he graduated with the diploma in June 1860, Markovnikov was retained by Butlerov to train for the professoriate. He began the study of a subject that would be one of the major focal points of his research for the next decade and longer: the relationship between isomeric or closely related organic compounds. His dissertation for the degree of kandidat was a study of the relationship between aldehydes, alcohols, and ketones.[19] The cover page of the dissertation is shown in Fig. 2.7. As originally submitted, only the right-hand side carried any text (*Al'degidy i ikh otnoshenie k alkogolyam i ketonam [Aldehydes and their relationship with alcohols and ketones]*). It was later overwritten by the Dean's request to submit, the dated affirmation by each of the examiners that they read it, and the examiners' report (by Butlerov).

In his kandidat dissertation, and in his early publications in *Liebigs Annalen der Chemie* and the *Zeitschroft für Chemie und Pharmacie*[20–23] and the *Zhurnal Russkago Khimicheskago Obshchestva*,[24,25] Markovnikov spelled his name with the first vowel as ***o***, not ***a***. He changed the spelling of his

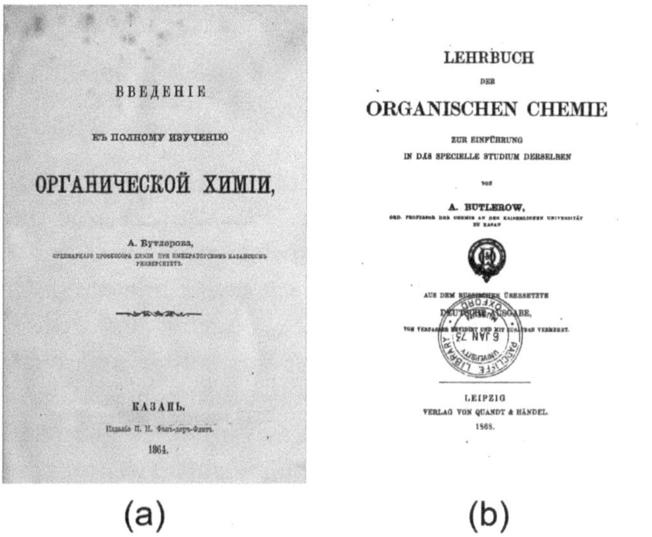

Figure 2.6 (A) The title page from Butlerov's Vvedenie k polnomu izucheniyu organicheskoi khimii [An introduction to the complete study of organic chemistry] by A. Butlerov. Kazan, 1864. (B) The title page from Butlerov's Lehrbuch der organischen Chemie zur Einführung in das specielle Studium demselben von A. Butlerow. Leipzig, 1867–68. *(N.d.) Downloaded June 2020 from (A) https://archive.org/stream/vvedeniekpolnomu008800#page/n5/mode/2up (B) https://books.google.com.ag/books/about/Lehrbuch_der_organischen_Chemie.html?hl=de&id=5-g-AAAAYAAJ&output=html_text.*

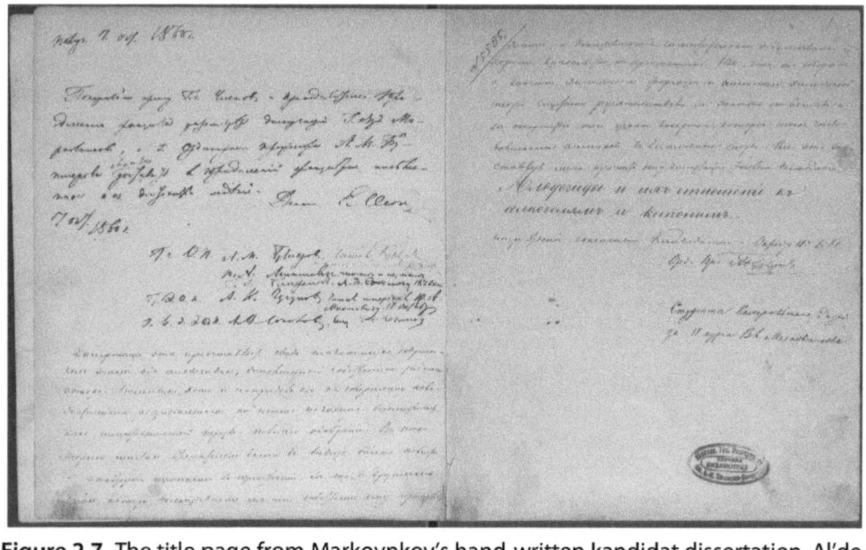

Figure 2.7 The title page from Markovnkov's hand-written kandidat dissertation, Al'degidy i ikh otnoshenie k alkogolyam i ketonam [Aldehydes and their relationship with alcohols and ketones], Kazan, 1860. The page is overwritten by notes from the Dean of the Faculty and the Professors on the Examining Committee, chaired by Butlerov. *Image from the rare manuscript collection of the Lobachevskii Scientific Library of Kazan Federal University. Reproduced by permission.*

name to Markovnikov while he was a student at Kazan University.[26] According to Shal'nov,[25,27] a local historian of the Dzherzinskii district (published online in the Dzherzinskii Vodokanal),[27] the vowel change was due to a slip of the pen, and Butlerov allowed him to retain the new spelling because "he liked it so much." A. V. Zakharov, on the other hand, provides the following alternative explanation:[26,28] "Writing his family name using *o*—Morkovnikov—was almost universal in official documents and journal articles in the 1860s, although it also occurs much later. Markovnikov himself explained that Butlerov kept writing the first syllable of his name using *o*, and that he was imitated by others. According to the original version by Klyuchevich and Bykov, "on admission to the university, he was named just Morkovnikov on the passport. He later corrected it to 'Markovnikov,' <but> Butlerov did not always remember about this."[29] Markovnikov was awarded the degree of kandidat in November 1860, and he began work on his Magistr Khimii (M. Khim.) research. The M. Khim. was the minimum degree required for an individual to occupy a position as a tenured Extraordinary (Associate) Professor in a Russian university. At the time of his kandidat research, it is clear that his mentor, Butlerov, still adhered to Type Theory. Had this situation changed, it is highly unlikely that Butlerov would not have imparted his new views to his favorite student, and nor would he have allowed his student to use what Butlerov himself would have viewed as an outdated theory, as the basis for his dissertation. Almost simultaneously with Markovnikov beginning his research for the M. Khim., Butlerov had the epiphany that would make him Structural Theory's most ardent advocate. The possible involvement of Markovnikov in Butlerov's transformation has been speculated upon,[30] but the question requires more research. In September 1861, Butlerov attended the XXXVI Versammlung deutscher Naturforscher und Ärzte (GDNÄ) zu Speyer. Here, in what has become known as, "the Speyer paper," he presented his version of structural theory. (Richard August Carl) Emil Erlenmeyer (1825–1909, Fig. 2.8), had been in the audience, and pressed Butlerov to publish the paper in his journal, *Zeitschrift für Chemie und Pharmacie*.[13]

Less than 3 years later, he had written and published the first organic chemistry textbook based exclusively on Structural Theory, the *Vvedenie k polnomu izucheniya organicheskoi khimii [An Introduction to the Complete Study of Organic Chemistry]*.[15] Butlerov instilled this same ardor into his student. Markovnikov's M. Khim. research concerned the question of isomers in

Figure 2.8 Emil Erlenmeyer (ca. 1863). *From (2) http://www.uni-heidelberg.de/institute/fak12/texte/histor2.engl.html (accessed Jun 2020).*

organic chemistry. Butlerov's version of Structural Theory had provided a tool for defining the differences between isomers of organic compounds that was superior to the versions of the theory that had gone before. In addition, Butlerov's presence at Kazan almost certainly allowed the two men to consult on problems as they arose. Toward the end of his M. Khim. research, one of the focuses of Markovnikov's research was the elucidation of the structure of the double bond. In 1864, he published a paper on the addition of bromine to allyl derivatives, as shown in Scheme 2.1.[31] This paper was followed in the same journal by a commentary by Butlerov.[32]

The dissertation itself (Fig. 2.9) resulted in five major conclusions[33]: 1. Not all polyatomic [polyvalent] elements are able to form compounds such as carbon, with connections of carbon affinities between each other. 2. The modern theory of atomicity [valence] does not provide a real basis for judging the true constitution of substances. 3. Structural formulas make redundant the laws of pairing of basicities [valences] in compounds (Basicitätsgesetze). 4. Substances of the same composition with different

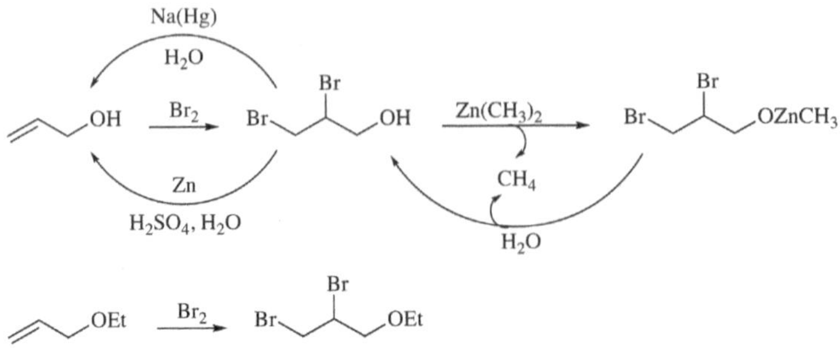

Scheme 2.1 Markovnikov's studies of the addition of bromine to allyl compounds.

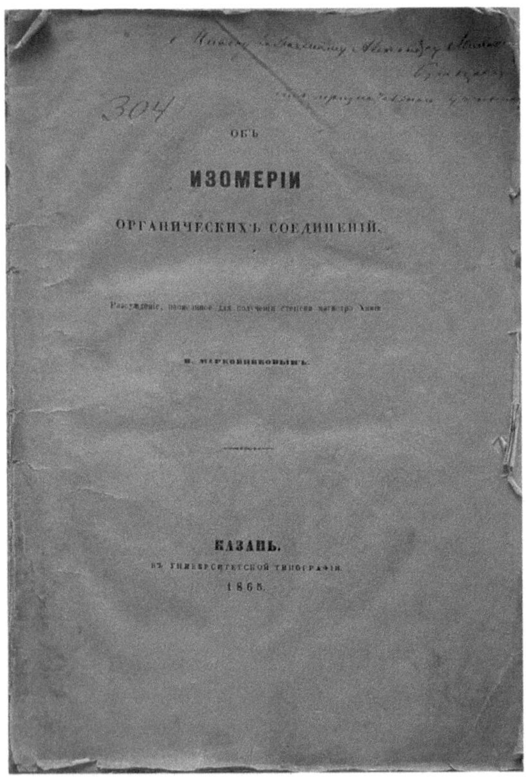

Figure 2.9 The cover of the examiners' copy of Markovnikov's M. Khim. dissertation (1865). The hand-written inscription in the upper right reads, "To much-esteemed Aleksandr Mikhailovich Butlerov from a grateful student."

circular polarization [optical rotation] are real chemical isomers. 5. The number of oxygen units in an acid radical cannot serve as a measure of its basicity.

The first of these conclusions refers to the property of catenation, which is exhibited by carbon to a staggering degree (although sulfur also exists as long polymeric chains). The property of catenation had been a foundational concept of the Couper,[34–37] Kekulé,[38,39] and Butlerov[13] forms of the Structural Theory of organic chemistry. Although sulfur also forms long chains of atoms, these chains merely correspond to different allotropes of the element and not to isomeric compounds. The second conclusion is referring to the shortcomings of empirical formulas and, by analogy, of Type formulas. The third is a direct challenge to Type Theory, in that it notes that Structural Formulas do not require the formal pairings of valences in compounds. This was an early explicit statement that optical isomers (enantiomers) are true chemical isomers. This was, of course, critical to the later work of van't Hoff[40–43] and Le Bel[44] and the founding of the subdiscipline of stereochemistry.[45] The final conclusion of this dissertation refers to Markovnikov's studies of the isomeric butyric acids and their hydroxylated analogues. He found that the hydroxybutyric acids (he mainly focused on α-hydroxyisobutyric acid) had the same overall acidity, but that they had differing numbers of hydroxyl groups. This clearly contradicted the conventional wisdom that the basicity of the acid (monobasic, dibasic, etc.) could be determined by simply dividing the number of oxygen atoms ($O = 16$) by 2.

With Kolbe at Leipzig

Markovnikov's M. Khim. degree was conferred in 1866, and the same year he was awarded a komandirovka (official study leave) to further his education abroad. He spent his first semester in Heidelberg, where he attended lectures by Erlenmeyer, Hermann Franz Moritz Kopp (1817–92, Fig. 2.10), and Gustav Robert Kirchhoff (1824–87, Fig. 2.10).

During his time at Heidelberg, Markovnikov published his paper on the synthesis of isobutyric acid and isopropyl ethyl ether from isopropyl iodide.[46] The pure secondary iodide was obtained from glycerol by the reaction with hydrogen iodide shown in Scheme 2.2.

Markovnikov next moved to Berlin, where he entered the laboratory of (Johann Friedrich Wilhelm) Adolph von Baeyer (1835–1917, Fig. 2.10). He was immediately confronted by Baeyer (a good friend of Kekulé)

(a) (b) (c)

Figure 2.10 (A) Hermann Kopp. (B) Gustav Kirchhoff. (C) Adolf von Baeyer in 1884. Image of Kopp from Universitätsbibliothek Heidelberg, and of Kirchhoff from the Smithsonian Libraries. Public domain image of Baeyer downloaded from Science History Institute. *Public domain images; (A) Universitätsbibliothek Heidelberg licensed under the Creative Commons Attribution-Share Alike 4.0 International license. (B) From Smithsonian Libraries https://commons.wikimedia.org/wiki/File:Gustav_Robert_Kirchhoff.jpg (C) https://digital.sciencehistory.org/catalog?f%5Bcreator_facet%5D%5B%5D=Muller%2C+Friedrich.*

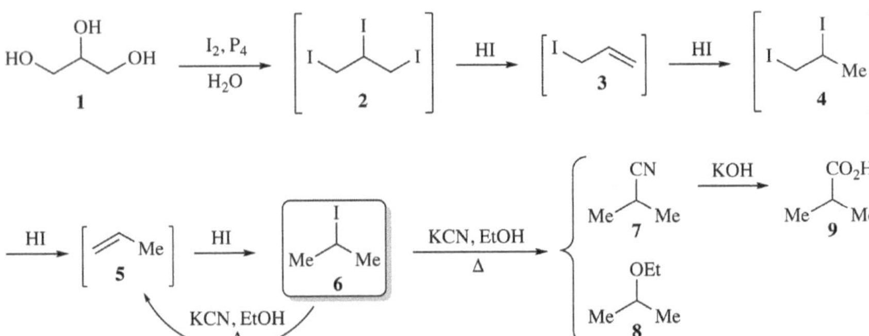

Scheme 2.2 Markovnikov's synthesis of isopropyl iodide and its conversion to isobutyric acid and ethyl isopropyl ether.

with the question of whether it was Markovnikov who had written a criticism of Kekulé's Lehrbuch, which Baeyer had obviously read with much displeasure. When Markovnikov replied in the affirmative, the usually courteous Baeyer became uncharacteristically confrontational, accusing him of attacking Kekulé's work "like a thief," and asserting that "had you dismantled Kolbe's book the same way, nothing would be left of it!"[47]

Although Markonikov's time in Baeyer's laboratory in Berlin was brief, it had long-term repercussions. As recounted by Zakharov[26,28,48]:

"The article on acetone acid was the first to explain the synthesis of hydroxy acids, obviously applicable to the synthesis of aldehydes (sic!). I then intended to carry out this reaction using other examples, but I was fascinated by other work that was obligatory to prove my views on the order of substitution, and which I found more urgently necessary and important for the science. "In the same article my views are outlined in more detail, and clarified further in my doctoral dissertation 'On the mutual influence of atoms, etc.' The laws that flowed from these views, controlling many important reactions, were then thoroughly novel, even conflicting with very authoritative opinions (Kekulé). Naturally, my views, as a young chemist and, moreover, a Russian, were ignored, and although later they were confirmed and accepted by everyone, the name of the first author was forgotten. Everyone accepts, for example, that in acids the hydrogen closest to the carboxyl is replaced, but no one mentions that I discovered this law. To what extent, even now, some prominent chemists are not completely clear about the mutual influence of atoms, prove the recent work of V. Meyer on the substitution order of hydrogen in monohaloid hydrocarbons. "At that time, the concepts of influence were so weak that once Graebe [Carl Graebe (1841–1927, Fig. 2.11)], with whom I worked with Bayer in 1866, asked me: why is the chlorine in acetyl chloride different from that in ethyl chloride? Schorlemmer [Carl Schorlemmer (1834–92, Fig. 2.11)], assured me in 1867 that when paraffins were chlorinated, one would get chlorides of normal alcohols, but I claimed that one would have a mixture with a predominant content of secondary chlorides. The latter has still not been clarified, although secondary chlorides have been obtained."[49]

These questions may well have instigated his thinking about the influence of structure on reactivity and properties.[48] Markovnikov left Berlin and moved to Leipzig to work with (Adolph Wilhelm) Hermann Kolbe (1818–84, Fig. 2.11). Because he already had an advanced degree, Markovnikov's position in Kolbe's laboratory was different from most of the Praktikanten, and he was allowed considerable latitude in the chemistry that he pursued. As he wrote to Butlerov:

"My position in Kolbe's laboratory was a little different from all the others, since I was a master [Magistr Khimii] and had worked on my own

Figure 2.11 (A) Carl Graebe in the 1890s. (B) Carl Schorlemmer in 1894. (C) Hermann Kolbe in the 1880s. *Public domain images; (A) https://commons.wikimedia.org/wiki/File:Carl_Graebe_1890s.jpg (B) https://commons.wikimedia.org/wiki/File:Schorlemmer_Carl.jpg (C) https://commons.wikimedia.org/wiki/File:Hermann_Kolbe_1880s.jpg.*

subject matter for 3 years. During this first year since my arrival in Germany, I have ascertained that, in theoretical matters, the Kazan laboratory has far outstripped all the laboratories of Germany."[50]

While in Leipzig, Markovnikov worked to find a German publisher for the German translation,[51] of Butlerov's *Vvedenie*. As well as extending the applications of Butlerov's version of Structural Theory. He also took advantage of the superior synthetic skills of the Praktikanten in the laboratory to improve is own. The main interaction between Kolbe and his senior Praktikant was, however, largely based on their opposing views on Structural Theory. Kolbe's position has been carefully explored by the eminent historian of chemistry, Alan J. Rocke of Case Western Reserve University.[52] Kolbe was both stubborn and irascible, and having been part of the struggle to establish organic chemistry as a reputable subdiscipline of science, he fought any changes in theory that were not firmly footed in experiment. It was this opposition to new ideas that led to his famous diatribe against the tetrahedral carbon hypothesis of van't Hoff, in which he used his Editor's pen at the Journal für Praktische Chemie to rail against van't Hoff's theory.[53] In a way, it is really unfortunate that this paper is one for which so many organic chemists remember him: His stubbornness did help to ensure the rigor of the science of organic chemistry, and he was actually far from the dogmatist he is often portrayed as, he did not insist on publishing his own ideas over those of his students (although he often attached lengthy "Bemerkungen" to those papers highlighting his disagreement with their conclusions.

Kolbe eschewed the term, "Chemical Structure," in favor of his own term, "Rational Constitution." Over time, his position did change, but he still refused to adopt the new terminology. Butlerov, who had first proposed the term, "Chemical Structure" (meaning the structure based on chemical not physical atoms),[13] could not have had a more able surrogate than Markovnikov to argue head to head with Kolbe. Kolbe was an unreconstructed Type Theorist, and he continued to use equivalent weights (C = 6, O = 8, S = 16, etc.) when many of his contemporaries had adopted the atomic weights of Cannizzarro. Markovnikov could be just as suborn as Kolbe, as is shown by his account of his final meeting with Kolbe in his memoir of Butlerov.[54] He wrote in this memoir:

"Before leaving Leipzig, I felt duty-bound to transfer my work to Kolbe, even though he did not take any part in it. After reading it, he invited me to his office and pointed to one of my explanations that was, in his opinion, wrong. The dispute came down to the old theory of oxygen. 'You do not understand me because you have not gotten used to my formulas—said Kolbe—let me express my thoughts in your formulas.' Well, I thought, now Herr Professor, you are trapped. For me, this was just what was needed. He began writing and stopped half-way through the formula. He thought for a moment and put down the pencil. 'Ja, Sie haben Recht.' Then he finished the formula, and repeated—Yes, yes, that's right, you are right!—and somewhat confusedly began to explain something. I quickly withdrew, sparing the feelings of an honored scholar."[55,56]

Markovnikov's return to Kazan

In May 1867, Markovnikov was due to leave Leipzig, but before returning to Kazan he sought and obtained a 6-month extension of his komandirovka from the Ministry of Education.[57] This permitted him to remain in Leipzig through the summer and then spend a month working in the Paris laboratory of one of the prominent French chemists (Wurtz or Berthelot) until mid-September. He then attended the 41 Versammlung deutscher Naturforscher und Ärzte in Frankfurt am Main (18–24 September 1867)[58] on his way back to Kazan. He was anxious to begin his teaching, so he returned to Kazan before the end of his extension. While abroad, he had been promoted to Docent (Assistant Professor) in the Chair of Chemistry, and on his return to Kazan, he began his independent teaching career by teaching the general course in chemistry. This act permitted Butlerov to take a komandirovka for the 1867–68 academic year, because Markovnikov would be training the students who would then be prepared to enter Butlerov's advanced course on his return. Within a year of his return to Kazan, Butlerov had accepted an appointment in St. Petersburg; in December 1868, he left Kazan for the last time. One of his last acts was to endorse Zaitsev for insertion into the professorial ranks at Kazan. On April 11, 1869, Markovnikov submitted a request to the Dean for the public defense of his Dr. Khimii dissertation, Materials on the question of the mutual influence of atoms in chemical compounds (Fig. 2.12). Glowing examiners' reports from Zaitsev and Butlerov paved the way for the conferral of the degree 16 days later. Butlerov's departure made room for Markovnikov to assume the Chair of Chemistry, and the Dr. Khim degree made Markovnikov eligible for promotion beyond the rank of Docent (and Extraordinary Professor, too). He was promoted to Extraordinary (Associate) Professor

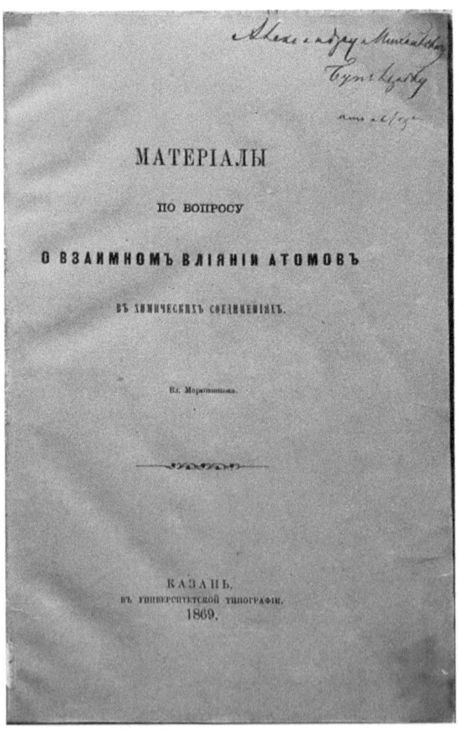

Figure 2.12 The cover of the examiners' copy of Markovnikov's Dr. Khim. dissertation. The onscription in the upper right-hand corner reads, "To Aleksandr Mikhailovich Butlerov from the author."

of Chemistry just 25 days after graduating (May 22, 1869) and, less than 10 months after that (March 6, 1870), to Ordinary (Full) Professor of Chemistry. In his 4 years as Head of the Chemistry Laboratory, Markovnikov accomplished the improvement of the teaching of analytical chemistry, he expanded the Chemistry Laboratory to an additional room, and he radically improved the gas plant, allowing the delivery of gas to all buildings and offices in the university. He did not, however, build a large or particularly active research group, and his research was an extension of his doctoral work. Thus in many ways, Markovnikov's first independent research work was, to a large extent, a continuation of Butlerov's work on Structural Theory. In all, alone and with his students (only three are coauthors), Markovnikov published only 12 papers[24,59–69] in this period of his career. Unlike his later years at Moscow, Markovnikov did not develop a "Markovnikov School" at Kazan. Nevertheless, his papers were important contributions to the Structural Theory of Organic Chemistry at a time when such contributions were key in establishing the validity of the new theory.

Markovnikov's resignation from Kazan

Markovnikov was an argumentative individual who had strongly held beliefs and views and who was not at all backward in expressing them. This often made him something of a thorn in the side of University administrations. When Butlerov finally left Kazan in 1869, the university was faced with the daunting possibility of having Markovnikov as the only Professor in chemistry and having to deal with him. There was a second Chair open at the time, and they initially planned to offer it to another Butlerov student, Aleksandr Nikiforovich Popov (ca. 1840–81, Fig. 2.13). However, by the time the University had tendered their offer, Popov had accepted a position at Warsaw University. This left Aleksandr Mikhailovich Zaitsev (1841–1910, Fig. 2.13), yet another Butlerov student, as the obvious choice, and he was appointed as Extraordinary Professor in 1869, immediately after successfully defending his M. Khim. dissertation. We will deal with Zaitsev in much more detail in the next chapter. Markovnikov and Zaitsev feuded for their entire careers,[24,59,61–70] and Zaitsev's appointment as his professorial colleague surely rankled Markovnikov. Certainly, Markovnikov never enjoyed with Zaitsev, the intimate relationship that he had enjoyed as Butlerov's junior colleague.

Markovnikov had a well-developed sense of fairness and ethics, and this led to his resignation from Kazan in support of Pyotr Frantsevich Lesgaft

Figure 2.13 A. M. Butlerov and his students, 1867–68. *Public domain image adapted from https://kazan.aif.ru/society/persona/velikiy_himik_kak_obidnoe_prozvishche_proslavilo_aleksandra_butlerova.*

(1837–1909, Fig. 2.14). Lesgaft was a popular, outspoken lecturer in Physiological Anatomy, but he was highly critical of the unscientific way that physiology was taught at the university. He was forbidden to teach his theories, and when he continued to do this, he was dismissed from the university. The faculty had voted to recommend Lesgaft for promotion to Ordinary (Full) Professor, but the University Council overrode this

Figure 2.14 The professors who resigned in protest at the unlawful treatment of Professor P. F. Lesgaft. Not pictured: Professor of Hygiene, A. I. Yakobii. *Public domain image from https://kpfu.ru/news-archive/k-jubileju-alma-mater-ob-uchenom-gigieniste-i.html.*

recommendation. In protest at the treatment of their colleague, Markovnikov and six of his colleagues—A. E. Gulobev (Extraordinary Professor of Histology), N. A. Golovkinskii (Professor of Geology and Paleontology), A. Ya. Danilevskii (Professor of Physiological Chemistry), L. I. Levitinskii (Privatdocent of Pathology and Therapy), V. G. Imshenetskii (Professor of Mathematics), and A. I Yakobii (Professor of Hygiene)—resigned to protest the university's actions. Lesgaft and six of the seven professors who resigned in protest of his dismissal are shown in Fig. 2.14. On December 18, 1871, Markovnikov's resignation was accepted.

Markovnikov was unemployed for just a few hours. The very same day, he was appointed Ordinary Professor of Chemistry at Novorossiisk University in Odessa (Fig. 2.15) as the successor to Nikolai Nikolaevich Sokolov (1826–77, Fig. 2.16). Sokolov had built the new chemistry laboratories at the university, so Markovnikov moved into an excellent facility.

Professor and head at Novorossiisk University

At Novorossiisk, Markovnikov delivered the lectures in organic chemistry, headed the chemistry laboratory, and led all the practical classes in organic chemistry. He and his students continued work on the Structural Theory of Organic Chemistry and the regiochemistry of organic reactions (Scheme 2.3). In his first paper from Odessa, Markovnikov showed that the reaction of epichlorohydrin (**10**) with hydrochloric acid gave two products (**11** and **12**), which he separated by repeated fractional distillation. When he subjected both to chromic acid oxidation, he found that the lower-boiling isomer gave a ketone (**13**) and that the higher-boiling isomer gave a carboxylic acid. (**14**) He demonstrated the formation of the ketone by

(a) (b)

Figure 2.15 (A) Novorossiisk University. (B) The Physics and Chemistry wing of the university building. *Public domain image from https://kpfu.ru/news-archive/k-jubileju-alma-mater-ob-uchenom-gigieniste-i.html.*

Figure 2.16 Nikolai Sokolov. *Public domain image; retrieved from СоколовНикола йНиколаевичхимикФайл https://ru.wikipedia.org/wiki/Соколов,_Николай_Николаевич_(химик)#/media/Файл:Sokolov_NN_1870.jpg.*

converting it to its bisulfite addition product and forming a semicarbazone derivative (**16**). In this way, he was able to show that the lower-boiling isomer, known at the time as "dichlorhydrin," was 1,3-dichloro-2-propanol (**11**) and that the higher-boiling isomer was the regioisomeric alcohol (**12**)[71]; he also carried out the dehydration of hydroxyacid **17** to **18**.[72] In an attempt to prepare isobutylidene chloride (**17**), Markovnikov's student Demchenko found[73] that the major product of treating isobutyraldehyde (**21**) with phosphorus pentachloride was the trioxane derivative, **23**. His student, Yelisafov, reported the synthesis and bromination of cetene (**24**).[74]

Markovnikov's move to Moscow

In the summer of 1872, Markovnikov attended the Polytechnic Exhibition in Moscow. The Dean of the Faculty of Physics and Mathematics of Moscow University, A. Yu. Davidov, invited him to take the vacant Chair of chemistry. Markovnikov accepted this proposal with some hesitation.

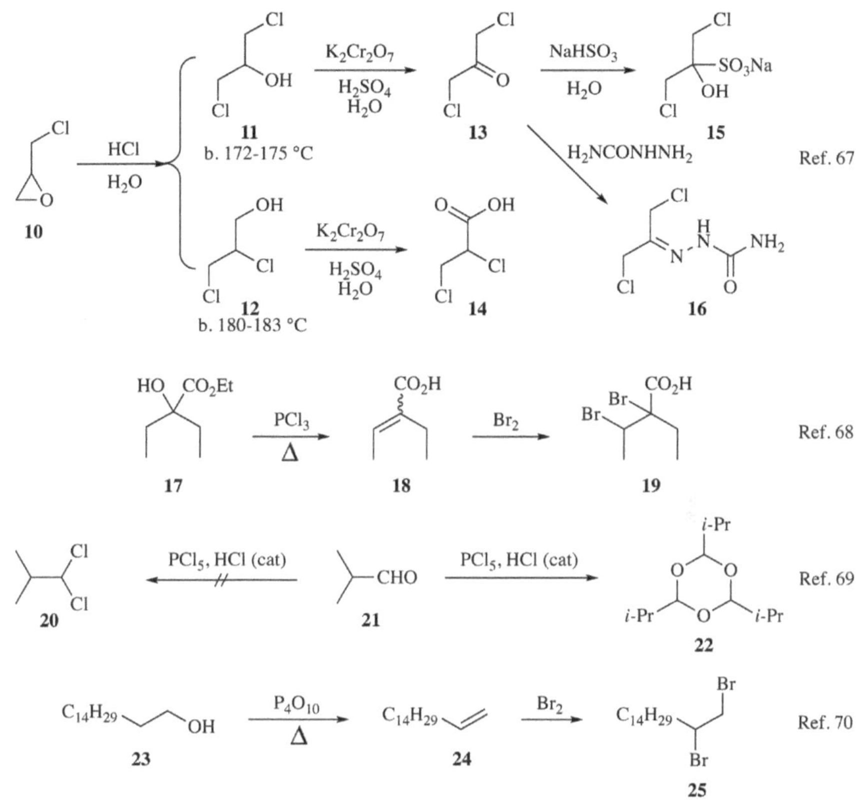

Scheme 2.3 Representative Markovnikov chemistry from Novorossiisk University.

That October, he wrote to Butlerov from Odessa, "You, of course, already know that they have chosen me at Moscow University. Here, the request has already been received without obstruction. I will probably stay here through the academic year and summer, living near Odessa. I don't know if I will have to regret it later. From the fact that I have decided to go to Moscow, it is obvious that I am counting on the best, but always there is always some involuntarily doubt. Of course, the local laboratory here is richer, but something can be arranged there."[75]

Initially, chemistry had been taught in the Faculty of Medicine, but with the establishment of the Faculty of Physical and Mathematical Sciences at the end of the 18th century, it had moved from the Faculty of Medicine to the new Faculty. In 1858, a small stone building was constructed for use as the chemistry laboratory (Fig. 2.17); the laboratory was renovated periodically over the next century and a half.

Figure 2.17 The "Red Building," original location of the chemistry laboratory at Moscow University on Mokhova Street. *(N.d.). Photograph by D. F. Sorokin (Archives of the Museum of MGU) retrieved June 2020, from http://www.chem.msu.su/rus/history/chemdept-history-1.html.*

Markovnikov's hesitation was understandable. As had been the case at the other major universities in Russia, chemistry at Moscow was initially taught by foreign professors—the inaugural Professor of Chemistry was Johann Christian Kerstens (1713–1802), who had obtained his M.D. at Halle in 1749. Kerstens remained in Moscow until 1770, when he left Russia. Under the foreign Professors of Chemistry, chemistry at Moscow failed to advance as it had at the other major chemistry centers of Russia: Kazan and St. Petersburg. Thus, the decision to leave Novorossiisk (and the opportunity to build his own organic chemistry program and the brand-new laboratories) was going to be very difficult. During the decade prior to his arrival, chemistry at Moscow had been taught by just two instructors: Nikolai Erastovich (Ernestovich) Lyaskovskii (1816–71, Fig. 2.18) and Dmitrii Kirilovich Kirilov (1833–77).

Lyaskovskii, who had submitted his M.D. dissertation, De cholere epidemici nonnullis causis atmosphericis [On some atmospheric causes of the cholera epidemic], in 1848, had been appointed to the Chemistry Department in 1854, following the elevation of Rodion Grigorievich Geiman (Rudolf von Heiman, 1802–65) to "Honored Professor of Moscow University." He had a strong reputation as a protein chemist, and in 1862, he wrote and defended his Dr. Khim. dissertation, Formuly proteinidov по (pervonachal'noi) teorii mul'dera [Formulas of proteinides in the (initial) theory of Mulder[76–81]]. This, however, was his last major scientific

Figure 2.18 Nikolai Lyaskovskii. *Public domain image; Retrieved from:* Лясковский *https://commons.wikimedia.org/wiki/File:Лясковский.png.*

research work. For a short time, he worked to modernize the department and had made significant progress by 1865. This, however, came to a halt in that year, and for the next 8 years, no further attention was paid to the condition of the laboratory.

From 1863 to 70, Lyaskovskii taught courses at the Alexander Military School, and from 1865 to 1869, he also taught inorganic and organic chemistry at the Petrovskii Academy of Agriculture and Forestry. This led to him turning over the teaching of organic chemistry and laboratory instruction in chemistry to teaching assistants and to Kirilov. On his death, the responsibility for all instruction in chemistry fell to Kirilov, who was totally unequipped to handle the task. By the time that Markovnikov arrived, there was no laboratory instruction in chemistry, and the discipline itself was no longer a mandatory subject for graduation. When Markovnikov took over, the Chemistry Department was in "a catastrophic condition."[82]

Markovnikov had made his move to Moscow contingent on the university consenting to the construction and renovation of the chemistry laboratory. The promise was not kept, however, despite the support of both the

Rector and the Physics-Mathematics Faculty. Markovnikov's plans were ambitious and called for a laboratory capable of serving all students requiring a course in general chemistry, including students in chemistry, medicine, pharmacy, agronomy, and so on—a total of over 200 students.

Markovnikov's initial work in Moscow

For his first decade in Moscow, Markovnikov continued his focus on the Structural Theory of Organic Chemistry. Among the earliest works from this period are studies of the oxidation of α-hydroxycarboxylic acid derivatives[83] and the structures of pyrotartaric acid and its derivatives.[84–86] He also carried out the synthesis of new tertiary alcohols obtained by the method of Frankland and Duppa[87–89] and then expanded this into the synthesis of crotonic acid derivatives.[90–92] This work included an extension of his investigations of the reactions of unsaturated compounds, including a restatement of his rule for addition.[93–95] In some cases, he took the occasion to respond to criticisms that he had incorrectly assigned the structure to either the alcohol or the alkene derived from it. A representative case is provided by the structure of 2,4-dimethyl-2-pentanol and its dehydration product. Butlerov's student, Pavlov, published the alcohol and alkene as new compounds,[96] but struck by the similarities in their boiling points and specific gravities, Markovnikov showed[97] that Pavlov's alcohol was identical to the tertiary heptyl alcohol Markovnikov had reported from Kazan in 1871.[64] Some representative examples from this stage of Markovnikov's work are gathered in Scheme 2.4. During this same period, Markovnikov also expanded his work into physiological chemistry and analytical organic chemistry, exemplified by his study of acetone in the urine of diabetics[98–100] and the theophylline content of tea.[101]

Markovnikov, the Sanitation Guru: The Russo-Turkish war

The Russo-Turkish war was fought from 1877 to 1888, and as a patriotic Russian, Markovnikov volunteered to help the war effort. He realized that sanitation and the availability of drugs (especially analgesics) were critical for a successful outcome of the war, so he immediately set about accumulating a stockpile of medicines for the front, in Moscow.[102–104] He was seconded to the front in Romania and Bulgaria, and here he oversaw the distributions of medicines and, more importantly, the sanitization of battlefields, ambulance trains, and hospitals, among others. It is worth remembering that during this period, half the loss of life in wars was due

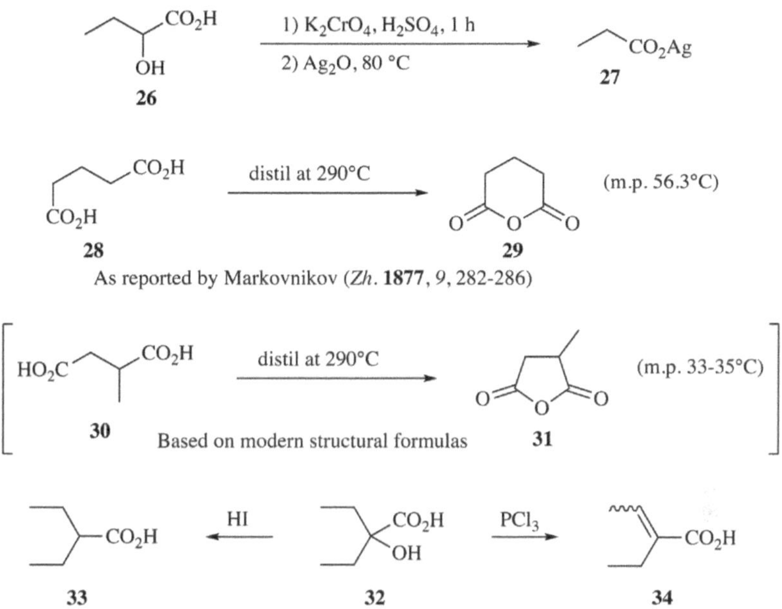

Scheme 2.4 Markovnikov's early work on Structural Theory in Moscow.

to disease, rather than enemy action, which made Markovnikov's work vital. Even so, over 80,000 Russians and between 60,000 and 90,000 Turks died of disease in this war.

For his services, Markovnikov was offered a stipend of 400 rubles in gold, but he declined this or any other payment while he was assigned to the Army of the Danube. He had two reasons for doing this: first, he was initially involved in the work under the auspices of the Red Cross, and second, he felt that he was merely doing his civic duty as a citizen of the Empire.[105] Immediately after the war, he revised his instructions for sanitation in wartime to encompass civilian healthcare workers as a 49-page booklet, A generally accessible, practical guide to disinfection for individuals, doctors, educational institutions, hospitals, factories, and railways.[106]

Markovnikov and cycloalkanes

Cycloalkanes became a focus of Markovnikov's research in Moscow at the same time that he began his seminal studies of Russian crude oils. At the time he began these studies, it was generally opined that cycloalkane derivatives with rings other than five- and six-membered would not be stable. Markovnikov's studies based on the Structural Theory of Organic Chemistry predisposed him to question all conventional wisdom. As a result of this, he attempted to make rings smaller than five-membered and larger than six-membered. He succeeded in both. In 1881, along with his student Krestovnikov, he synthesized cyclobutane-1,3-dicarboxylic acid (Scheme 2.5).[107–109]

A seven-membered structure for suberone (cycloheptanone) had been proposed in 1874 by Dale and Schorlemmer, when they obtained it by the dry distillation of calcium carbonate and suberic acid.[110] They obtained evidence in support of their structure by the permanganate oxidation of suberone to pimelic (heptanedioic) acid. In the second paper in the series,[111] these authors used known chemistry to suggest structures for azelaic acid. In 1884, a collaborator of these two chemists, A. Spiegel, published a paper on the results of his studies on suberone,[112] in which he suggested that the seven-membered structure was not the only possible one for suberone, but that alkylcycloalkanones with fewer members in the ring might be more appropriate. This was the status of the question when Markovnikov began his work.

Oddly enough, none of these workers sought to convert suberone to the corresponding saturated hydrocarbon. Markovnikov did. Around the same time, Markovnikov began his seminal studies of the Caucasus oils. These studies will be discussed in substantially greater detail in the next section; As a result of these studies, cycloalkanes had become a focus of his research in Moscow. By the sequence of reactions in Scheme 2.6, Markovnikov converted suberone to cycloheptane.[113,114] Markovnikov made the point that the sealed tube reduction with hydrogen iodide—Berthelot's method[115,116] —often gave rise to a complex mixture of products, especially if the reaction was carried out for extended periods of time.

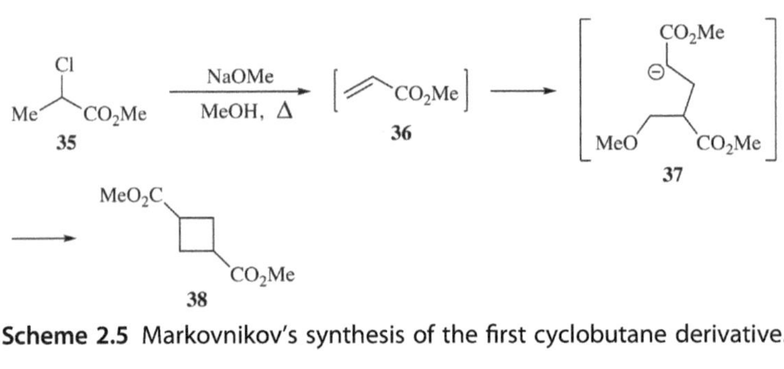

Scheme 2.5 Markovnikov's synthesis of the first cyclobutane derivative.

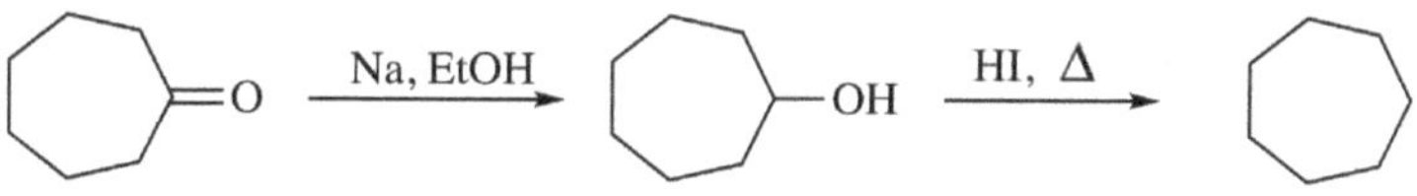

39 40 41

Scheme 2.6 Markovnikov's synthesis of cycloheptane.

A new direction: petroleum chemistry

It was Markovnikov's firmly held conviction that Russian scientists should apply their talents to the solution of problems in the public interest. In his view, the crude oils of Russia fell under that mantle, so following the end of the Russo-Turkish war, he again allowed his patriotic instincts to take over and began a decade-long study of the composition of Russian crude oils. With his student, Vladimir Oglobin, Markovnikov began the first of the long series of papers on the Caucasus crude oils[117–119] that built on the early work of Beilstein and Kurbatov on the low-boiling fractions of the Caucasus crude oil.[120] The work of Beilstein and Kurbatov showed that the compound was unaffected by concentrated sulfuric acid, nor oxidized by chromic anhydride or potassium permanganate, and therefore was saturated. They tentatively identified it as methylcyclohexane (hexahydrotoluene).

The analysis of a number of fractions of Caucasus crude oil boiling between 220 and 300°C showed that all had the same empirical formula (CH_2). Their lack of reactivity toward hot, concentrated sulfuric acid and toward chlorine made it clear that they were saturated hydrocarbons.

To characterize the compounds that he had called octonaphthene (C_8H_{16}, **42**) and nonanaphthene (C_9H_{18}, **46**), Markovnikov first tried to oxidize them with potassium permanganate. However, he found that only a trace of oxidation products was formed in either acid or neutral solution. There was no more success with a nitrating mixture under reflux: octonaphthene gave just a trace of trinitro-m-xylene and nonanaphthene gave a trace of trinitromesitylene. To obtain more evidence for his proposed structures of the naphthenes as alkylated cyclohexanes, he turned to aromatization with sulfur. On heating octonaphthene with sulfur at 220°C in a sealed tube, and subsequently treating the crude product with a nitrating mixture, trinitro-m-xylene (**44**) was obtained. The question of the intermediacy of m-xylene (**43**) itself, however, was not resolved because Markovnikov confirmed an earlier observation that xylene does not react with a nitrating mixture. Markovnikov did find, however, that when octonaphthene was mixed with 10% fuming sulfuric acid, the temperature rose to 50–60°C.

When this process was repeated, eventually all the octonaphthenes dissolved. The product formed a barium salt that was difficultly soluble in water. Heating the disulfonic acid with superheated steam gave a hydrocarbon that could be nitrated at elevated temperature to give **44**. Similar observations were made with nonanaphthene (**46**), leading to the production of trinitromesitylene (**48**). These studies[121,122] are summarized in Scheme 2.7.

Over the next decade, Markovnikov and his students systematically studied the composition of the Russian crude oils, and they found a relative paucity of aromatic hydrocarbons relative to their American counterparts.

Markovnikov's ouster from his professorship

Throughout his career, Markovnikov had a reputation as being very blunt-spoken, demanding, and for holding people to a high set of (especially ethical) standards. This seldom sat well with university administrators and governing boards, who viewed the presence of this brilliant scientist and teacher as a mixed blessing. An example of this attitude can be traced to his departure from Kazan. With the six colleagues who followed his suit, he rebelled against what he saw as the unjust and unethical actions against P. F. Lesgaft. This characteristic became more difficult to contain as he rose in rank: one can tell an entry-level person how to behave toward his colleagues and superiors, but this is a much more difficult task to accomplish with a senior professor.

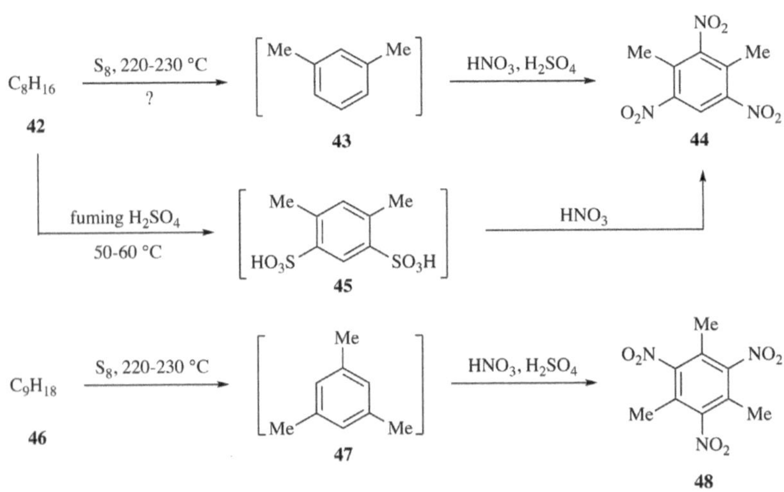

Scheme 2.7 Markovnikov's aromatization studies

In 1881, a new University Statute replaced the statute of 1863 that had granted the universities a large measure of autonomy. This new statute implemented a stricter government supervision in the universities that Markovnikov refused to accept, and his habitual outspokenness almost certainly made him a marked man. The new Statute carried an arcane provision that a professor could be forcibly retired from his post after 25 years of service. This happened to Markovnikov. His initial appointment to faculty rank was in 1868, at Kazan, so in 1893 his quarter century was up. In the summer of 1893, the trustee of the Moscow school district ordered Markovnikov to hand over the laboratory to Nikolai Dmitrievich Zelinskii (1861–1953, Fig. 2.19), a young Extraordinary Professor appointed by the Ministry, and a former privat-docent of Novorossiisk University, where Markovnikov had, himself, served as Head. This was a hard blow for Markovnikov, the world-renowned, distinguished Professor, to take. Moreover, the loss of his Professorship meant that he was also ordered to vacate the state-owned apartment in the building, so he had to move his family out.

Figure 2.19 Nikolai Zelinskii. *Public domain image from https://ru.wikipedia.org/wiki/Файл:Зелинский_Николай_Дмитриевич.jpg.*

There are some fascinating descriptions of Markovnikov's final decade in the biography by Platé, Bykov and Eventova,[6] and also in the article in Vestnik Moskovskogo Universiteta, Ser. 2, Khimiya, by O. N. Zefirova and N. S. Zefirov.[123] What follows here is largely the Author's translation of pages 140–143 of the Platé book, with the first paragraph translated from the Zefirov paper:

As he wrote in his diary, "They said to me: 'Get out! You are not needed!' ... It's a shame not only for myself, but for science in Russia and for its unfortunate representatives who are turned into slaves." It was obvious to Markovnikov that their attitude toward him was not unintentional, but was a consequence of his statements in opposition to the new Statute. " ... it was always thus, that the authorities consider all beliefs other than their own to be harmful, and now this view is the basis of the government system. Thus, Mendeleev and Sechenov survived from St. Petersburg University, and Novalevsky from Moscow University." Judging from his other entries and noted, Markovnikov remained upset at this treatment (the "catastrophe" at Moscow University) for the remainder of his life.

The situation surrounding Markovnikov's dismissal was also clear to outsiders. On July 10, 1893, Gavriil Gavrilovich Gustavson (1842–1908, Fig. 2.20) wrote to Nikolai Yakovlevich Dem'yanov (1861–1938, Fig. 2.20)[124] in Moscow, "You, of course, already know that Markovnikov has been outrageously expelled from the University. Lvov and I are thinking about how to convey it to the honorable members of the Chemical Society."

Markovnikov was offered an honorary membership in the University and the chance to retain his research laboratory and continue teaching practical courses in organic chemistry. After a period of bitter reflection, Markovnikov decided to remain at the university. Part of the laboratory of organic chemistry and one laboratory assistant were allocated in his order. In this way, the "Organic Department of Honored Professor V. V. Markovnikov" was formed; it lasted until his death in 1904. Still, there is a visible decrease in his scientific productivity following his ouster. Toward the end of his life, he became much more of a social commentator, responding to societal and political events. In this regard, his article "From Carlsbad on China," a response to the beginning of international intervention in China, is especially interesting. Contrary to the official policy of the tsarist government, Markovnikov wrote in it, "It can be safely shown that most Russians consider it completely unfair to forcefully impose on China the civilization that European missionaries and merchants bring there. Each state, and even

Figure 2.20 (A) G. G. Gustavson. (B) M. I. Konovalov. (C) N. Ya. Dem'yanov. (D) N. M. Kizhner. (E) A. Ye. Chichibabin. *(N.d.). Retrieved June 2020, from (A) http://publ.lib.ru/ARCHIVES/G/GUSTAVSON_Gavriil_Gavriilovich/_Gustavson_G.G.html (B) https://kpi.ua/ru/konovalov-about (C) ДемьяновНиколайЯковлевичhttps://en.wikipedia.org/wiki/Nikolay_Demyanov#/media/File:Демьянов_Николай_Яковлевич.jpg (D) https://za-kadry.tpu.ru/article/1521/10598.htm (E) screen capture from https://library.bmstu.ru/static/Presentations/WW1/ww1c.pdf.*

more so the one that has a two-thousand-year-old culture, has the right to settle down and live the way it wants. This is the main point of view that we can point out to Europe regarding China."

From 1898 to 1901, Markovnikov was involved in arrangements for the celebration of the sesquicentennial of the Chemistry Laboratory at Moscow. An incident that occurred on the evening of February 25, 1901 illustrates his relationship with his students during this era. On that evening, a dinner was held in the Hermitage, which was described by one correspondent (denoted "A Chemist" in the paper) as follows: "All the most prominent representatives of the Faculty of Physics and Mathematics were there, as well as some professors of the Agricultural Institute (now the Agricultural Academy named after K. A. Timiryazev—Platé.), students of Markovnikov—Konovalov [Mikhail Ivanovich Konovalov, 1858—1906, Fig. 2.20] (who had come specially from Kiev—Platé.), Kizhner, [Nikolai Matveevich Kizhner, 1867—1935, Fig. 2.20],[125—128] Kablukov [Ivan Alekseevich Kablukov, 1857—1942], and younger chemists, whose names are still unknown in science, (there were many "factory" chemists from industry)—[V.] Oglobin and others. After the first toast to the health of the hero of the day, a deputation of students, who came to congratulate their teacher, was received. In his reply, Markovnikov, pointing out that perhaps he was sometimes harsh with them because of great nervousness, emphasized the fact that he always saw his most important duty and favorite work as working with students and arousing interest in scientific research in them."[129]

Following his ouster, Markovnikov mentored three graduate students who were to become organic chemists with their own eponymous reactions: Dem'yanov, Kizhner, and Aleksei Yevgen'evich Chichibabin (1871—1945, Fig. 2.20).[130,131] Each of these chemists was elected to Corresponding or Full Membership of the Academy of Sciences of the USSR.

Zelinskii's antipathy toward Markovnikov was shown quite clearly by his treatment of Aleksei Yevgen'evich Chichibabin (1871—1945), Markovnikov's last graduate student. Zelinskii's treatment of Chichibabin, who worked with Markovnikov, his assistant, Konovalov, and his student, Kablukov, was, in this author's opinion, unconscionable. When Markovnikov lost his position, Zelinskii refused to have anything to do with his predecessor's student, and when Chichibabin presented his M. Khim. dissertation for examination, Zelinskii cast one of two negative votes (his colleague, Sabaneev, disparaged Chichibabin as "a self-made man" during the examination). Zelinskii and Sabaneev notwithstanding, Chichibabin's degree was awarded.

In December 1903, Markovnikov attended a meeting of the Russian Physical Chemical Society in St. Petersburg, where he presented an overview of his work. During his return to Moscow, he caught a chill that gradually worsened. Sixteen days later (January 29/February 11, 1901) Markovnikov, the founder of organic chemistry at Moscow University, died.

References

1. Lewis, D. E. Aleksei Yevgen'evich Chichibabin (1871—1945): A Century of Pyridine Chemistry. *Angewandte Chemie - International Edition* **2017,** *56* (33), 9660—9668. https://doi.org/10.1002/anie.201611724.

2. Kablukov, I. A. *Vladimir Vasil'evich Markovnikov (Biograficheskaya Svedeniya i Kratkii Ocherk Nauchnikh Rabot) [Vladimir Vasil'evich Markovnikov (Biographical Information and a Short Account of His Scientific Works;* V. Demakova-Naslend: St. Petersburg, 1905.
3. Kizhner, N. *Lomonosov Collection: Moscow,* 1901; p 114.
4. Markovnikov, V. V. Autographical Data. In *Biographical Dictionary of the Professors and Teachers of the Imperial Kazan University: For the Hundred Years (1804–1904);* Zagoskin, N. P., Ed.; Tipo-lit. Imp. Univ.: Kazan, 1904; pp 424–428.
5. Markovnikov, Vladimir Vasil'evich. In Entsiklopehicheskii slovar Brokgauza i Efrona [The Encyclopedic dictionary of Brockhaus and Efron]; St. Petersburg.
6. Platé, E. F.; Bykov, G. V.; Eventova, M. S. *Vladimir Vasil'evich Markovnikov: Ocherk Zhizni i Deyatel'nosti, 1837–1904 [Vladimir Vasil'evich Markovnikov: An Essay on His Life and Work 1837–1904];* Izd-vo Akademii nauk SSSR: Moscow, 1968.
7. Manolov, K. Vladimir Vasil'evich Markovnikov (1837-1904). In *Velikie Khimiki. V Dvukh tomakh [Great Chemists. In Two Volumes]; Izd-vo,* Vol. 2; "Mir": Moscow, 1986.
8. Wakefield, A. Books, Bureaus, and the Historiography of Cameralism. *Eur. J. Law Econ.* **2005,** *19* (3), 311–320. https://doi.org/10.1007/s10657-005-6640-z.
9. Dmitriev, A. A. Pervaya Seriya Biograficheskikh Ocherkov [The first series of biographical sketches]. In *Biograficheskii ukazatel pamyatnykh deyatelei Permskogo kraya [Biographical index of memorable figures of the Perm region];* Vol. 1; Tipo-Litogr. Pravleniya: Perm Gub., 1902, pp 17–21.
10. *Markovnikov, Vl. Organicheskaya Khimiya, Sistavlennaya Po Lektsiyam Ord. Prof. A. M. Butlerova Stud. Vl. Morkovnikovym [Organic Chemistry Compiled from the Lectures of Ord. Prof. A. M. Butlerov by Student Vl. Morkovnikov] 1859–1860;* Kazan Litografiya: Kazan, 1860.
11. Markovnikov, V. Vospominaniya i Cherty Iz Zhizni i Deyatel'nosti A. M. Butlerova [Memories and traits from the life and work of A. M. Butlerov]. *Zh. Russ. Fiz.-Khim. O-va.* **1887,** *19,* 69–96.
12. Platé, E. F.; Bykov, G. V.; Eventova, M. S. *Vladimir Vasil'evich Markovnikov: Ocherk Zhizni i Deyatel'nosti, 1837–1904 [Vladimir Vasil'evich Markovnikov: An Essay on His Life and Work 1837–1904];* Izd-vo Akademii nauk SSSR: Moscow, 1968; p 10.
13. Butlerow, A. Einiges Uber Die Chemische Structur Der Körper. *Z. Chem. Pharm* **1861,** *4,* 549–560.
14. Butlerow, A. Ueber Die Verwandtschaft Der Mehraffinen Atome. *Z. Chem. Pharm* **1862,** *5,* 297–304.
15. Butlerov, A. *Vvedenie k Polnomu Izucheniyu Organicheaskoi Khimii [Introduction to the Complete Study of Organic Chemistry];* Fan-der-Flit: Kazan, 1864.
16. Butlerow, A. Ueber Die Verschiedenen Erklärungsweisen Einiger Fälle von Isomerie. *Z. Chem. Pharm* **1863,** *6,* 500–534.
17. Boutlerow, A. Sur Les Explications Différents de Quelques Cas d'isomerie. *Bull. Soc. Chim.* **1864,** *1,* 100–128. Paris, N. S.
18. Lewis, D. E. *Introduction to an English Translation, "On the Different Explanations of Certain Cases of Isomerism" by Aleksandr Butlerov, and Primary Documents. "On the Different Explanations of Certain Cases of Isomerism" Mr. A. Boutlerow, Bulletin de La Société Chimique de Paris, Nouvelle Série* **1864,** *Vol. 1; pp 100–128 [Translated by David E. Lewis]. Bull. Hist. Chem. 2015, 40, 9–28.*
19. Morkovnikov, V. *Al'degidy i Ikh Otnoshenie k Alkogolyam i Ketonam [Aldehydes and Their Relationship to Alcohols and Ketones];* Kazan, 1860.
20. Morkownikoff, W. Zur Geschichte Des Acetaldehyds. *Ann. Chem. Pharm.* **1860,** *115,* 327–329.
21. Morkownikoff, W. Ueber Das Allylen. *Z. Chem. Pharm* **1861,** *4,* 466–467.
22. Morkownikoff, W. Ueber Die Acetonsäure. *Ann. Chem. Pharm* **1868,** *146* (3), 339–351.

23. Morkownikoff, W. Ueber Das Quecksilberacetamide. *Z. Chem. Pharm.* **1863,** *6,* 534–537.
24. Morkovnikov, V. Prevrashenie Izobutil'nago Alkogolya v Tetichnyi Psevdobutil'nyi Alkogol [On the question of the mutual influence of atoms in chemical compounds. The conversion of isobutyl alcohol into tertiary pseudobutyl alcohol]. *Zh. Russ. Khim. O-va* **1869,** *1,* 242–247.
25. Lewis, D. E. *Introduction to an English Translation of Markovnikov's First Paper Describing "Markovnikov's Rule," and, Primary Documents. "On the Question of the Mutual Influence of Atoms in Chemical Compounds." V. Morkovnikov. Zhurnal Russkago Fiziko-Khimicheskago Obshchestva* **1869,** *Vol. 1; pp 242–247. Translated by David E. Lewis. Bull. Hist. Chem. 2018, 43, 21-26.*
26. Zakharov, A. V. *Kazanskii Universitet. Khronologiya Stanovleniya Khimicheskoi Laboratorii i Kazanskoi Khimicheskoi Shkoly. Chast I, 1806–1872. [Kazan University: Chronology of the Formation and Development of the Chemical Laboratory and the Kazan School of Chemistry. Part I. 1806–1872];* Kazan University: Kazan, 2011; p 755.
27. Shal'nov, S. Lyudi Rodom Iz Chernorech'ya [People Originally from Chernorechye]. Otkrytoe aktzionernoe obshchestvo Dzherzinskii Vodokanal. Rozhdyonnyi dlya zhizni gorod! [Open Joint Stock Company Dzerzhinsky Vodokanal. Born for the life of the city!].
28. Zakharov, A. V. Khronologiya Stanovleniya Khimicheskoi Laboratorii i Kazanskoi Khimicheskoi Shkoly. Chast I, 1806–1872. In *[Kazan University: Chronology of the Formation and Development of the Chemical Laboratory and the Kazan School of Chemistry. Part I. 1806–1872];* Kazan University: Kazan, 2011; pp 406–410.
29. Klyuchevich, A. S.; Bykov, G. V. *Aleksandr Mikhailovich Zaitzev. 1841–1910;* Kazan University Press: Kazan, 1980; p 18.
30. Lewis, D. E. Magic years in the development of the structural theory of organic chemistry. *Bull. Hist. Chem* **2019,** *44,* 77–91.
31. Morkownikoff, W. Ueber Allylalkoholbromid Und Die Isomerie Einiger Substanzen Der Propylreihe. *Z. Chem. Pharm* **1864,** *7,* 68–77.
32. Butlerow, A. Bemerkungen Zu Den Vorstehenden Abhandlungen. *Z. Chem. Pharm* **1864,** *7,* 79–82.
33. Markovnikov, V. *Ob Izomerii v Organicheskikh Soedineniyakh;* Universitetskaya tipografiya: Kazan, 1865.
34. Couper, A.-S. Sur une nouvelle théorie chimique. *Ann. Chim. Phys. Sér.* **1858,** *53* (3), 469–489.
35. Couper, A. S. On a new chemical theory. *Phil. Trans. R. Soc.* **1858,** 104–116.
36. Couper, A. S. Ueber eine neue chemische Theorie. *Justus Liebigs Ann. Chem.* **1859,** *110* (1), 46–51. https://doi.org/10.1002/jlac.18591100105.
37. Couper, A.-S. Sur Une Nouvelle Théorie Chimique. *C. R. Hebd. Séances Acad. Sci.* **1858,** *46,* 1157–1160.
38. Kekulé, A. Ueber die Constitution und die Metamorphosen der chemischen Verbindungen und über die chemische Natur des Kohlenstoffs. *Justus Liebigs Ann. Chem.* **1858,** *106* (2), 129–159. https://doi.org/10.1002/jlac.18581060202.
39. Kekulé, A. Ueber die s. g. gepaarten Verbindungen und die Theorie der mehratomigen Radicale. *Justus Liebigs Ann. Chem.* **1857,** *104* (2), 129–150. https://doi.org/10.1002/jlac.18571040202.
40. van't Hoff, J.-H. Sur Les Formules de Structure Dans l'espace. *Bull. Soc. Chim.* **1875,** *23,* 295–301. France, N. S.
41. van't Hoff, J. H. *Voorstel Tot Uitbreiding Der Tegenwoordig in de Scheikunde Gebruitke Structuur-Formules in Die Ruimte; Benevens Een Daarmeê Samenhangende Opmerkung Omtrent Het Verband Tusschen Optisch Actief Vermogen En Chemische Constitutie van Organsiche Verbindingen;* Greven: Utrecht, 1874.

42. van't Hoff, J. H. *La Chimie Dans l'espace;* Bazendijk: Rotterdam, 1875.
43. van't Hoff, J. H.; Hermann, F. *Die Lagerung Der Atome Im Raume* (translator); Vieweg: Braunschweig, 1875.
44. Le Bel, J.-A. Sur Les Relations Qui Existent Entre Les Formules Atomiques Des Corps Organiques et Le Pouvoir Rotatoire de Leurs Dissolutions. *Bull. Soc. Chim.* **1874,** *22,* 337–347. Paris.
45. Ramberg, P. J. *Chemical Structure, Spatial Arrangement. The Early History of Stereochemistry, 1874–1914;* Routledge Publishing: New York, 2017.
46. Morkownikoff, W. Ueber die Isobuttersäure und den Pseudopropyl-Aethyl-Aether. *Justus Liebigs Ann. Chem.* **1866,** *138* (3), 361–375. https://doi.org/10.1002/jlac.18661380309.
47. Platé, E. F.; Bykov, G. V.; Eventova, M. S. *Vladimir Vasil'evich Markovnikov: Ocherk Zhizni i Deyatel'nosti, 1837–1904 [Vladimir Vasil'evich Markovnikov: An Essay on His Life and Work 1837–1904];* Izd-vo Akademii nauk SSSR: Moscow, 1968; p 27.
48. Markovnikov, V. V. *Izbrannye Trudy [Selected Works]. Editorial, Articles and Notes by E. F. Platé and G. V. Bykov;* Izd-vo Akademii nauk SSSR: Moscow, 1955; pp 790–791.
49. Markovnikov, V. V. *Izbrannye Trudy [Selected Works]. Editorial, Articles and Notes by E. F. Platé and G. V. Bykov;* Izd-vo Akademii nauk SSSR: Moscow, 1955; pp 830–831.
50. Leicester, H. M. Vladimir Vasil'evich Markovnikov. *J. Chem. Educ.* **1941,** 53–57. https://doi.org/10.1021/ed018p53.
51. Butlerow, A. *Lehrbuch Der Organischen Chemie: Zur Einführung in Das Specielle Studium Derselben;* Quandt & Handel: Leipzig, 1867.
52. Rocke, A. J. *The Quiet Revolution: Hermann Kolbe and the Science of Organic Chemistry;* University of California Press: Berkeley, CA, 1993.
53. Kolbe, H. Zeichen Der Zeit. *J. Prakt. Chem.* **1877,** *15,* 473–477.
54. Markovnikov, V.; Vospominanie, I.; Cherty, I.; Zhizni, I.; Deyatel'nosti, A. M. Butlerova [Reminiscences and traits from the life and work of A. M. Butlerov]. *Zh. Russ. Fiz.-Khim. O-va.* **1887,** *19,* 69–96 (p. 87).
55. Arbuzov, A. E. *Kratkii Ocherk Razvitiya Organicheskoi Khimii v Rossii [A Brief Account of the Development of Organic Chemistry in Russia]; Izd-Vo;* Izd-vo Akademii nauk SSSR: Moscow, 1948; p 113.
56. Lewis, D. E. Static to dynamic: Kazan chemists and the transformation from chemical structure to reaction regiochemistry. *Angew. Chem. Int. Ed.* **2019,** *58* (12), 3694–3705. https://doi.org/10.1002/anie.201804036.
57. Zakharov, A. V. *Kazanskii Universitet. Khronologiya Stanovleniya Khimicheskoi Laboratorii i Kazanskoi Khimicheskoi Shkoly. Chast I, 1806–1872. [Kazan University: Chronology of the Formation and Development of the Chemical Laboratory and the Kazan School of Chemistry. Part I. 1806–1872];* Kazan University: Kazan, 2011; p 757.
58. *Tageblatt Der 41. Versammlung Deutscher Naturforscher Und Ärzte in Frankfurt Am Main Vom 18. Bis 24;* C. Naumann's Press: Frankfurt a./M, 1867.
59. Sorokin, V. Stud. O Deistvii Iodovodoroda Na Nekotorie Galoidnye Soedinenie Efilena i Propilena [Concerning the action of iodine hydrogen on some halide compounds of ethylene and propylene]. *Zh. Russ. Khim. O-va* **1870,** *2,* 212–219.
60. Markovnikov, V. Tupolev. Izomer Pirovinnoi Kisloty [An isomer of pyrotartaric acid]. *Zh. Russ. Khim. O-va.* **1872,** *4,* 45–45.
61. Markownikoff, W. Die Dritter Isomere Pyroweinsaüre. *Ber. Dtsch. Chem. Ges.* **1872,** *6,* 1440–1440.
62. Markovnikov, V. Ob Uglevodorode C7H14 i Nekotorykh Ego Prevrashcheniyakh [On the hydrocarbon C7H14 and some of its derivatives]. *Zh. Russ. Khim. O-va* **1871,** *3,* 206–212.

63. Markovnikov, V. Predvaritel'noe Soobshchenie o Produktakh Dikhlorgidrina [Preliminary report on the products of dichlorohydrin]. *Zh. Russ. Khim. O-va* **1871,** *3,* 212–213.
64. Sorokin, V. O. Stroenii Khloriodistago Propilena [On the structure of chloroiodopropane]. *Zh. Russ. Khim. O-va.* **1871,** *3,* 192–195.
65. Krivaksin, V. K. Voprosu o Stroenii Efilena [Towards the question of the structure of ethylene]. *Zh. Russ. Khim. O-va.* **1871,** *3,* 196–199.
66. Krivaksin, V. O. Tak Nazyvaemom Dikhloratsetone [Concerning so-called dichloroacetone]. *Zh. Russ. Khim. O-va.* **1871,** *3,* 200–204.
67. Krivaksin, V. O. Razlozhenii Bromistago Efilena s Vodoyu [On the decomposition of ethylene bromide with water]. *Zh. Russ. Khim. O-va.* **1871,** *3,* 204–206.
68. Markovnikov, V. O. Voprosu o Russkoi Khimicheskoi Nomenklature [On the question of Russian chemical nomenclature]. *Zh. Russ. Khim. O-va.* **1871,** *3,* 65–69.
69. Markovnikov, V. Kisloty Oksiizocaprilovaya, Oksiizovaler'yanovaya i Novyi Predstavitel Ryada Ketonov—Dipsevdopropil'nyi Keton [Hydroxyisocaprylic acid, hydroxyvaleric acid and a new representative of a class of ketones—diisopropyl ketone]. *Zh. Russ. Khim. O-va* **1870,** *2,* 92–100.
70. Lewis, D. E. Feuding Rule-Makers: Vladimir Vasil'evich Markovnikov (1838–1904) and Aleksandr Mikhailovich Zaitsev (1841–1910). A commentary on the origins of Zaitsev's rule. *Bull. Hist. Chem.* **2010,** *35,* 115–124.
71. Markovnikov, V. Dikhlorgidrin i Produkty Ego Okisleniya [Dichlorohydrin and the products of its oxidation]. *Zh. Russ. Khim. O-va. i Fiz. O-va.* **1873,** *5,* 309–320.
72. Markownikoff, W. Mittheilung Aus Dem Chemisches Universitäts-Laboratorium in Odessa. *Ber. Dtsch. Chem. Ges* **1873,** *6,* 1175–1178.
73. Markovnikov, V.; Demchenko, S.; Soobshcheniya, V. Morkovnikova [Reports of V. Morkovnikov]. Ob Usloviyakh Polimerizatsii Izobutirinago al'degida [On the conditions for the polymerization of isobutyraldehyde]. *Zh. Russ. Fiz.-Khim. O-va.* **1874**, *6,* 83–85.
74. Markovnikov, V.; Yelisafov, S.; Soobshcheniya, V. Morkovnikova [Reports of V. Morkovnikov]. O Tsetene [On cetene]. *Zh. Russ. Fiz.-Khim. O-va.* **1874**, *6,* 85–88.
75. Platé, E. F.; Bykov, G. V.; Eventova, M. S. *Vladimir Vasil'evich Markovnikov: Ocherk Zhizni i Deyatel'nosti, 1837–1904 [Vladimir Vasil'evich Markovnikov: An Essay on His Life and Work 1837–1904];* Izd-vo Akademii nauk SSSR: Moscow, 1968.
76. Mulder, C. J. Acide Xantho-Protéique. *Bull. Sci. Phys. Nat. Néerlande* **1838,** 152–155.
77. Mulder, C. J. Sur La Décomposition Des Matières Animales Par Les Alcalis. *Bull. Sci. Phys. Nat. Néerlande* **1838,** 166–168.
78. Mulder, C. J. Sur la chondrine. *Bull. Sci. Phys. Nat. Néerlande* **1838,** 77–79.
79. Mulder, C. J. Sur La Composition de Quelques Substances Animales. *Bull. Sci. Phys. Nat. Néerlande* **1838,** 105–119.
80. Mulder, C. J. Sur Le Sucre de Gélatine et La Leucine. *Bull. Sci. Phys. Nat. Néerlande* **1838,** 145–149.
81. Mulder, C. J. Action de l'acide Hydrochlorique Sur La Protéine. *Bull. Sci. Phys. Nat. Néerlande* **1838,** 163–166.
82. Figurovskii, N. A.; Bykov, G. V.; Komanova, T. A. *Khimiya v Moskovskom Universitete Za 200 Let. Let [Chemistry at Moscow University to 200 Years];* Izd-vo Mosk. Un-ta.: Moscow, 1955.
83. Markovnikov, V. Produkty Okisleniya α-Oksibutirinovoi Kisloty [Products of the Oxidation of α-Hydroxybutyric Acid. *Zh. Russ. Fiz.-Khim. O-va.* **1875,** *7,* 75–80.
84. Markovnikov, V. Angidrid Itakonovoi Kisloty [The anhydride of itaconic acid]. *Zh. Russ. Fiz.-Khim. O-va* **1880,** *11,* 447–449.

85. Markovnikov, V. Normal'naya Pirovinnaya Kislota. (Predvaritel'naya Soobshchenie) [Normal pyrotartaric acid (preliminary communication)]. *Zh. Russ. Khim. O-va. i Fiz. O-va* **1876,** *8,* 233–233.
86. Markovnikov, V. Nekotoriya Proizvodniya Normal'noi Pirovinnoi Kisloty [Some derivatives of normal pyrotartaric acid]. *Zh. Russ. Khim. O-va. i Fiz. O-va.* **1877,** *9,* 283–286.
87. Frankland, E.; Duppa, B. F. Ueber Die Einwirkungvon Zink Auf Eine Mischung von Jodmethyl Und Oxalsäurem Methyl. *Ann. Chem. Pharm* **1865,** 80–84.
88. Frankland, E.; Duppa, B. F. Notizen Aus Untersuchungen Über Die Säuren Der Milchsäure-Reihe. II. Ueber Die Einwirkung Des Zinks Auf Eine Mischung von Jodäthyl Und Oxalsaurem Methyl. *Ann. Chem. Pharm.* **1865,** 25–29. https://doi.org/10.1002/jlac.18651350103.
89. Frankland, E. Vorläufige Notiz über die Synthese der Leucinsäure. *Justus Liebigs Ann. Chem.* **1863,** *126* (1), 109–113. https://doi.org/10.1002/jlac.18631260115.
90. Markovnikov, V.; Drobyazgin Soobshcheniya, V. Markovnikova [Communications from V. Markovnikov].1. O Nekotorykh Prevrashcheniyakh Etilkrotonovoi Kisloty. [On some transformations of ethylcrotonic acid]. *Drobyagin. Zh. Russ. Khim. O-va. i Fiz. O-va* **1874,** *6,* 47–51.
91. Markovnikov, V.; Egiz, Y.; Soobshcheniya, V. Markovnikova [Communications from V. Markovnikov]. 2. Nepreryvnyi Sposob Polucheniya Nekotorykh Slozhnykh Efirov [A continuous method for producing certain esters]. *Zh. Russ. Khim. O-va. i Fiz. O-va* **1874,** *6,* 51–54.
92. Markovnikoff, V. Sur Les Lois Qui Régissent Les Réactions de l'addition Directe. *C. R. Hebd. Séances Acad. Sci* **1875,** 776–779.
93. Markovnikoff, V. Sur Les Lois Qui Régissent Les Réactions de l'addition Directe. *C. R. Hebd. Séances Acad. Sci* **1875,** *81,* 668–671.
94. Markovnikoff, V. Sur Les Lois Qui Régissent Les Réactions de l'addition Directe. *C. R. Hebd. Séances Acad. Sci* **1875,** *87581,* 728–730.
95. Markovnikov, V. O. Zakonakh Obrazovaniya Pryamikh Soedinenii Nepredel'nymi Organicheskimi Chastitsami [On the laws of the direct formation of compounds with unsaturated organic groups]. *Zh. Russ. Khim. O-va. i Fiz. O-va* **1876,** *8,* 16–29.
96. Pavlov, D. O. Dimetilizobutilkarbinole i Novom Geptilene Iz Nego Poluchaemom [On dimethyl isobutyl carbinol and a new heptylene derived from it]. *Zh. Russ. Khim. O-va. i Fiz. O-va.* **1874,** *6,* 170–172.
97. Markovnikov, V. Po Povodu Novykh Dimetil-Izobutil-Karbinola i Proiskhodyashchago Iz Nego Geptilena [Regarding the new dimethyl isobutyl carbinol and the heptylene drived from it]. *Zh. Russ. Khim. O-va. i Fiz. O-va.* **1875,** *7,* 32–33.
98. Markovnikov, V. Ob Atsetone v Moche Diabetikov [On acetone in the urine of diabetics]. *Zh. Russ. Khim. O-va. i Fiz. O-va.* **1876,** *8,* 14–16.
99. Markovnikov, V. Ob Atsetone v Moche Diabetikov [On acetone in the urine of diabetics]. *Zh. Russ. Khim. O-va. i Fiz. O-va.* **1876,** *8,* 286–287.
100. Markownikoff, W. Das Aceton Im Harne Der Diabetiker. *Justus Liebigs Ann. Chem.* **1876,** *182,* 362–364.
101. Markovnikov, V. Ob Opredelenii Teina v Chae [On the determination of theine in tea]. *Zh. Russ. Khim. O-va. i Fiz. O-va* **1876,** *8,* 226–232.
102. Markovnikov, V. Desinfektsiya i Rumynii i Bolgarii (Otchet Obshchestvu Popecheniya o Bol'nykh i Ranenykh Voinakh) [Disinfection of Both Romania and Bulgaria (Report of the Society for the Care of Sick and Wounded Warriors)]. *Zdorov'e* **1878,** *4* (No. 79), 5–8.
103. Markovnikov, V. Desinfektsiya i Rumynii i Bolgarii (Otchet Obshchestvu Popecheniya o Bol'nykh i Ranenykh Voinakh) [Disinfection of Both Romania and Bulgaria

(Report of the Society for the Care of Sick and Wounded Warriors)]. *Zdorov'e* **1878,** *4* (30), 31–33.
104. Markovnikov, V. Desinfektsiya i Rumynii i Bolgarii (Otchet Obshchestvu Popecheniya o Bol'nykh i Ranenykh Voinakh) [Disinfection of Both Romania and Bulgaria (Report of the Society for the Care of Sick and Wounded Warriors)]. *Zdorov'e* **1878,** *4* (No. 81), 54–55.
105. Markovnikov, V. V. *Izbrannye Trudy [Selected Works]. Editorial, Articles and Notes by E. F. Platé and G. V. Bykov;* Izd-vo Akademii nauk SSSR: Moscow, 1955; p 850.
106. Markovnikov, V. V. *Obshchedostupnoe Praktiheskoe Rukovodstvo k Desinfekvii Dlya Chastykh Lits, Vrachei, Uchebnykh Zavedenii, Gospitalei, Fabrik i Zheleznykh Dorog. [A Generally Accessible, Practical Guide to Disinfection for Individuals, Doctors, Educational Institutions, Hospitals, Factories, and Railways]; Moscow,* 1879.
107. Markovnikov, V.; Krestovnikov, G. Gomoitakonovaya Kislota [Homoitaconic acid]. *Zh. Russ. Fiz.-Khim. O-va.* **1879,** *11,* 255–256.
108. Markovnikov, G. Krestovnikov. Tetrilendikarbonovaya Kislota (Gomoitakon-Ovaya) [Tetrylenedicarboxylic (homoitaconic) acid (cyclobutane-1,1-dicarboxylic acid—del)]. *Zh. Russ. Fiz.-Khim. O-va* **1880,** *12,* 449–461.
109. Markownikoff, W.; Krestownikoff, A. Aus dem chemischen Universitätslaboratorium Moscau von Prof. Markownikoff. Tetrylendicarbonsäure (Homoïtakonsäure). *Justus Liebigs Ann. Chem.* **1881,** *208* (3), 333–349. https://doi.org/10.1002/jlac.18812080306.
110. Schorlemmer, C.; Dale, R. S. Ueber Das Suberon. *Ber. Dtsch. Chem. Ges* **1874,** *7,* 806–808.
111. Dale, R. S.; Schorlemmer, C. Ueber Suberinsaüre Und Azelaïnsäure. *Justus Liebigs Ann. Chem.* **1879,** *199,* 144–150.
112. Spiegel, A. Ueber das Suberon. *Justus Liebigs Ann. Chem.* **1882,** *211* (1), 117–120. https://doi.org/10.1002/jlac.18822110106.
113. Markownikoff, W. Derivés de l'heptaméthylène. *C. R. Hebd. Séances Acad. Sci.* **1890,** *110,* 466–468.
114. Markownikoff, W. Derivés de l'heptaméthylène. *C. R. Hebd. Séances Acad. Sci* **1892,** *115,* 462–464.
115. Berthelot, M. Nouvelles applications des méthodes de réduction en chimie organique. *Bull. Soc. Chim. Paris* **1867,** *7,* 53–65.
116. Berthelot, M. Méthode Universelle Pour Réduire et Saturer d'hydrogène Les Composées Organiques. *Bull. Soc. Chim.* **1868,** *9,* 8–31. Paris.
117. Markovnikov, V.; Oglobin, V. Issledovanie Kavkaskoi Nefti [An investigation of the caucasus oils]. *Zh. Russ. Fiz.-Khim. O-va.* **1881,** *13,* 179–183.
118. Markovnikov, V.; Oglobin, V. Issledovanie Kavkaskoi Nefti [An investigation of the caucasus oils]. *Zh. Russ. Fiz.-Khim. O-va.* **1882,** *14,* 36–40.
119. Markovnikov, V.; Oglobin, V. Issledovanie Kavkaskoi Nefti [An investigation of the caucasus oils]. *Zh. Russ. Fiz.-Khim. O-va* **1883,** *15,* 307–354.
120. Beilstein, F.; Kurbatow, A. Ueber die Natur des kaukasischen Petroleums. *Ber. Dtsch. Chem. Ges.* **1880,** *13* (2), 1818–1821. https://doi.org/10.1002/cber.188001302143.
121. Markovnikov, V. Issledovaniya Iz Ryada Naftenov Ili Polymetilenov [Studies from a series of naphthenes and polymethylenes]. *Zh. Russ. Fiz.-Khim. O-va.* **1890,** *22,* 275–279.
122. Markownikoff, W.; Spady, J. Zur Constitution Der Kohlenwasserstoffe, CnH2n, Des Kaukasischen Petroleums. *Ber. Dtsch. Chem. Ges.* **1887,** *20,* 1850–1853. https://doi.org/10.1002/cber.188702001417.
123. Zefirova, O. N.; Zefirov, N. S. Vladimir Vasilievich Markovnikov and Moscow University. *Vestn. Mosk. Univ.* **2004,** *45* (5), 350–356. Seriya 2 Khimiya.

124. Lewis, D. E. *Early Russian Organic Chemists and Their Legacy;* Springer: Heidelberg, 2012; pp 103–105.
125. Lewis, D. E. Disability, Despotism, Deoxygenation - From Exile to Academy Member: Nikolai Matveevich Kizhner. *Angew. Chem. Int. Ed.* **2013,** *52* (45), 11704–11712. https://doi.org/10.1002/anie.201303165.
126. Suntsov, V.; Lewis, D. E. A Century of Base-Promoted Decomposition of Hydrazones: The Tomsk Years of Nikolai Matveevich Kizhner (1867–1935). *Bull. Hist. Chem* **2014,** *39,* 43–52.
127. Suntsov, V.; Lewis, D. E. After the Revolution: Nikolai Matveevich Kizhner (1867–1935) in Soviet Moscow. *Bull. Hist. Chem.* **2017,** *42,* 46–56.
128. Lewis, D. E. *The Wolff-Kishner and Related Reactions;* Discovery and Development, 2019.
129. Chemist, A. *Vestnik. khimiheskoi tekhnologii i khimicheskoi praktiki* **1901,** 689.
130. Cerkovnikov, E. Aleksei Evgen'evich Chichibabin. *J. Chem. Educ.* **1961,** *38* (12), 622–624.
131. Lewis, D. E. Aleksei Yevgen'evich Chichibabin (1871–1945): A Century of Pyridine Chemistry. *Angew. Chem. Int. Ed.* **2017,** *56* (33), 9660–9668. https://doi.org/10.1002/anie.201611724.

CHAPTER THREE

Markovnikov's rule: history and development

The first decades: early challenges to the validity of the rule

Two accounts of the history of Markovnikov's Rule were published in 2019, the sesquicentennial year of Markovnikov's initial publication.[1] The paper by Lewis[2] describes how Markovnikov's Rule originated from his studies in Organic Structural Theory during his master's and doctoral research. The paper by Beletskaya and Nenajdenko[3] describes its application (and misapplication) in the period following its initial disclosure.

Following its initial publication in 1869,[1] Markovnikov's Rule slowly came under general examination. At Markovnikov's insistence and over Butlerov's objections, it had been published only in Russian, so the rule remained largely unknown in the West. To confuse things further, Markovnikov's German paper containing the Rule,[4] which he published in the *Annalen* in 1870, was largely devoted to a discussion of his research into the nature of the isomeric hydroxybutyric acids. The part of the paper devoted to the regiochemistry of the addition to alkenes does not appear until the 29th page of this 32-page paper, which makes it appear to be an afterthought. This was really a simplistic evaluation of a much more complex situation. Markovnikov's mentor and close friend, Butlerov, had published his clarification of Structural Theory in 1861, but he was not receiving what Markovnikov viewed as fair credit for his contribution to the theory; Markovnikov saw this as a mark of disrespect for his mentor and his work. Consequently, Markovnikov's German publication was focused on advocating for his mentor. Markovnikov's first Russian-language publication, in the *Zhurnal Russkago Khimicheshago Obshchestva*, targeted a readership that was already well aware of Butlerov's contributions to the theory. With this audience, he did not need to advocate for Butlerov, so he published his Rule as the sole subject of the paper.[5]

Shortly after the appearance of Markovnikov's *Annalen* paper of 1870, the Belgian chemist Louis Henry (1834–1913, Fig. 3.1) published a paper challenging his result in the addition of hypochlorous acid to propylene.[6]

Addition, Elimination and Substitution: Markovnikov, Hofmann, Zaitsev and Walden
ISBN: 978-0-12-821027-7
https://doi.org/10.1016/B978-0-12-821027-7.00001-3

Figure 3.1 (left) Louis Henry. (right) Laurent-Guillaume de Koninck. *(N.d.). Image of Henry retrieved July 2020 from https://commons.wikimedia.org/wiki/File:Louis_Henry_(1834-1913).jpg Image of de Koninck retrieved July 2020 from https://commons.wikimedia.org/wiki/File:De_Koninck_Laurent-Guillaume_(1809-1887).jpg*

It was partly in response to Henry's paper that Markovnikov made a final clarification of his rule in three papers in the *Comptes rendus*.[7–9] This set off a brief series of intense polemical exchanges between the two men[10–12] over the period 1874–76. The question was finally resolved when American chemist Arthur Michael (1852–1942, Fig. 3.2) verified Markovnikov's observations.[13,14]

Louis Henry

Louis Henry[15] was born in Marche-en-Famenne in the Walloon region of Belgium. In October 1851, at the age of 16 years and 11 months, he entered the Université de Louvain. In August 1853, he graduated as Candidate of Natural Sciences (at the head of a class of 23), and just 2 years later, he received the degree of Docteur en Sciences Naturelles. Following his success in the doctoral examinations (no dissertation was required at the time), the Chair of the Examining Committee, Laurent-Guillaume de Koninck (1809–87, Fig. 3.1), invited him to carry out postdoctoral study in his private laboratory. There, Henry learned laboratory techniques in organic chemistry and wrote his first paper, on organic radicals, which was presented

Figure 3.2 Arthur Michael, 1896. *Courtesy of Digital Collections and Archives, Tufts University. (n.d.). Retrieved from https://dl.tufts.edu/concern/images/ws859r46q*

before the Académie Royale de Belgique, in 1857 just before his departure for Germany. This paper was subsequently published in the *Mémoires Couronnés* in 1859.[16]

At de Koninck's urging, Henry traveled to Giessen in 1857 to study in the Liebig laboratory, now under the direction of Heinrich Will. He spent his time at Giessen learning the techniques of combustion analysis. He spent three semesters at Giessen and then returned to Louvain in June 1858, bringing with him the first paper based on his work in Giessen. This work was published in the *Bulletins de l'Académie Royale*.[17] In 1863, Henry was appointed to the Chair of General Chemistry at Louvain, and in 1867, he completed his three-volume *Précis de Chimie Générale élémentaire*,[18] the first Belgian chemistry textbook based on Structural

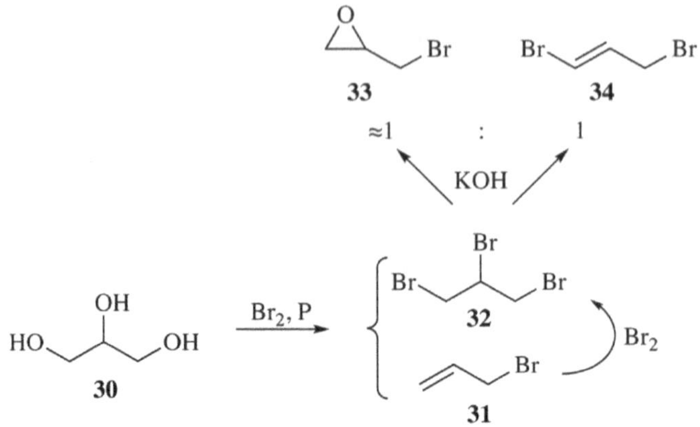

Scheme 3.1 Henry's research on glycerol derivatives.

Theory. In his career, Henry published some 200 works, which included systematic studies of glycerol and allyl halide derivatives (Scheme 3.1).[19] He recognized that the vapor pressure of organic liquids tended to decrease with increasing molecular weight, but also noted the exception for "associated liquids."

Arthur Michael

Arthur Michael (Fig. 3.2)[20–25] was born to an affluent real estate investor in Buffalo, New York, and attended the Classical School there. Here he had his first encounters with chemistry at his home, where he had a personal laboratory, and at the Classical School in the form of private instruction in the discipline. In his late teens, he contracted what was then called "brain fever," which may correspond to today's scarlet fever or meningitis. This illness prevented his entry into Harvard College, but the time off did allow him to spend time with literature and art (he was an avid art collector, of oriental art, in particular), before his return to chemistry. His career also shows that he was an inveterate traveler.

Although he had neither a degree nor the formal chemical training required, he was nevertheless admitted to Berlin University in 1871 and permitted to enter the laboratory of A. W. Hofmann (1818–92, Fig. 3.3). A year later, he transferred to the Heidelberg laboratory of R. W. E. Bunsen (1811–99, Fig. 3.3), where he honed his synthetic skills and published the first synthesis of helicin, a natural glycoside.[26] The method he developed, which he again exploited 3 years later[27] in the synthesis of other glycosides, is still used today. In 1875, he returned to Hofmann's laboratory in Berlin,

(a) (b) (c) (d)

Figure 3.3 (A) A. W. Hofmann late-19th century. (B) R. W. E. Bunsen, image dated as 1860, but almost certainly 2–3 decades later. (C) D. I. Mendeleev in 1897 and (D) C. A. Wurtz ca. 1870. *(N.d.). (A) August Wilhelm von Hofmann. Process print, 1902. Credit: Wellcome Collection.Attribution 4.0 International (CC BY 4.0). (B) Portrait of R.W.E. Bunsen, half-length to right in oval. Credit: Wellcome Collection. Attribution 4.0 International (CC BY 4.0). (C) Image of D. I. Mendeleev retrieved from https://commons.wikimedia.org/wiki/File:DIMendeleevCab.jpg not subject to any known restrictions and (D) Image of C. A. Wurtz retrieved from https://commons.wikimedia.org/wiki/File:Charles-Adolphe_Wurtz.jpg not subject to any known restrictions.*

then in 1879, the last year of his first sojourn in Europe, he spent time with Adolphe Wurtz (1817–84, Fig. 3.3) in Paris and Dmitrii Mendeleev (1834–1907, Fig. 3.3) in St. Petersburg.

Michael again attained a position for which he did not hold the formal qualifications when he returned to America in 1880: despite still having no formal degree, he was appointed as an Assistant in Chemistry at Tufts College. A year later, he was promoted to Professor. Tufts University did award him two honorary degrees: the A.M. in 1882 and the Ph.D. in 1890. In 1889, he married one of his students, Helen Abbot, who was one of the few women in the field and her husband's intellectual equal. They traveled the world for the next 18 months.

Michael became Head of the Chemical Laboratory at Clark University in Worcester, Massachusetts, when the couple returned to the United States. A year later, the Michaels moved to the Isle of Wight, where Michael built a private laboratory. They worked here for 3 years before returning to the United States, where Michael resumed his Professorship at Tufts. He retired in 1904, after his wife's death from influenza. While in retirement, Michael continued his research at a new private laboratory that he set up near Boston. In 1912, Harvard University persuaded him to reemerge from retirement to become Professor without lecturing duties. He supervised graduate research at Harvard until 1936, when he retired a second time.

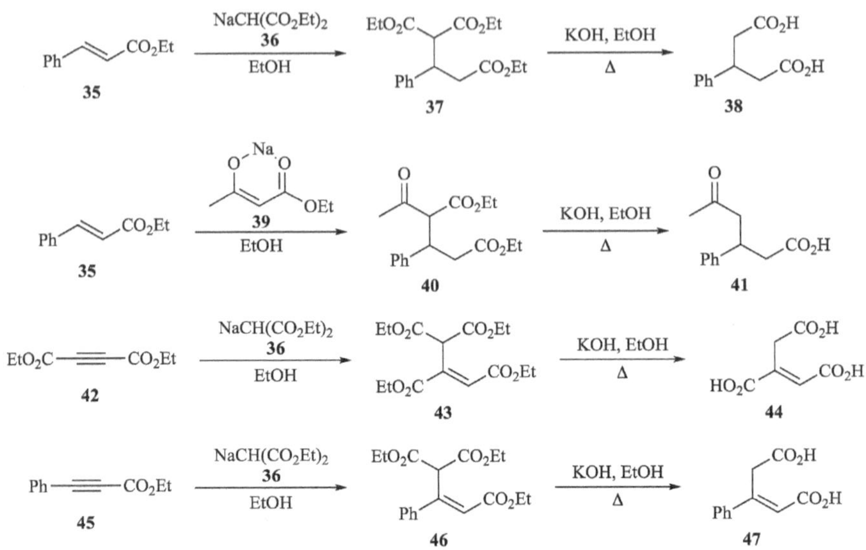

Scheme 3.2 The earliest nucleophilic addition reactions of malonate and acetoacetate ester anions to conjugated esters carried out by Michael.

Michael's lasting contribution to organic chemistry is his eponymous reaction (Scheme 3.2), which he first reported in a series of four papers published between 1887 and 1894.[28–31] The value of this reaction as a synthetic method is clear from the number and length of review articles on the subject.[32–40] His credibility on the international stage was cemented by his critique of van't Hoff's stereochemical theory.[41] For the purposes of this chapter, his reputation for insightful and incisive logic almost certainly lent additional weight to his two lengthy theoretical papers on laws and their application in organic chemistry.[13,14] In these two papers, which run to more than 180 pages total, Michael came down firmly on the side of Markovnikov in the controversy with Henry.

Markovnikov's predictions encouraged other organic chemists to investigate additions of hydrogen halides to unsymmetrical alkenes under a variety of conditions. The addition of hydrogen bromide to allyl bromide became particularly popular subject for investigation, partly because it gave two products, but in inconsistent ratios. Systematic work by a number of investigators[42–54] began to delineate the factors affecting the regiochemistry of the addition reaction. As we see in Scheme 3.3, the amounts of the two dibromides vary with the reaction temperature, with higher temperatures favoring the formation of 1,3-dibromopropane.

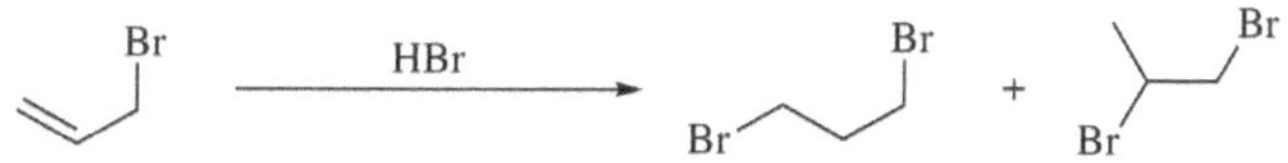

31 **35** **36**

Germont	conc.	100 °C	2:1
Lermontoff	conc.	100 °C	1:3
	conc.	165-170 °C	≈1:0
Reboul	conc.	100 °C	>3:1
	dil.	100 °C	≤1:1

Scheme 3.3 Additions of hydrogen bromide to allyl bromide.

Yuliya (Julia) Vsevolodovna Lermontova

Among the students of Markovnikov who made contributions to this effort[42–54] was Yuliya Vsevolodovna Lermontova (1846–1919, Fig. 3.4), the first Russian woman to obtain a doctoral degree in chemistry.[55] She had received a good education at home, but her efforts to enter the Petrovskii Agricultural Institute were unsuccessful, despite her acquaintance with Mendeleev. Consequently, she moved to Berlin in the company of the mathematician, Sof'ya Kovalevskaya (1850–91), who, as a married woman (although her marriage was actually a "marriage of convenience"), could chaperone her. Lermontova entered the laboratory of Bunsen and Kirchoff

Figure 3.4 (Left) Yuliya Lermontova and (Right) Morris Kharasch. *(N.d.). Public domain image of Lermontova retrieved July 2020 from https://en.wikipedia.org/wiki/Julia_-Lermontova#/media/File:Julia_Lermontova_(1846-1919).jpg. Image of Kharasch is courtesy of the Special Collections Research Center, University of Chicago Library. https://photoarchive.lib.uchicago.edu/db.xqy?show=browse8.xml|1560*

in 1869 where she studied in Bunsen's laboratory, learning his methods for the analysis of metals in ores and joining his research effort in the separation of the platinum metals. In 1871, she moved to Hofmann's laboratory in Berlin, where she carried out her dissertation research on methylene halides and also worked on aromatic amines; the subject of her first publication[56] was the chemistry of 4,4′-diaminoazobenzene.

Neither Heidelberg nor Berlin would confer a doctoral degree on a woman, so she submitted her Ph.D. dissertation, "Zur Kenntnis der Methylenverbindungen," to Göttingen, a more liberal university where the degree was awarded in 1874. On her return to Russia, she first visited St. Petersburg, where she was feted by Mendeleev, Butlerov, and the chemistry community. She then moved on to Moscow, where she joined Markovnikov's laboratory. In 1878, she moved to St. Petersburg, where she accepted Butlerov's invitation to join his research group. Here, she worked on the chemistry of branched olefins. The same year and the following year, she published three papers on the alkylation of isobutylene with tert-butyl iodide[57–59] to give dimeric and trimeric alkenes by heating the mixture with lime (Scheme 3.4), and, at the same time, became Godmother to Sof'ya Kovalevskaya's daughter, Fufa. Markovnikov began his work on the Caucasus crude oil in 1880, and he was able to attract Lermontova back to Moscow at that time. She worked until 1886, at which time she retired to her family estate. She survived World War I, and when the new Soviet government tried to nationalize her estate, she was able to prevent it. She died on her estate in 1919.

Morris Selig Kharasch

The addition of hydrogen bromide to propylene was well known as being unpredictable. The origin of this problem for synthetic chemists was first satisfactorily explained by Russian-American organic chemist Morris Selig Kharasch (1895–1957, Fig. 3.4)[60] in 1933,[61] which was followed by eight more papers on the same topic within a year.[62–68]

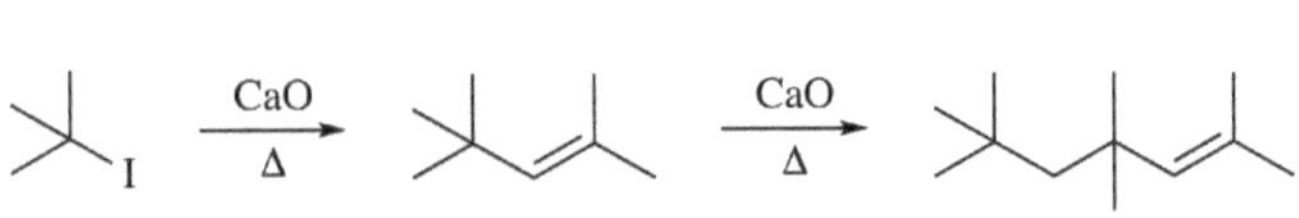

37 38 39

Scheme 3.4 Lermontova's studies of the oligomerization of isobutylene with tert-butyl iodide and metal oxides.

Kharasch was born in Kremenetz, in the Ukraine, and despite his family's well-off status, he moved to the United States in 1908 to take advantages of the educational opportunities there. An older brother, who had emigrated earlier, took care of the 13-year-old émigré when he arrived in Chicago. He entered the University of Chicago in 1913 and graduated with his B.S. in 1917. During World War I, he spent time in the U.S. Army, working in the Chemical Warfare Service, while working on his Ph.D. research concurrently. He earned his Ph.D. in 1919 for a dissertation in organomercury synthesis[69] under the mentorship of Swiss-born American chemist, Jean Felix Piccard (1884–1963).

Following his graduation, Kharasch remained in Chicago on a National Research Council Fellowship until 1922, when he accepted an appointment as Associate Professor of Chemistry at Maryland. Two years later, he was appointed Professor in the same department. In 1928, he returned to his alma mater as Associate Professor; once again, he received rapid promotion to Professor (in 1930). From 1935 until the end of his life, he held endowed professorships at Chicago.

Much of Kharasch's work was involved in the study of free radicals. Beginning with his Ph.D work, he studied organomercury compounds, regularly publishing papers on the topic for the period 1921–36.[70–78] His work with organometallic compounds expanded to another series of papers on the mechanism of the addition of Grignard reagents and organolithiums to carbonyl compounds, in which he proposed a single-electron transfer mechanism.

The work for which Kharasch is best remembered in his work on the "peroxide effect" in the additions of hydrogen halides to alkenes. He discovered that in the absence of oxygen or peroxides, the addition of hydrogen bromide proceeded with normal regiochemistry, as predicted by Markovnikov's Rule, whereas their presence dramatically accelerated the rate of the reaction, as well as reversing the "normal" regiochemistry. The addition of hydrogen bromide to propylene under a wide range of experimental conditions (Scheme 3.5) provided the first systematic evidence for the effects of peroxides on the regiochemistry of the addition.[62]

It was Kharasch who also proposed a mechanism for the "abnormal" addition reaction based on free radicals as an offshoot of his observations on the peroxide effect.

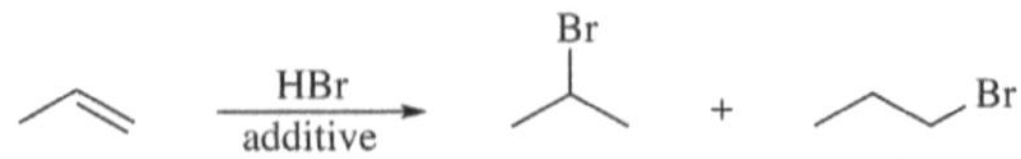

40 41 42

	Additive	41		42
	Ph_2NH	100	:	0
	BuNC	100	:	0
	MeC_6H_4SH, $MnCl_2$	100	:	0
(room temp.)............................	$(PhCO_2)_2$	87	:	13
(−78 °C)................................	$(PhCO_2)_2$	30	:	70
(room temp.; HBr added v. slowly)	ascaridole	2	:	98
(−78 °C; HBr added rapidly).........	ascaridole	0	:	100

Scheme 3.5 The effect of peroxides and antioxidants on the regiochemistry of the addition of hydrogen bromide to propylene.

What is in a name? Markovnikov and anti-Markovnikov

Much of the controversy that has arisen in the last two decades within the chemical education community over Markovnikov's Rule has been focused on its pedagogical value or lack thereof. The origins of this problem can be traced, in this author's view, to the two mechanisms for addition of hydrogen bromide to propylene.

In 1919, Julius Cohen made it clear that Markovnikov's interpretation was in full accord with the current state of theory at that time[79]:

"This principle, taken in conjunction with that of maximum neutralisation, will determine the course of the additive process. The latter may take the form of what is termed by Michael the negative-positive rule, where the maximum neutralization is attained by the electronegative atom or group of the addendum attaching itself to the more electropositive atom of the unsaturated molecule, and the more electropositive atom to the more electronegative part of the molecule.[80] For example, in propylene, $CH_3.CH:CH_2$, the electropositive radical CH_3 will influence the central more than the end ethylenoid carbon by rendering it more electropositive. Consequently, on the addition of hydrogen iodide, the electronegative iodine atom will be mainly directed to the central carbon. This proves to be the case."

" … From the above considerations, the rule laid down by Markownikoff (3)[4] that the halogen of a halogen acid attaches itself to the least hydrogenated carbon, though by no means free from exceptions, will be readily understood; for the least hydrogenated carbon will usually be situated next to the strongest hydrocarbon radical."

In his 1935 textbook, Howard Lucas described the addition in terms that are exactly the same as those used by Markovnikov.[81] A decade later, Edgar Wertheim also used the "abnormal addition" vocabulary for what today is "anti-Markovnikov addition" in the second edition of his textbook.[82] Even as late as 1956, Royals used the Markovnikov language in his *Advanced Organic Chemistry*.[83] Lucas' language, which is typical, is given here:

"In general, unsymmetrical addition to olefins produces largely that isomer in which the negative radical (Cl, Br, SO_4H) is attached to the carbon joined to the least number of hydrogen atoms (Markownikoff's rule). This statement is qualitative only, for the proportion is sometimes reversed by the use of different solvents, or by the presence of oxygen."

In the early 1930s, Frank Whitmore at Pennsylvania State College (Fig. 3.5) was carrying out his seminal researches into the nature of the intermediates in the Wagner–Meerwein rearrangement. As part of this research, he established the ionic mechanism for addition and proposed the currently accepted mechanism for the ionic addition of hydrogen halides to alkenes by initial protonation of the alkene to give a cationic intermediate.[84–86] At the same time, he inverted the way that Markovnikov's rule was taught. In his textbook, published 2 years after Lucas' book, Whitmore provides the first version of Markovnikov's rule based on where the hydrogen attaches to the alkene[87]:

"When the addend has two unlike parts, two different addition products are possible since the double bond is not symmetrically placed. Both possible

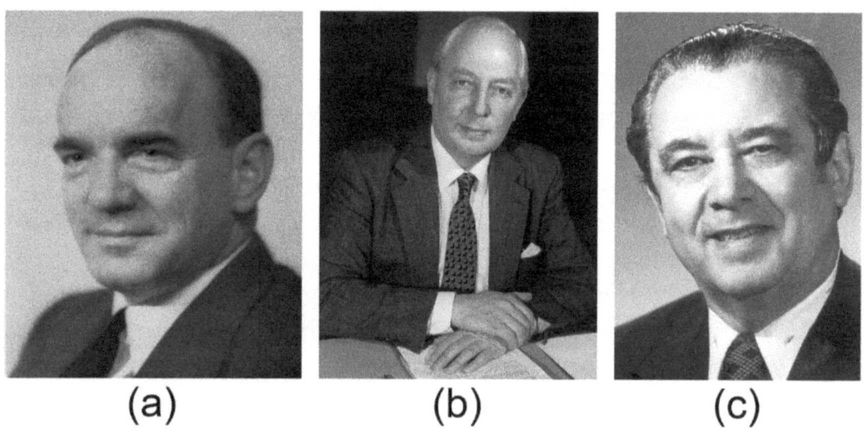

Figure 3.5 (A) Frank C. Whitmore, (B) B. C. L. Weedon, and (C) Herbert C. Brown. *(N.d.). (A) Image of Frank C. Whitmore from Marvel, C. S., Frank Clifford Whitmore (1887–1947), Biographical Memoirs of the National Academy of Sciences; National Academy of Sciences: Washington, DC, 1954; pp. 289–311. (B) Image of B. C. L. Weedon, FRS, from Pattenden, G. Basil Charles Leicester Weedon July 18, 1923–October 10, 2003.* Biogr. Mem. Fell. Roy. Soc. ***2005,*** 51, *425–436. (C) Image of Herbert C. Brown courtesy of the Nobel Foundation.*

products are formed but one usually predominates to the extent of 90% or more. This chief product is the one in which the hydrogen or "positive" part adds to the end carbon, Markownikoff's Rule (1870). Thus propylene gives mainly isopropyl compounds with halide acids and sulfuric acid ... "

This statement was based on Whitmore's own research into the role of carbocations (a term he assiduously avoided) in the Wagner—Meerwein rearrangement. It took Markovnikov's original empirical observation a step further by placing it in a mechanistic framework. Whitmore's statement eventually became the standard wording of Markovnikov's Rule (although not until the end of the 1950s, as revealed by the books by Roberts and Caserio,[88] Morrison and Boyd,[89] and other major textbooks of this era). Since its rewriting in terms of the protonation-first mechanism, the origins of the regiochemistry of electrophilic additions to alkenes have continued to intrigue chemists. In May 2020, there were over 4000 hits in a Google Scholar search for "Markovnikov's Rule," and 158 explicit and over 30,000 concept references returned by SciFinder Scholar.

In 1949, Batty and B. C. L. Weedon (Fig. 3.5) were first to use the term "anti-Markownikoff" to replace the term "abnormal," when describing addition regiochemistry.[90] In many ways, this has been an unfortunate development because the term implies that the addition does not obey Markovnikov's Rule. The term may have been an oddity initially, but it moved squarely into the mainstream when H. C. Brown (Fig. 3.5) used it to describe the regiochemistry of alkene hydration by the hydroboration—oxidation sequence.[91] With so many chemists using the method because of its regio- and stereospecificity, it was to be expected that Brown's usage would prevail and it has.

From a pedagogical standpoint, the term "anti-Markovnikov" has long been a source of confusion for beginning students in organic chemistry. This stems—in part, at least—from students now being taught the rule from a mechanistic perspective. Early in the course, however, students are often not ready to appreciate the changes in the course of the reaction when the mechanism changes. Additions that do not fall neatly under the electrophilic addition mechanistic purview (e.g., the addition of HBr under free-radical conditions or the two-stage hydroboration-oxidation) lead to confusion.

There is a growing tendency among educators in organic chemistry to call for the elimination of the Markovnikov Rule from the curriculum,[92–95] while there is also a group who equally avidly oppose this.[96–99] In this latter group, Pedrag-Peter Ilich and his coworkers make an excellent case for the wrongness of "anti-Markovnikov," but also make the important point that this terminology is now firmly entrenched as part of the vocabulary of the discipline, and that students need to be introduced to it. This is the very problem in achieving the elimination of Markovnikov's Rule from introductory textbooks.

Some of these problems come from the fact that Markovnikov's original work was published in Russian, but derivative works by Western authors almost invariably rely on the German versions of the papers. Often, the German version does not correspond exactly with the Russian version. It is such a situation that led Hughes, in his widely cited paper in the *Journal of Chemical Education*,[100] to pose the question, "Was Markovnikov's Rule an Inspired Guess?" As one who has read Markovnikov's M. Khim.[101] and Dr. Khim.[102] dissertations in the original Russian, I can answer this question with an emphatic, "NO!"[5] His rule was the result of a highly analytical mind working to elucidate the structure of the double bond in propylene and higher 1-alkenes using the regiochemistry of the addition of hydrogen halides and water to bolster his arguments. He then saw the generalization possible from his observations of those additions. The logic in the dissertations is clear and devoid of any, " ... then a miracle occurs ...," moments: his Rule emerges naturally from his arguments.

The most recent efforts of chemists in rationalizing Markovnikov's Rule have been more focused on computational research.[103–106]

This author believes that it is unlikely that Markovnikov's Rule will be eliminated from the curriculum in the near future, but it is also doubtful that it will not remain under siege.

References

1. Morkovnikov, V. K. Voprosu o vzaimnom vlianii atomov v khimicheskikh soedineniyakh [On the question of the mutual influence of atoms in chemical compounds]. *Zh. Russ. Fiz-Khim. O-va.* **1869,** *1,* 242–247.
2. Lewis, D. E. Static to dynamic: Kazan chemists and the transformation from chemical structure to reaction regiochemistry. *Angew. Chem. Int. Ed.* **2019,** *58* (12), 3694–3705. https://doi.org/10.1002/anie.201804036.
3. Beletskaya, I. P.; Nenajdenko, V. G. Towards the 150th anniversary of the Markovnikov Rule. *Angew. Chem. Int. Ed.* **2019,** *58* (15), 4778–4789. https://doi.org/10.1002/anie.201810035.
4. Markownikoff, W. Ueber Die Abhängigkeit Der Verschiedenen Vertretbarkeit Des Radikalwasserstoffs in Den Isomieren Buttersäuren. *Ann. Chem. Pharm* **1870,** *153,* 228–259. https://doi.org/10.1002/jlac.18701530204.
5. Lewis, D. E. The Logic behind Markovnikov's Rule. Was it an Inspired Guess? ...No!. *Angew. Chem. Int. Ed.* **2021,** *60* (9), 4412–4421. https://doi.org/10.1002/anie.202008228.

6. Henry, L. Sur Le Produit d'addition Du Propylene à l'acide Hypochloreux. *C. R. Hebd. Séances Acad. Sci.* **1874,** *79,* 1203–1207.
7. Markovnikoff, V. Sur Les Lois Qui Régissent Les Réactions de l'addition Directe. *C. R. Hebd. Séances Acad. Sci.* **1875,** *81,* 668–671.
8. Markovnikoff, V. Sur Les Lois Qui Régissent Les Réactions de l'addition Directe. *C. R. Hebd. Séances Acad. Sci.* **1875,** *81,* 728–730.
9. Markovnikoff, V. Sur Les Lois Qui Régissent Les Réactions de l'addition Directe. *C. R. Hebd. Séances Acad. Sci.* **1875,** *81,* 776–779.
10. Henry, L. Sur La Constitution Des Monochlorohydrines Propyléniques et La Loi d'addition de l'acide Hypochloreux. *C. R. Hebd. Séances Acad. Sci.* **1876,** *82,* 1266–1268.
11. Henry, L. Sur Les Chlorohydrines Propyléniques et La Loi d'addition de l'acide Hypochloreux. *C. R. Hebd. Séances Acad. Sci.* **1876,** *82,* 1390–1392.
12. Markovnikov, V. O. Zakonakh Obrazovaniya Pryamikh Soedinenii Nepredel'nymi Organicheskimi Chastitsami [On the laws of of forming saturated compounds from unsaturated organic particles]. *Zh. Russ. Fiz-Khim. O-va.* **1876,** *8,* 16–29.
13. Michael, A. Ueber einige Gesetzte und deren Anwendung in der organischen Chemie. I. *J. Prakt. Chem.* **1899,** *60,* 286–384. https://doi.org/10.1002/prac.18990600131.
14. Michael, A. Ueber einige Gesetzte und deren Anwendung in der organischen Chemie. II. *J. Prakt. Chem.* **1899,** *60,* 409–486. https://doi.org/10.1002/prac.18990600137.
15. Bruylants, A. Henry (Louis). *Biographie Nationale publiée par l'Académie Royale des Sciences des Lettres et des Beaux-Arts de Belgique* **1979,** *41,* 414–427.
16. Henry, L. Considérations sur quelques classes de composés organiques et sur les radicaux en général. *Mém. Couronn. Autres Mém.* **1859,** *8,* 1–42. Collection in-8°.
17. Henry, L. Sur Le Berbérine et Ses Sels. *Bull. Acad. Roy. Sci. Lett.* **1859,** *7,* 579–610. Beaux-Arts Belg.
18. Henry, L. *Précis de Chimie Générale Élémentaire;* Ch. Peeters: Louvain, 1867.
19. Henry, L. Recherches Sur Les Combinaisons Glycériques. *Ann. Chem. Phys.* **1870,** *20,* 341–350.
20. Lewis, D.; Dieringer, R. Arthur Michael (1853–1942): The Michael Addition Reaction. *Synform* **2017,** A142–A144. https://doi.org/10.1055/s-0036-1590305. No. 2017/09.
21. Costa, A. B. Arthur Michael (1853–1942): The meeting of thermodynamics and organic chemistry. *J. Chem. Educ.* **1971,** *48* (4), 243–246. https://doi.org/10.1021/ed048p243.
22. Fieser, L. F. Arthur Michael. August 2, 1853–February 8, 1942. *Biogr. Mem. Natl. Acad. Sci.* **1975,** *46,* 330–366.
23. Gillispie, C. C.; Arthur, M. *In Complete Dictionary of Scientific Biography;* Charles Scribner & Sons: Detroit, 2008.
24. Ramberg, P. J. Arthur Michael's critique of stereochemistry, 1887–1899. *Hist. Stud. Phys. Biol. Sci.* **1995,** *26* (1), 89–138. https://doi.org/10.2307/27757757.
25. Takashi, T. Discovery of the Michael reaction. *Eur. J. Org Chem.* **2010,** *2010,* 2009–2016. https://doi.org/10.1002/ejoc.200901130.
26. Michael, A. Sur La Synthèse du Phenylglucoside et de l'orthoformylglucoside ou Hélicine. *C. R. Hebd. Séances Acad. Sci.* **1879,** *89,* 355–358.
27. Michael, A. Ueber die Synthese des Salicins und des Anhydrosalicylglucosids. *Ber. Dtsch. Chem. Ges.* **1882,** *15* (2), 1922–1925. https://doi.org/10.1002/cber.188201502105.
28. Michael, A. Ueber die Addition von Natriumacetessig- und Natriummalonsäureäthern zu den Aethern ungesättigter Säuren. *J. Prakt. Chem.* **1887,** *35,* 349–356. https://doi.org/10.1002/prac.18870350136.

29. Michael, A. Ueber die Addition von Natriumacetessig- und Natriummalonsäureäther zu den Aethern ungesättigter Säuren. *J. Prakt. Chem.* **1894,** *49,* 20–25. https://doi.org/10.1002/prac.18940490103.
30. Michael, A.; Freer, P. C. Ueber die Addition von Natriumacetessig- und Natriummalonsäureäthern zu den Aethern ungesättigter Säuren. Zweite Mitteilung. *J. Prakt. Chem.* **1891,** 390–395. https://doi.org/10.1002/prac.18910430140.
31. Michael, A.; Schulthess, O. Ueber die Addition von Natriumacetessig- und Natriummalonsäureäthern zu den Aethern ungesättigter Säuren. *J. Prakt. Chem.* **1892,** 55–63. https://doi.org/10.1002/prac.18910450104.
32. Bergmann, E. D.; Ginsburg, D.; Pappo, R. The Michael reaction. *Org. React.* **1959,** *10,* 179–556. https://doi.org/10.1002/0471264180.or010.03.
33. Jung, M. Stabilized nucleophiles with electron deficient alkenes and alkynes. In *Comprehensive Organic Synthesis,* Vol. 4, Semmelhack, M., Ed.; Pergamon: London, 1991; pp 1–67. https://doi.org/10.1016/B978-0-08-052349-1.00090-1.
34. Schmaltz, H.-G. Asymmetric nucleophilic addition to electron deficient alkenes. In *Comprehensive Organic Synthesis,* Vol. 4, Semmelhack, M., Ed.; Pergamon: London, 1991; pp 199–236. https://doi.org/10.1016/B978-0-08-052349-1.00094-9.
35. Little, R. D.; Masjedizadeh, M. R.; Wallquist, O.; Mcloughlin, J. I. The intramolecular Michael reaction. *Org. React.* **1995,** *47,* 315–552. https://doi.org/10.1002/0471264180.or047.02.
36. Mather, B. D.; Viswanathan, K.; Miller, K. M.; Long, T. E. Michael addition reactions in macromolecular design for emerging technologies. *Prog. Polym. Sci.* **2006,** *31* (5), 487–531. https://doi.org/10.1016/j.progpolymsci.2006.03.001.
37. Pellissier, H. Asymmetric organocatalysis. *Tetrahedron* **2007,** *63* (38), 9267–9331. https://doi.org/10.1016/j.tet.2007.06.024.
38. Yong, Z.; Wei, W. Recent advances in organocatalytic asymmetric Michael reactions. *Catal. Sci. Technol.* **2012,** 42–53. https://doi.org/10.1039/c1cy00334h.
39. Reyes, E.; Uria, U.; Vicario, J. L.; Carrillo, L. The catalytic, enantioselective Michael reaction. In *Organic Reactions,* Vol. 90; John Wiley and Sons Inc.: Spain, 2016; pp 1–898. https://doi.org/10.1002/0471264180.or090.01.
40. Lee, V. J. Conjugate additions of reactive carbanions to activated alkenes and alkynes. In *Comprehensive Organic Synthesis,* Vol. 4, Semmelhack, M., Ed.; Pergamon: London, 1991; pp 69–137. https://doi.org/10.1016/B978-0-08-052349-1.00091-3.
41. Michael, A. Untersuchungen Über Alloïsomerie. *J. Prakt. Chem.* **1895,** 289–325. https://doi.org/10.1002/prac.18950520123.
42. Geromont, F. Zur Constitution der Allylverbindungen. *Justus Liebigs Ann. Chem.* **1871,** *158* (3), 369–371. https://doi.org/10.1002/jlac.18711580312.
43. Geromont, F. Zur Constitution der Allylverbindungen. *Ber. Dtsch. Chem. Ges.* **1871,** *4* (2), 548–550. https://doi.org/10.1002/cber.18710040206.
44. Reboul, E. Combinaisons Des Hydracides Avec l'éthylène et Le Propylène Bromés. *C. R. Hebd Séances Acad. Sci.* **1870,** *70,* 398–401.
45. Reboul, E. Iodhydrates et Chlorhydrates d'éthylène et de Propylène Monobromés. *C. R. Hebd, Séances Acad. Sci.* **1870,** *70,* 853–856.
46. Reboul, E. Sur Deux Nouveaux Isomères du Bromure de Propylène. *C. R. Hebd Séances Acad. Sci.* **1872,** *74,* 613–616.
47. Reboul, E. Sur les Divers Chlorures de Propylène. *C. R. Hebd Séances Acad. Sci.* **1873,** *76,* 1270–1272.
48. Reboul, E. Recherches Sur L'isomérie Dans La Série Du Propylène. *Ann. Chem. Phys.* **1878,** *14,* 470–473.
49. Erlenmeyer, E. Ueber die beiden isomeren Bromüre $C_3H_6Br_2$. *Ann. Chem. Pharm.* **1873,** *197,* 169–185. https://doi.org/10.1002/jlac.18791970203.

50. Lermontoff, J. Ueber die Darstellung von Trimethylenbromid. *Justus Liebigs Ann. Chem.* **1876,** *182* (3), 358–362. https://doi.org/10.1002/jlac.18761820308.
51. Lermontoff, J. Société Chimique Russe. – Séance Du 4/16 Mai 1878. – correspondance Russe. *C. R. Hebd Séances Acad. Sci.* **1878,** 20–28.
52. Lermontova, Y. O. Poluchenii normal'nago bromistago propilena (trimetilenbromida) [On obtaining normal propylene bromide (trimethylene bromide)]. *Zh. Russ. Fiz.-Khim. O-va.* **1876,** *8,* 281–283.
53. Bogomolets, I. O. prigotovlenii bromistago trimetilena [On the preparation of trimethylene bromide]. *Zh. Russ. Fiz.-Khim. O-va.* **1878,** *10,* 258–259.
54. Bogomolez, I. P. 1257 in G. Wagner, Aus Sitzung der russischen chemischen Gesellschaft am 4/16 Mai 1878. *Ber. Dtsch. Chem. Ges.* **1878,** *11,* 1251–1263. https://doi.org/10.1002/cber.187801101345.
55. Creese, M. R. S. Early Women Chemists in Russia: Anna Volkova, Iuliia Lermontova and Nadezhda Ziber-Shumova. *Bull. Hist. Chem.* **1998,** *21,* 19–24.
56. Lermontoff, J. Ueber die Zusammensetzung des Diphenins. *Ber. Dtsch. Chem. Ges.* **1872,** *5,* 231–236. https://doi.org/10.1002/cber.18720050172.
57. Lermontova, Yu, O. deistvie tretichnogo iodistogo butila na izobutilen v prisustvii metallicheskikh okislov [On the action of tertiary butyl iodide on isobutylene in the presence of metal oxides. *Zh. Russ. Fiz.-Khim. O-va* **1878,** *10,* 238–244.
58. Lermontoff, J. Sur l'action de l'iodure de Butyle Tertiare Sur l'isobutylène En Présence d'oxydes Métalliques. *Bull. Acad. Imp. Sci.* **1879,** *25,* 203–209. St. Pétersb.
59. Lermontoff, J. Ueber die Einwirkung des tertiären Butyljodürs auf Isobutylen bei Gegenwart von Metalloxyden. *Justus Liebigs Ann. Chem.* **1879,** *196* (1), 116–122. https://doi.org/10.1002/jlac.18791960109.
60. Westheimer, F. H. Morris Selig Kharasch (1895–1957): A biographical memoir. In *Biographical Memoirs of the Academy of Sciences;* National Academy of Sciences: Washington, DC, 1960; pp 123–152.
61. Kharasch, M. S.; Mayo, F. R. The peroxide effect in the addition of reagents to unsaturated compounds. I. The addition of hydrogen bromide to allyl bromide. *J. Am. Chem. Soc.* **1933,** *55* (6), 2468–2496. https://doi.org/10.1021/ja01333a041.
62. Kharasch, M. S.; McNab, M. C.; Mayo, F. R. The peroxide effect in the addition of reagents to unsaturated compounds. II. The addition of hydrogen bromide to vinyl bromide. *J. Am. Chem. Soc.* **1933,** *55* (6), 2521–2530. https://doi.org/10.1021/ja01333a048.
63. Kharasch, M. S.; Hannum, C. W. The peroxide effect in the addition of reagents to unsaturated compounds. IV. The addition of halogen acids to vinyl chloride. *J. Am. Chem. Soc.* **1934,** *56* (3), 712–714. https://doi.org/10.1021/ja01318a054.
64. Kharasch, M. S.; Hinckley, J. A. The peroxide effect in the addition of reagents to unsaturated compounds. V. The addition of hydrogen bromide to butene-1. *J. Am. Chem. Soc.* **1934,** *56* (5), 1212–1214. https://doi.org/10.1021/ja01320a068.
65. Kharasch, M. S.; Hinckley, J. A., Jr. The peroxide effect in the addition of reagents to unsaturated compounds. VI. The addition of hydrogen bromide to isobutylene. *J. Am. Chem. Soc.* **1934,** *56,* 2521–2530. https://doi.org/10.1021/ja01320a068.
66. Kharasch, M. S.; McNab, M. C. The peroxide effect in the addition of reagents to unsaturated compounds. *J. Am. Chem. Soc.* **1934,** *56.* https://doi.org/10.1021/ja01321a508, 1425–1425.
67. Kharasch, M. S.; Hinckley, J. A., Jr.; Gladstone, M. M. The peroxide effect in the addition of reagents to unsaturated compounds. *J. Am. Chem. Soc.* **1934,** *56* (7), 1642–1644. https://doi.org/10.1021/ja01322a509.
68. Kharasch, M. S.; Hannum, C. The peroxide effect in the addition of reagents to unsaturated compounds. VIII. The addition of hydrogen iodide to ethylene compounds. *J. Am. Chem. Soc.* **1934,** *56* (8), 1782–1784. https://doi.org/10.1021/ja01323a040.

69. Kharasch, M. S.; Piccard, J. F. Aromatic mercuri-organic derivatives. *J. Am. Chem. Soc.* **1920,** *42* (9), 1855–1864. https://doi.org/10.1021/ja01454a014.
70. Kharasch, M. S.; Jacobsohn, I. M. Mercuri-organic derivatives. The mercurization of aromatic amines and its relation to the theory of substitution. *J. Am. Chem. Soc.* **1921,** *43* (8), 1894–1903. https://doi.org/10.1021/ja01441a017.
71. Stieglitz, J.; Kharasch, M.; Hanke, M. Preparation of 5, 5′-mercuri-bis-3-nitro-4-hydroxy-phenyl-arsonic acid. *J. Am. Chem. Soc.* **1921,** *43* (5), 1185–1193. https://doi.org/10.1021/ja01438a018.
72. Kharasch, M. S.; Chalkley, L. Mercuri-organic derivatives: II. nitrobenzene mercury compounds; an indirect method of mercurizing organic compounds. *J. Am. Chem. Soc.* **1921,** *43* (3), 607–612. https://doi.org/10.1021/ja01436a026.
73. Kharasch, M. S. Aromatic mercuri-organic derivatives. the Hofmann rearrangement and the nature of valences of mercury in mercuri-organic derivatives. *J. Am. Chem. Soc.* **1921,** *43* (8), 1888–1894. https://doi.org/10.1021/ja01441a016.
74. Kharasch, M. S. An indirect method of preparation of organic mercuric derivatives and a method of linking carbon to carbon. Preliminary report. *J. Am. Chem. Soc.* **1921,** *43* (10), 2238–2243. https://doi.org/10.1021/ja01443a017.
75. Kharasch, M. S.; Staveley, F. W. The behavior of mercuric salts of organic acids toward heat. *J. Am. Chem. Soc.* **1923,** *45* (12), 2961–2972. https://doi.org/10.1021/ja01665a023.
76. Kharasch, M. S.; Chalkley, L. Factors affecting the stability of mercurials and the mercuration of substituted phenylammonium salts. *J. Am. Chem. Soc.* **1924,** *46* (5), 1211–1223. https://doi.org/10.1021/ja01670a014.
77. Kharasch, M. S.; Grafflin, M. W. The preparation and decomposition of unsymmetrical mercuri-organic compounds. A method of establishing the relative degree of negativity of organic radicals. *J. Am. Chem. Soc.* **1925,** *47* (7), 1948–1954. https://doi.org/10.1021/ja01684a024.
78. Kharasch, M. S.; Marker, R. The decomposition of unsymmetrical mercuriorganic compounds: A method of establishing the relative degree of electronegativity of organic radicals. *J. Am. Chem. Soc.* **1926,** *48* (12), 3130–3143. https://doi.org/10.1021/ja01691a018.
79. Cohen, J. B. *Organic Chemistry for Advanced Students. Part I. Reactions;* Edward Arnold: London, 1919; p 114.
80. Michael, A. Zur Constitution des Natriumacetessigäthers. *J. Prakt. Chem.* **1892,** *46,* 189–208. https://doi.org/10.1002/prac.18920460114.
81. Lucas, H. J. *Organic Chemistry;* American Book Company: New York, 1935; p 72.
82. Wertheim, E. *Textbook of Organic Chemistry;* The Blackiston Company: Philadelphia, 1945; p 56.
83. Royals, E. E. *Advanced Organic Chemistry;* Prentice-Hall: Englewood Cliffs, NJ, 1954; pp 214–222.
84. Whitmore, F. C.; Johnston, F. Secondary isoamyl chloride, 3-chloro-2-methylbutane. *J. Am. Chem. Soc.* **1933,** *55* (12), 5020–5022. https://doi.org/10.1021/ja01339a053.
85. Whitmore, F. C. The common basis of intramolecular rearrangements. *J. Am. Chem. Soc.* **1932,** *54* (8), 3274–3283. https://doi.org/10.1021/ja01347a037.
86. Whitmore, F. C. Mechanism of the polymerization of olefins by acid catalysts. *Ind. Eng. Chem.* **1934,** *26* (1), 94–95. https://doi.org/10.1021/ie50289a023.
87. Whitmore, F. C. *Organic Chemistry;* D. Van Nostrand Company, Inc.: New York, 1937; pp 37–38.
88. Roberts, J. D.; Caserio, M. C. *Basic Principles of Organic Chemistry,* 1st ed.; W. A. Benjamin: New York, 1964.
89. Morrison, R. T.; Boyd, R. T. *Organic Chemistry,* 1st ed.; Allyn and Bacon: Boston, 1959.

90. Batty, J. W.; Weedon, B. C. L. Acetylene reactions. Part III. Reaction of aminobutynes with carbon disulphide. *J. Chem. Soc.* **1949,** 786–789. https://doi.org/10.1039/JR9490000786.
91. Brown, H. C.; Subba Rao, B. C.; Hydroboration, I. The reaction of olefins with sodium borohydride-aluminum chloride. A convenient route to organoboranes and the anti-markownikoff hydration of olefins. *J. Am. Chem. Soc.* **1959,** *81* (24), 6423–6428. https://doi.org/10.1021/ja01533a023.
92. Gooch, E. E. Moving past Markovnikov's Rule. *J. Chem. Educ.* **2001,** *78,* 1358. https://doi.org/10.1021/ed078p1358.
93. Kerber, R. C. Markovnikov's rule. *J. Chem. Educ.* **2007,** *1109.* https://doi.org/10.1021/ed084p1109.1.
94. Graulich, N.; Bhattacharyya, G. Investigating students' similarity judgments in organic chemistry. *Chem. Educ. Res. Pract.* **2017,** *18* (4), 774–784. https://doi.org/10.1039/c7rp00055c.
95. Kerber, R. Markovnikov's rule in history and pedagogy. *Found. Chem.* **2002,** *4,* 61–72. https://doi.org/10.1023/A:1014479921278.
96. Jones, G. The Markovnikov rule. *J. Chem. Educ.* **1961,** *38* (6), 297–300. https://doi.org/10.1021/ed038p297.
97. Isenberg, N.; Grdinic, M. A modern look at Markovnikov's rule and the peroxide effect. *J. Chem. Educ.* **1969,** *46* (9), 601–605. https://doi.org/10.1021/ed046p601.
98. Ilich, P. P.; Rickertsen, L. S.; Becker, E. Polar addition to C=C group: Why is Anti-Markovnikov hydroboration-oxidation of alkenes not "anti-". *J. Chem. Educ.* **2006,** *83* (11), 1681–1685 https://doi.org/10.1021/ed083p1681.
99. Ilich, P.-P. Markovnikov's rule. *J. Chem. Educ.* **2007,** *84* (7), 1109–1109. https://doi.org/10.1021/ed084p1109.2.
100. Hughes, P. Was Markovnikov's rule an inspired guess? *J. Chem. Educ.* **2006,** *83* (8), 1152–1154 https://doi.org/10.1021/ed083p1152.
101. Markovnikov, V. *Ob Izomerii V Organicheskikh Soenineniyakh [On the Isomerism of Organic Compounds];* Kazan, 1865.
102. Markovnikov, V. *Materialy Po Voprosu Vzaimnom Vliyanii Atomov V Khimicheskikh Soenineniyakh [Materials on the Question of the Mutual Influence of Atoms in Chemical Compounds];* Kazan, 1869.
103. Sæthre, L. J.; Thomas, T. D.; Svensson, S. Markovnikov addition to alkenes. A different view from core-electron spectroscopy and theory. *J. Chem. Soc. Perkin Trans.* **1997,** *2* (4), 749–755. https://doi.org/10.1039/a606562g.
104. Aizman, A.; Contreras, R.; Galván, M.; Cedillo, A.; Santos, J. C.; Chamorro, E. The Markovnikov regioselectivity rule in the light of site activation models. *J. Phys. Chem.* **2002,** *106* (34), 7844–7849. https://doi.org/10.1021/jp020214y.
105. Yang, Z. Z.; Ding, Y. L.; Zhao, D. X. Insight into Markovnikov reactions of alkenes in terms of ab initio and molecular face theory. *ChemPhysChem* **2008,** *9* (16), 2379–2389. https://doi.org/10.1002/cphc.200800364.
106. Franzen, S.; Cochran, K. H.; Weng, J.; Bartolotti, L.; Delley, B. The quadrapolar character of the Markovnikov reaction transition state. *Chem. Phys.* **2016,** *464,* 46–54. https://doi.org/10.1016/j.chemphys.2015.11.001.

CHAPTER FOUR

Aleksandr Mikhailovich Zaitsev and his empirical rule for elimination

In Chapter 2, we looked at Vladimir Vasil'evich Markovnikov (1837–1904), and his contributions to the advancement of organic chemistry, among which his eponymous rule for addition to alkenes[1–7] has been one of the most important and lasting. In this chapter, we will meet his contemporary and fellow student at Kazan, Aleksandr Mikhailovich Zaitsev (1841–1910, Fig. 4.1). Biographies of Zaitsev are available in Russian,[8–11] English,[12,13] and German.[14]

Aleksandr Mikhailovich Zaitsev (1841–1910)

Zaitsev was born into a prominent merchant family who had lived in Kazan and the surrounding area at least since the time of Ivan the Terrible (reigned 1547–84). His father, Mikhail Savvich Zaitsev, was a member of the Guilds and held control over the tea and sugar trade in the area; Kazan's location, as a gateway between east and west, made this a very influential position. His grandfather, Savva Stepanovich Zaitsev (1780–1844), was a wealthy and prominent businessman in the city and a Warden of the Cathedral of Saints Peter and Paul (Fig. 4.2), built in 1723 in commemoration of the 1722 visit to Kazan of Tsar Peter the Great on his way to the Russo-Persian War of 1722–23. The Cathedral was damaged by fire periodically over the next two centuries; the fire of 1815 was particularly destructive.[15] The reconstruction of the cathedral after this fire was carried out at the expense of Savva Stepanovich between 1824 and 1825.

After the death of his first wife (known only by her initials, N. N.), Zaitsev's father, Mikhail Savvich, married Natal'ya Vasil'evna Lyapunova (1819–47). Aleksandr Mikhailovich was their middle son. Mikhail Savvich was determined that his sons would follow him into the Guilds, but his

Addition, Elimination and Substitution: Markovnikov, Hofmann, Zaitsev and Walden
ISBN: 978-0-12-821027-7
https://doi.org/10.1016/B978-0-12-821027-7.00007-4

Figure 4.1 A. M. Zaitsev, 1880s. *(N.d.). Retrieved from https://en.wikipedia.org/wikl/Alexander_Mikhaylovich_Zaytsev#/media/File:Alexander_Mikhaylovich_Zaytsev.jpg.*

brother-in-law, Mikhail Vasil'evich Lyapunov (1820–68, Fig. 4.3), the Professor of Astronomy at Kazan University, worked hard to convince him that the times demanded a more educated citizenry for Russia to advance in the modern world.

Lyapunov played a critical role in shaping Zaitsev's career because he persuaded Mikhail Savvich to permit the young Aleksandr to enroll in the Kazan Gymnasium and then Kazan University, instead of being enrolled in the Guilds. Mikhail Savvich's assent came at a price: he demanded that his son enroll in the Cameral Science Division of the Juridical Faculty at Kazan University. In 1858, following his graduation from the Gymnasium, Zaitsev enrolled at Kazan University as a Cameral student; we discussed cameralism in Chapter 2.

At the time, a reading knowledge of Latin was required for any student to enter the Juridical Faculty at Kazan University. Latin was not taught at the

Figure 4.2 The Kazan Cathedral of SS. Peter and Paul, 1839. *From https://commons.wikimedia.org/wiki/File:А._Дюран._Кафедральный_собор_святых_Петра_и_Павла.jpg.*

Kazan Gymnasium at the time, so his uncle and aunts taught Zaitsev the Latin that he would need to be permitted to enroll. (Unlike Markovnikov before him, Zaitsev was not a member of the nobility, so the fact that he had graduated from the Gymnasium did not suffice for admission to the university).

As agreed, once he had passed the entrance examinations, Zaitsev entered the Cameral Science Division. Like Markovnikov before him, the 2 years of chemistry for graduation led to Zaitsev also coming under the influence of Aleksandr Mikhailovich Butlerov (1828–86, Fig. 4.3). Before he graduated in economic science, Zaitsev had become a committed chemist

Figure 4.3 Mikhail Vasil'evich Lyapunov (Left), Aleksandr Mikhailovich Butlerov (Right). *(N.d.). Public domain images downloaded from (Lyapunov) https://commons.wikimedia.org/wiki/File:Liapounov.jpg and (Butlerov) https://commons.wikimedia.org/wiki/File:Butlerov,_A._M._1828-1886.jpg.*

thanks to Butlerov's mentorship. Unlike Markovnikov, who always made it a point of pride that he had graduated in economics, Zaitsev felt no such connection.

Zaitsev entered Butlerov's class in 1860. We have notes taken by Markovnikov during Butlerov's lecture course for the 1859–60 academic year[16] that make it clear that Butlerov was, at that time, a staunch Type Theorist; Markovnikov's kandidat dissertation of October 1860[17] reaffirms this. By 1861, however, he was a committed Structuralist. Lecture notes taken in his course in organic chemistry for the 1862–63 academic year (the year after Zaitsev graduated) by student Ivan Bukhvostov,[18] on the other hand, make it clear that by this time, he had become a confirmed Structural Theorist and an active opponent of Kolbe's Theory of Rational Constitution.

In Western Europe

Shortly before Aleksandr Mikhailovich graduated in 1862, his father died, and the family business was liquidated. At this time, the progression through an academic career in Russia was fairly clearly established.

Immediately after graduation, the student would carry out a small research project that would be summarized in a thesis for the degree of kandidat. In the 1870s, this degree was approximately the equivalent of a modern Honors degree of Bachelor in a university in the British Commonwealth and somewhere between a Senior Thesis and a Master's dissertation in an American University. Today, the kandidat is the full equivalent of a Western Ph.D. degree. In the 1870s, the kandidat was the minimum qualification for holding a salaried Assistantship in a Russian university. Zaitsev's next actions placed his entire career in jeopardy.

Rather than staying in Kazan and following the traditional course for students seeking to pursue an academic career, Zaitsev took his inheritance and left for Western Europe, intending to study with the most eminent organic chemists there. He first attended the World's Exhibition in London and then followed the example of his older brother, Konstantin Mikhailovich (1840–1919, Fig. 4.4), by going to Marburg and entering the laboratory of (Adolph Wilhelm) Hermann Kolbe (1818–84, Fig. 4.4). He spent the next four semesters at Marburg, before moving to Paris to study with (Charles) Adolphe Wurtz (1817–84, Fig. 4.4) for another two semesters. He returned to Marburg for one last semester.

Zaitsev had studied with Butlerov before his graduation, before Butlerov had completed the development of his version of Structural Theory. During this time, Butlerov had decided to groom him for the professoriate. But it appears that, unlike his contemporary, Markovnikov, Zaitsev did not closely follow Butlerov's subsequent work on chemical structure.

Figure 4.4 (L–R): Konstantin Mikhailovich Zaitsev, A. W. Hermann Kolbe, C. Adolphe Wurtz. *(N.d.). Images downloaded from: (Zaitsev) http://www.ksu.museum.ru/chmku/images/65b.jpg, (Kolbe) https://commons.wikimedia.org/wiki/File:Hermann_Kolbe.jpg and (Wurtz) https://commons.wikimedia.org/wiki/File:Charles-Adolphe_Wurtz.jpg.*

1863: faux pas and recovery

In February 1863, after a year in Kolbe's laboratory, Zaitsev was confident that he was ready to write his kandidat dissertation on a theoretical topic. He was sure that Kolbe's doctrine on the relationship of organic compounds (carboxylic acids) to inorganic carbon dioxide[19] "shed a great deal of light on the genesis of chemical compounds and provide fundamental proof of Kolbe's view on the rational composition ... " of a number of aliphatic compounds.

Zaitsev had maintained contact with Butlerov, who had agreed to the focus of the project, but Zaitsev did not take the simple step of clearing the title and content of the dissertation itself with Butlerov before submitting it to Kazan. He submitted the dissertation to Butlerov from Marburg. Like Markovnikov's dissertation, this 75-page document was hand-written. The dissertation[20] —The theoretical views of Kolbe on the rational constitution of organic compounds and their relationship with inorganic ones—was submitted to Butlerov at Kazan for examination.

The following account of this episode is a translation* from Zakharov's *Kazanskii Universitet. Khronologiya stanovleniya khimicheskoi laboratorii i Kazanskoi khimicheskoi Shkoly*:[21]

In February 1863, without prior agreement with Butlerov, Zaitsev wrote his first kandidat dissertation "The theoretical views of Kolbe on the rational constitution of organic compounds and their relationship with inorganic [compounds]," which is now stored in the N. I. Lobachevsky Scientific Library Kazan University, Russian sector of the department of manuscripts and rare books, No. 5651. As you know, Kolbe derived the "constitution" of organic compounds "from inorganic" carbon dioxide, which he gave the formula C_2O_4, using half the actual atomic weights of carbon and oxygen. This method of formal modeling of the constitution of organic compounds led him to numerous errors, although it sometimes made it possible to make successful predictions, as well as to bring experimental material into the system. On this basis, Kolbe considered his theoretical system to be the only correct one, and Zaitsev fully adopted this point of view, writing that Kolbe's teachings and his new (1861) discoveries "spread a great deal of light on the principles of chemical compounds and provide fundamental proof of the validity of Kolbe's view of the rational composition ... " of aliphatic compounds. Not only did Zaitsev fail to critically compare the views of

* All English translations from the Russian are translated by the Author.

Kolbe and Butlerov, but he also included the term—unacceptable to Butlerov— "constitution of organic compounds" in the title of the dissertation.

The manuscript was delivered to Butlerov through the office of the Dean of the Juridical Faculty, as shown by the inscription on the cover: "I have read [this]. A. Butlerov. and returned it on April 16th. I will give you a special report." Some members of the Faculty Council also signed up for the reading. Unfortunately, Butlerov's report has not survived, and it may not even have been written, but everything is clear from several dozen of his notes and comments on the manuscript itself—"hardly?" "probably years," "in German," "?? not clear," "not in Russian," "purely German translation," "oh, German slang!" etc. in the margin of Zaitsev's dissertation. Butlerov criticizes Kolbe's views, of course, from the standpoint of his theory of chemical structure. So when Zaitsev expressed the idea that hydrobenzamide "is a true diamine, this probably comes from the position of its atoms," in touching on the work of A.P. Borodin's "Investigation of the chemical structure of hydrobenzamide and amarin" (1858), Butlerov immediately noted ironically: "After all, Kolbe rejects the possibility of determining the raumliche Lagerung < spatial arrangement> of atoms." It is not surprising, therefore, that Butlerov rejected this work of Zaitsev, concluding on the last page of the dissertation: "A bad translation from German!"

"How could it have happened," asked A. S. Klyuchevich "that by the beginning of 1863, Zaitsev did not yet know the full extent of Butlerov's doctrine and write a work glorifying Kolbe?" He then suggested: "It could be that A. M. Zaitsev had attended the course of Butlerov's lectures in 1860, but did not attend them later, and did not know about Butlerov's famous report in Speyer on October 19, 1861[22] and the article that reflected it, published in 1862 in Kazan.[23] But this also leads us to the idea that A.M. Zaitsev was not yet a close student of Butlerov at that time and overestimated his strength, taking on a theoretical theme."

Although Butlerov appears to have given Zaitsev's original plan his approval, the dissertation that arrived on his office was clearly nothing like he had expected or approved. He was ruthless in dissecting the dissertation: his comments are all negative and unusually acerbic for the normally equanimous professor, who criticized Zaitsev's logic without mercy. Overall, it is clear that Butlerov was not impressed with Zaitsev's command of either

Kolbe's theoretical stance or his own, so that the dissertation came across as substandard and unconvincing. Is it surprising, therefore, that the degree was not awarded?

Thus, Zaitsev's move to submit was impolitic, to say the least, if not foolhardy. One fact is indisputable: he had sent a dissertation advocating the theories of Structural Theory's most intransigent opponent to be examined by its most ardent protagonist. Whether Zaitsev was a committed member of Butlerov's "inner circle" at the time has been questioned, but him submitting this dissertation would seem to imply that he was not. So, what motivated this ill-advised move? One rationale that the young Zaitsev could not control his youthful overconfidence, which degenerated into youthful hubris. Certainly, it was not out of character for Zaitsev to overestimate his own ability, and therefore to act without thinking which, of course, left him facing the resultant mess he found himself in. He believed that his best chance to write an outstanding kandidat dissertation would come from working with the best chemists in Western Europe and then writing the dissertation and submitting it to Kazan. This, at least, is Zelinskii's interpretation of the likely reasons for Zaitsev's faux pas.[24] In this instance, Zaitsev was very close to being tragically wrong.

Butlerov may well have viewed Zaitsev's 1863 thesis as an indirect affront to himself and Russian organic chemistry in general, but his disposition suggests that he found it hard to hold a grudge (except against Beilstein, following Mendeleev's failed nomination to the Academy of Sciences[25]). On the other hand, there is little doubt that Markovnikov, whose dislike of Zaitsev was already well known, did. Fortunately, Butlerov also looked at what Zaitsev had accomplished with Kolbe and saw that here was a young synthetic organic chemist of the first water (as contrasted with his mediocre ability as a theoretician). He may not have disseminated Butlerov's views on Structural Theory as enthusiastically as Markovnikov, but Butlerov saw in him someone who had the potential to make important contributions to the development of synthetic organic chemistry in Russia.

Theoretician no more: the synthetic organic chemist emerges

It did not take long for Zaitsev to realize his mistake and recognize the import of Butlerov's evaluation; his mentor's reaction had eliminated the chance for submitting another theory-based dissertation. One might expect that this irreparably damaged Zaitsev's relationship with Butlerov, but actually this was not the case. It appears that Butlerov saw in Zaitsev the chemist worth salvaging as a Professor for Kazan University. Equally, it is perhaps fortunate for organic chemistry that this was the case, because Zaitsev's later career showed him to be a gifted synthetic organic chemist (as will become obvious later in this chapter, he was, in fact, a synthetic organic chemist of rare skill).

Zaitsev had kept Butlerov informed of progress on his projects in Kolbe's laboratory, so Butlerov knew that he had also begun synthetic work on two projects, studying: (1) the reactivity of diaminosalicylic acid[26–28] and (2) the reaction of chloroacetate esters with potassium isocyanate.[29–32] Here Zaitsev was playing to his own strengths. Butlerov viewed the latter project as being a suitable topic for a kandidat dissertation and encouraged Zaitsev to continue working on it until it was ready to submit.

Zaitsev's synthesis of diaminosalicylic acid began with nitration of salicylic acid (**43**) under the standard conditions to obtain the dinitro compound **44**, and then reduction of the nitro groups with hydrogen iodide generated from phosphorus triiodide and water to give the di(hydroiodide) of the product (**45**). This salt is quite air-stable, but on neutralization with sodium carbonate, the air-sensitive diamino acid (**46**) is released. When treated with excess base, the final nitrogen is deprotonated to give an extremely air-sensitive product (**47**) that reacted with air to give an intractable brown material (Scheme 4.1). This work was carried out in Kolbe's laboratory as the papers in the *Bulletin de la Société chimique de Paris*[26,29] point out explicitly.

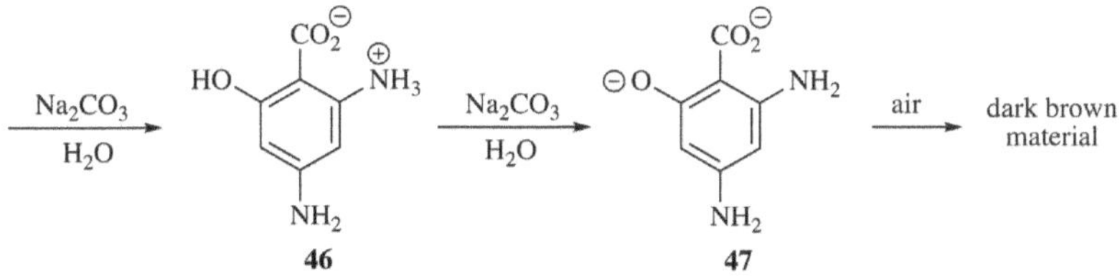

Scheme 4.1 Zaitsev's studies of diaminosalicylic acid.

Given that he did not appear to initiate any new research in Paris, his work in Wurtz' laboratory in Paris may have consisted of attending lectures and seminars.

According to the biography by Klyuchevich and Bykov,[8] Zaitsev submitted his new dissertation in August 1864. This time, Butlerov noted that, although Zaitsev had not achieved all his goals for the dissertation, he nevertheless had written a dissertation[28] that completely satisfied his mentor. Not only did Butlerov give it his full support, he did so to the extent that Zaitsev was excused reading the (not mandatory, but customary) two associated colloquia. Zaitsev was awarded the degree in November 1864. Zaitsev published papers on both topics in French and German journals in 1866, following his return to Kazan.

In Zaitsev's work on the reaction of potassium isocyanate with ethyl chloroacetate, he showed that the reaction occurred with 2:1 stoichiometry to give the derivative **52**, related to biuret, as the final product (Scheme 4.2).

Sulfur compounds in Marburg

While in Kolbe's laboratory, Zaitsev began working with organic sulfur compounds, so when he returned to Marburg from Paris, he moved to complete the project. The key discovery in this project was the discovery of the sulfoxides,[33,34] which Butlerov recognized as a major achievement in synthesis.

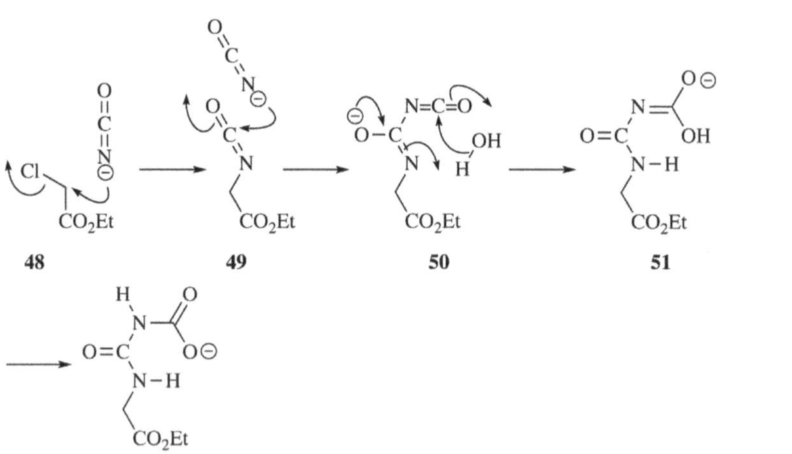

48 **49** **50** **51** **52**

Scheme 4.2 Zaitsev's studies of the reaction of potassium cyanate with ethyl chloroacetate.

Zaitsev began the project in the traditional way: by repeating literature syntheses. In this case, he set about reproducing work by Adolf von Oefele, who had reported the synthesis of diethylsulfone[35,36] by the nitric acid oxidation of diethyl sulfide. An example of the synthesis of unsymmetrical sulfides by the same strategy was disclosed by his brother, Mikhail Mikhailovich, who reported synthesis of ethyl isoamyl sulfide from isoamyl mercaptan and ethyl iodide in 1870.[37] However, when Zaitsev oxidized the dialkylsulfides (**53**) with boiling nitric acid, he did not obtain the sulfone as had von Oefele, but obtained instead the sulfoxides (**54**),[33,34] as shown in Scheme 4.3.

In 1864, von Oefele had also described the reaction between diethyl sulfide and ethyl iodide to give what we now know as triethylsulfonium iodide (**56a**).[38] He did not pursue the reaction further, but Zaitsev recognized the inherent basicity (nucleophilicity) in the sulfur atom of the sulfide, and tested his theory by the reaction between ethyl isoamyl sulfide and methyl iodide to give the sulfonium salt **56b**;[39] these reactions are shown in Scheme 4.4.

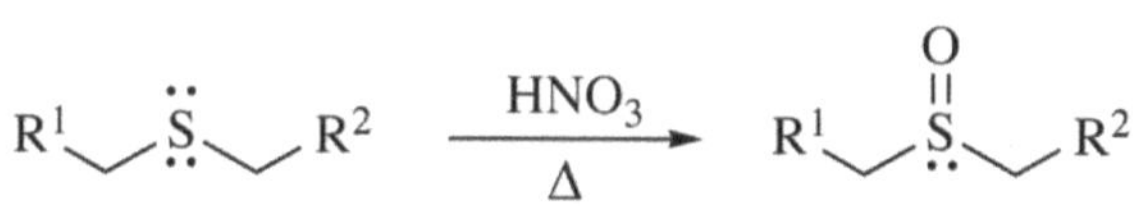

53a: $R^1 = R^2 = i$-Bu
53b: $R^1 = R^2 = i$-Pr
53c: $R^1 = i$-Bu, R^2 = Me
53d: $R^1 = R^2$ = Me
53e: $R^1 = R^2$ = H

54a: $R^1 = R^2 = i$-Bu
54b: $R^1 = R^2 = i$-Pr
54c: $R^1 = i$-Bu, R^2 = Me
54d: $R^1 = R^2$ = Me
54e: $R^1 = R^2$ = H

Scheme 4.3 Zaitsev's syntheses of sulfoxides.

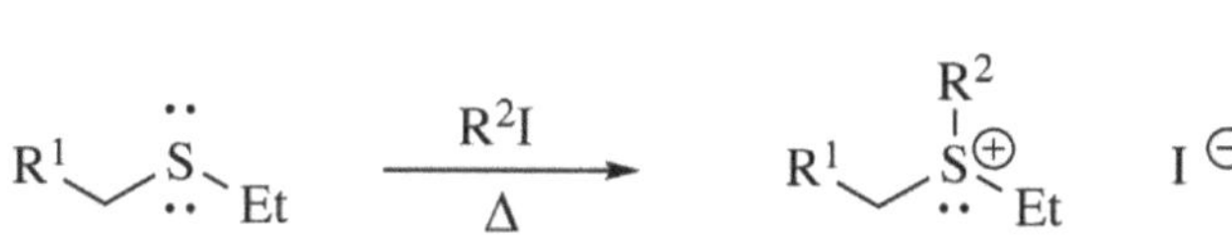

55d
55c

56a: R^1 = Me, R^2 = Et
56b: $R^1 = i$-Bu, R^2 = Me

Scheme 4.4 Syntheses of sulfonium salts by von Oefele and Zaitsev.

The return to Kazan

When Kolbe moved to Leipzig in 1865, Zaitsev's funds had finally run out, so he returned to Russia in September that year. Although he had remediated himself with his new kandidat thesis, his kandidat degree was in cameral science, not natural sciences. Unfortunately for him, the kandidat degree in chemistry was administered by the Physics–Mathematics Faculty, so this kandidat degree in cameral sciences did not qualify him for entry into the graduate program in chemistry. With this degree, he could qualify for a low-level, salaried position in the sciences at Kazan, but he could not ascend the academic ladder any further.

In other words, almost devoid of funds, Zaitsev had returned to Russia with no real prospects of salaried employment and certainly not employment as an Aspirant (graduate assistant). His solution to this problem was typical of him. He knew that Butlerov had sufficient influence to obtain a position for him at Kazan—provided that there was a position available. There was not. So, while Butlerov was attempting to secure him a position, Zaitsev began working for him "as a private person," i.e., without a stipend (although he did not have to pay Butlerov's fees for supervising the research). In February 1866, a position did open up in the agronomy laboratory, and so here Zaitsev began his academic career.

Zaitsev's ultimate goal was to become a Professor of Chemistry, so his kandidat degree in cameral science was a real impediment to his plans. In order to become an Associate Professor, Zaitsev needed the M. Khim. degree, but his kandidat degree did not allow him entry into the Master's program in chemistry, which was administered by the Physics–Mathematics Faculty. The one solution to this problem that Zaitsev chose was to earn a doctoral degree at a Western university by submitting a research-based dissertation and earning the degree of Ph.D. He could then petition Kazan to be allowed to submit for the M. Khim. degree. Consequently, Zaitsev submitted his work on sulfur compounds to Kolbe at Leipzig in 1866.[40] Thanks, no doubt, to Kolbe's influence, he earned the Ph.D. in absentia, without ever having stepped on the Leipzig campus. Following his graduation, more stringent requirements were placed on dissertations for the Ph.D. degree in Germany.

With his Ph.D. in hand, and with Butlerov's help, Zaitsev received permission to submit for the degree of M. Khim. in the Physics–Mathematics Faculty. He submitted his dissertation,[41] which described his research results in organosulfur chemistry, in 1867. Butlerov and Markovnikov served as examiners of the dissertation, and in their joint report, they recommended its approval. In October 1868, Zaitsev successfully defended his dissertation and 2 months later was admitted to the degree. Although the University Charter required the reading of two sample lectures, this requirement was waived.

On April 27, 1868. Mendeleev submitted Butlerov's name as a candidate for the vacant Professorship in chemistry. Sixteen days later, on May 13, 1868, Butlerov accepted an appointment as Professor at large (with no Department associated with the chair) at St. Petersburg University.[42–44] At the request of Kazan University, Butlerov's departure was delayed until the end of December. In an ironic twist, Butlerov was replaced by the transfer of Pyotr Frantsevich Lesgaft (whom we encountered in Chapter 2).

Butlerov's departure posed a problem for the administration at Kazan. It was obvious that Markovnikov, as the senior student and Butlerov's close associate, would take one of the Chairs of Chemistry on Butlerov's departure. Since the move of Kittary to Moscow, Butlerov had assumed the duties of both chairs.

For a time, Butlerov alone had held a Chair of Chemistry, but the University Charter specified that there should be two Chairs, so with Butlerov's departure, holders of two Chairs needed to be appointed. Based on seniority and ability, Markovnikov was the obvious choice to succeed Butlerov, especially given that he had already substituted for his mentor while Butlerov was abroad in Western Europe. But his reputation as a fiery, argumentative individual meant that the last thing that the university Administration wanted was to appoint him as the sole successor to Butlerov. Fortunately, the need to appoint two Professors gave them an opportunity to choose a more moderate colleague for him.

The first candidate considered for the second professorship was Aleksandr Nikiforovich Popov (c. 1840–81, Fig. 4.5). Popov was another Butlerov student who graduated with the degree of kandidat in 1865 for a thesis, "On the affinity of the carbon atom."[45] In this work, Popov had shown experimentally that the four valences of a carbon atom are equivalent, contravening the views of Hermann Kolbe and providing strong support for Butlerov's version of Structural Theory.

He remained at Kazan for his doctoral work, and in 1869, he defended his M. Khim. dissertation, "On the oxidation of monoketones" under the direction of Markovnikov.[45] This dissertation summarized the results of his studies of the oxidation of ketones under forcing conditions (Scheme 4.5).

Figure 4.5 A. M. Butlerov with his research students and Laboratory Assistant Loman, 1868–69. *(N.d.). Image adapted from www.ksu.museum.ru/chmku/images/14b.jpg.*

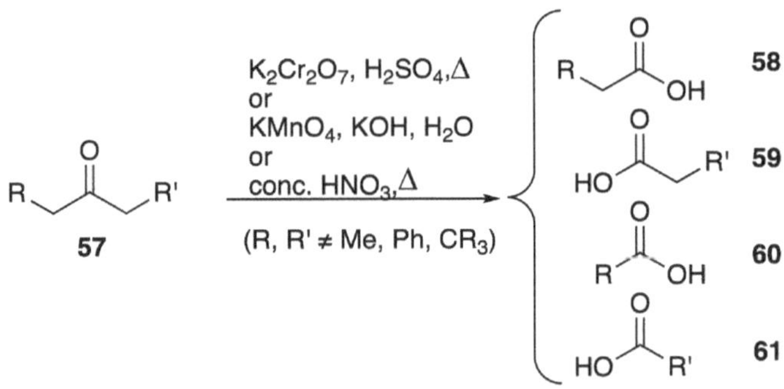

Scheme 4.5 Popov's doctoral research.

The University of Warsaw was founded the year that Popov defended his M. Khim. dissertation, and he was invited to the university as an Extraordinary Professor that same year. In some ways, this was fortunate for Kazan because, although Popov could count on Markovnikov's support, he was difficult to get along with and was never in robust health (he was barely more than 40 years old when he died). In 1872, he submitted his Dr. Khim. dissertation, *O zakonnosti okisleniya ketonov i o primenenii ee k opredeleniyu stroeniya alkogolej i kislot [On the law of oxidation of ketones and its application to the determination of the structure of alcohols and acids]* to Warsaw Imperial University.[46] This work extended his M. Khim. work, but he did little else of note after his move to Warsaw.

Moving into leadership of the Kazan Chemical School

This left Zaitsev as the obvious candidate for the second Professorship, and his much less confrontational nature was almost certainly a comfort to the Administration. Markovnikov thought little of (actually, detested might be more appropriate) his junior colleague, but Butlerov's strong support had allowed Zaitsev to complete his M. Khim. degree in 1869[41] and thus qualify for a faculty position. In a rare display of unity, the Council unanimously elected him Docent after his successful defense.

Zaitsev continued his work toward the Dr. Khim. degree, and in 1870, he presented a dissertation, *A new way to convert fatty acids to their corresponding alcohols. Normal butyl alcohol (propyl carbinol) and its conversion to secondary butyl alcohol (methyl ethyl carbinol)* (Fig. 4.6).[47] As part of this work, Zaitsev reported the reduction of acid chlorides to primary alcohols and their esters by sodium amalgam in buffered ether (Scheme 4.6).[48–50] During this work, he also reported the first synthesis of γ-butyrolactone.[51,52]

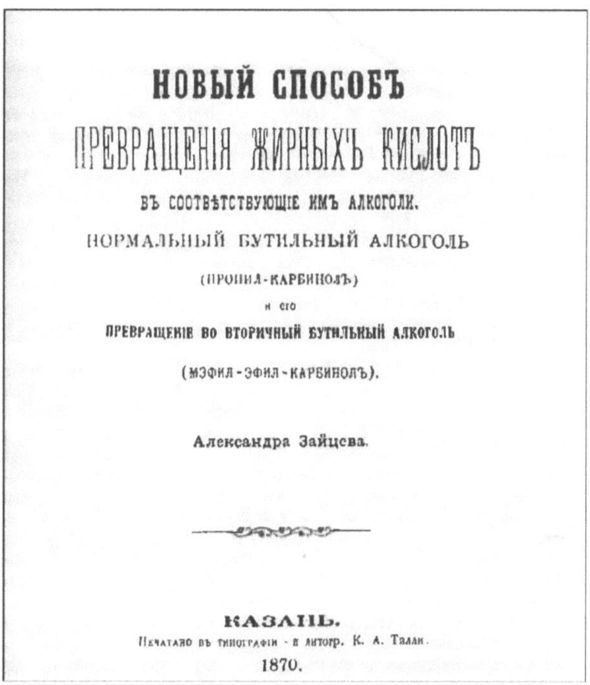
НОВЫЙ СПОСОБЪ
ПРЕВРАЩЕНІЯ ЖИРНЫХЪ КИСЛОТЪ
ВЪ СООТВѢТСТВУЮЩІЕ ИМЪ АЛКОГОЛИ.
НОРМАЛЬНЫЙ БУТИЛЬНЫЙ АЛКОГОЛЬ
(ПРОПИЛ-КАРБИНОЛЪ)
и его
ПРЕВРАЩЕНІЕ ВО ВТОРИЧНЫЙ БУТИЛЬНЫЙ АЛКОГОЛЬ
(МЭФИЛ-ЭФИЛ-КАРБИНОЛЪ).

Александра Зайцева.

КАЗАНЬ.
Печатано въ типографіи и литогр. К. А. Тили.
1870.

Figure 4.6 The title page from Zaitsev's *Dr. Khim.* dissertation. *Taken from Klyuchevich, A. S.; Bykov, G. V. Aleksandr Mikhailovich Zaitsev 1841–1910. Moscow: Izd-vo, "Nauka," 1980; p. 77.*

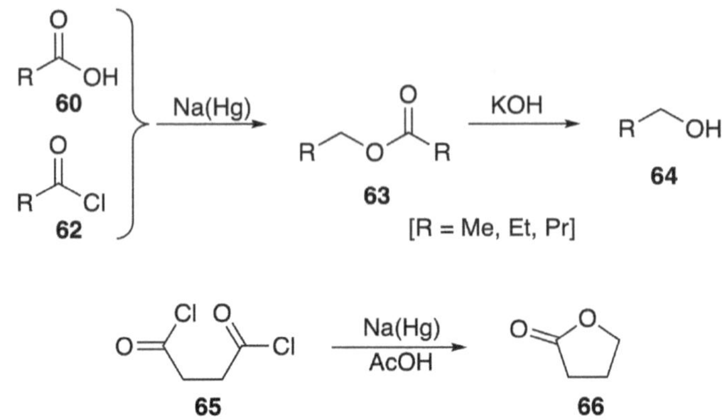

Scheme 4.6 Zaitsev's syntheses of primary alcohols and g-butyrolactone by reduction of acid chlorides.

The Dr. Khim. degree permitted Zaitsev to be appointed to a Professorial Chair should one become available. Six weeks after his successful defense, Zaitsev was appointed Extraordinary Professor of Chemistry. In 1871, a Chair in Chemistry did open up, and the Dean of the Physics–Mathematics faculty, physicist Iosif Antonovich Bol'tsani (1818–76), suggested Zaitsev as a candidate to fill the position. The vote was 19–12 in favor of Zaitsev's appointment, and he was duly appointed Professor (with a 50% boost in salary) in early November 1871.

He had little time to enjoy being the coequal of Markovnikov—before the end of the year, Markovnikov had resigned from the University, leaving Zaitsev responsible for heading the chemistry laboratory and teaching the general and specialist courses in chemistry (Fig. 4.7).

This was a difficult position to sustain, but his options for a junior colleague were severely limited. The best option, Popov had left Kazan 2 years earlier and was not interested in returning. Zaitsev also approached two of the chemists from St. Petersburg who had been founders of the Russian Chemical Society: Nikolai Aleksandrovich Menshutkin (1842–1907, Fig. 4.8) and Viktor Yur'evich Richter (1841–91, Fig. 4.8). Unfortunately, neither accepted his offers. Eventually, the University settled on Grigorii Nikolaevich Glinskii (1842–84, Fig. 4.5), who was a student at Kazan and had done research under Zaitsev's brother, Konstantin Mikhailovich. Glinskii had submitted his kandidat dissertation, which had been passed—but only just. The choice was not a good one for Zaitsev because Glinskii's lack of in-depth knowledge meant that the brunt of the work still fell on him. Following Glinskii's death at the relatively young age of 42, a Butlerov student, Flavian Mikhailovich Flavitskii (1848–1917, Fig. 4.8) was appointed to the second Chair at Kazan.

Figure 4.7 Zaitsev reading a lecture in the Butlerov Auditorium, 1890s. *(N.d.). Retrieved from www.ksu.museum.ru/chmku/foto/00233.jpg.*

Figure 4.8 (L—R) Menshutkin, Richter, Flavitskii. *(N.d.). Retrieved from (Menshutkin) https://commons.wikimedia.org/wiki/File:Nikolai_Menshutkin.jpg, (Richter) https://commons.wikimedia.org/wiki/File:%D0%92%D0%B8%D0%BA%D1%82%D0%BE%D1%80_%D0%AE%D0%BB%D1%8C%D0%B5%D0%B2%D0%B8%D1%87_%D0%A0%D0%B8%D1%85%D1%82%D0%B5%D1%80_(1841-1891).jpg and (Flavitskii) https://commons.wikimedia.org/wiki/File:Flavian_Flavitskij_(1848-1917).jpg.*

Following his appointment as Docent of Chemistry at Kazan in 1869, Zaitsev proceeded to build a four-decade long career at the University. This was by far the longest career of any chemist at Kazan to that point and was only surpassed by his student, A. Ye. Arbuzov (1877–1968), in the 20th century. To celebrate the 40th anniversary of his first appointment at Kazan, his students organized a Jubilee and presented him with bound copies of all his works.

Scientific work as professor at Kazan

Under Butlerov and Markovnikov, the Kazan School had focused on the Structural Theory of Organic Chemistry and obtaining evidence for Butlerov's version of it. This entailed the synthesis of the compounds predicted by this theory, which was essential to its confirmation. Butlerov had accomplished the first synthesis of *tert*-butyl alcohol in 1863[53–55] by the reaction between dimethylzinc and phosgene and, a year later, by the reaction of dimethylzinc with acetyl chloride[56,57] (Scheme 4.7).

Zaitsev subsequently modified his mentor's procedure by replacing the dialkylzinc reagent with a mixture of the alkyl iodide and zinc metal (usually with a small amount of sodium as an activating agent). The resulting alkylzinc iodide was treated with an acid chloride to give the corresponding tertiary alcohols where two of the alkyl groups were derived from the alkyl iodide. With his students, Zaitsev made his organozinc synthesis the state of the art[58] until its displacement by the Grignard reaction.[59–66]

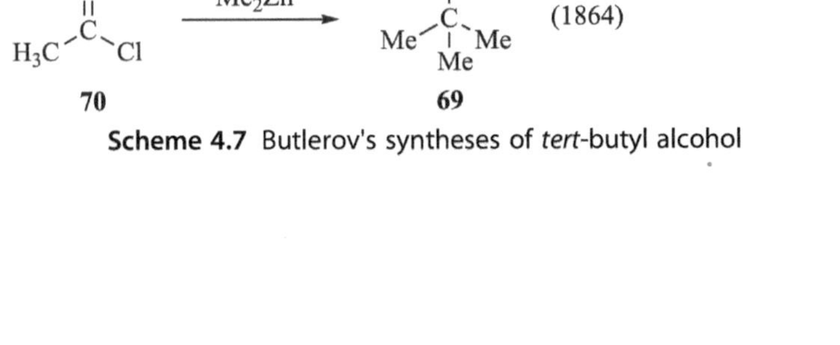

Scheme 4.7 Butlerov's syntheses of *tert*-butyl alcohol

Zaitsev's advances in this synthesis were in two major areas: (1) he expanded the types of carbonyl compounds to include formate esters; and (2) he developed an excellent method for the addition of allyl nucleophiles to carbonyl compounds.

With his students, Innokenti Ivanovich Kanonnikov (1854–1902, Fig. 4.9)[67–70] and Yegor Yegorovich Vagner (Georg Wagner 1849–1903, Fig. 4.9),[71,72] Zaitsev extended the organozinc approach to the synthesis of symmetrical and unsymmetrical secondary alcohols. Later, Wagner would independently extend this to the reaction of aldehydes with organozinc reagents.[73,74] Zaitsev's brother (and student), Mikhail Mikhailovich (1845–1904, Fig. 4.9), was involved in the development of the allylation reaction for the synthesis of tertiary homoallylic alcohols,[75] a reaction later exploited by Wagner for the synthesis of homoallylic secondary alcohols.[76,77] These reactions are summarized in Scheme 4.8.

Zaitsev's synthesis of alcohols fell into disuse for nearly a century after the disclosure of the Grignard reaction, but its use was resurrected by Ryoji Noyori in the 1980s, when he found that excellent asymmetric induction in the formation of secondary alcohols with zialkylzinc reagents could be

Figure 4.9 (L–R): Innokenti Kanonnikov, Yegor Vagner, Mikhail Zaitsev. *(N.d.). Images retrieved from (Kanonnikov) https://commons.wikimedia.org/wiki/File:Inocêncio_Ivanovich_Kanonikof_(1854–1902).jpg, (Wagner) https://commons.wikimedia.org/wiki/File:Georg_Wagner.jpg, and (M. M. Zaitsev) www.ksu.museum.ru/chmku/images/68b.jpg.*

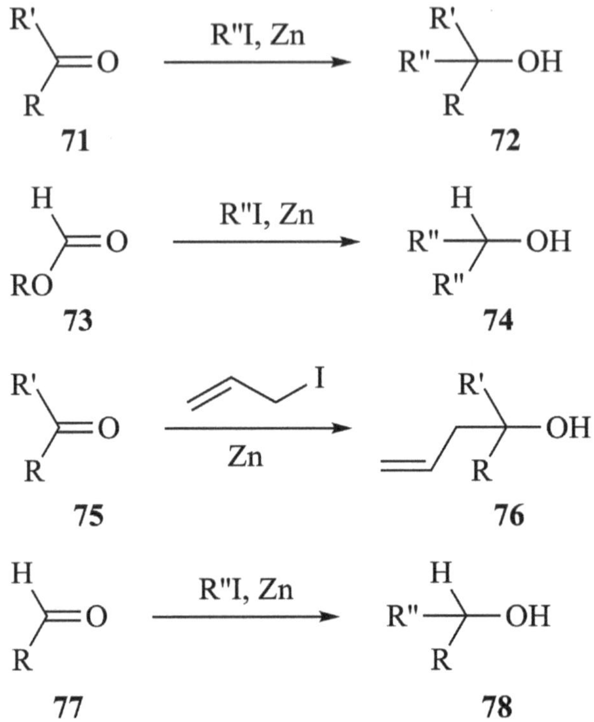

Scheme 4.8 Zaitsev alcohol syntheses.

accomplished by using a chiral 3-aminoisoborneol adjuvant as a catalyst[78–82] (Scheme 4.9). This work, for which Noyori shared the Nobel Prize, was one of the first demonstrations of nonlinear effects of asymmetric induction.

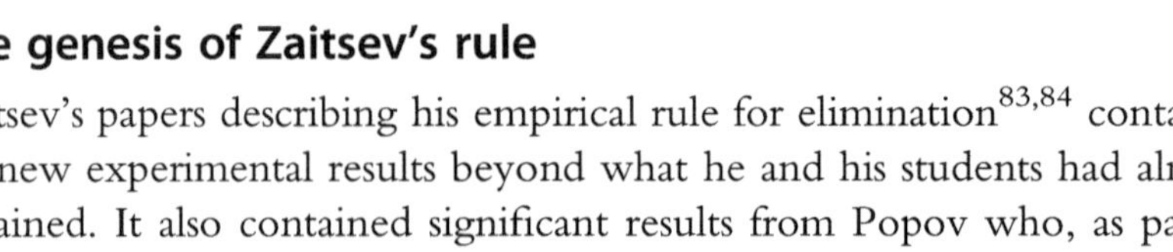

Scheme 4.9 Noyori's asymmetric Zaitsev–Wagner synthesis of secondary alcohols.

The genesis of Zaitsev's rule

Zaitsev's papers describing his empirical rule for elimination[83,84] contained no new experimental results beyond what he and his students had already obtained. It also contained significant results from Popov who, as part of his doctoral work, had formulated what became known as "Popov's Rule." This rule states that when an unsymmetrical ketone is oxidized, one of the bonds to the carbonyl group is broken. In a series of nine papers in German and Russian between 1868 and 1872,[85–93] Popov reported the results from an extensive study of the oxidation of ketones. Except for noting that bonds to methyl, phenyl, and tertiary alkyl groups do not break, Popov reported that a mixture of all possible products is obtained. This observation is consistent with the enol being an intermediate in the reaction.

In addition to the work of Popov, Zaitsev also relied heavily on the work of his students, Nikolai I[vanovich?] Grabovskii,[94,95] and Wagner.[72,96,97] In these students' studies, the 1,2-transposition of a hydroxyl group was accomplished by the sequence of conversion of the starting alcohol (**82**) to the iodide (**83**) with iodine and phosphorus, dehydrohalogenation with base, and rehydration with aqueous sulfuric acid according to Markovnikov's rule. Using alcoholic potassium hydroxide, 1-iodobutane gave 23%–28% of the alkene (**84**); the rest of the product as 1-ethoxybutane (**85**). The alkene intermediates were also treated with bromine to obtain the dibromides (**86**); treatment with sodium hypochlorite and sodium hydroxide gave the corresponding glycols (**87**) (Scheme 4.10).

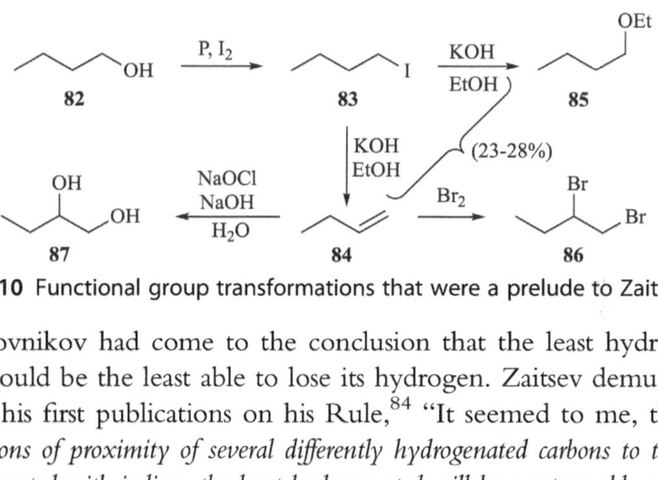

Scheme 4.10 Functional group transformations that were a prelude to Zaitsev's Rule.

Markovnikov had come to the conclusion that the least hydrogenated carbon would be the least able to lose its hydrogen. Zaitsev demurred, and wrote in his first publications on his Rule,[84] "It seemed to me, that *under the conditions of proximity of several differently hydrogenated carbons to the unit of carbon connected with iodine, the least hydrogenated will be most capable of losing its hydrogen.*" This conclusion is exactly the opposite to that of Markovnikov. In more modern terms, Markovnikov had proposed that base elimination of alkyl iodides should occur with Hofmann regiochemistry; the experimental work of Zaitsev and his students showed that this was not, in fact, the case.

The feud

There was no love lost between Markovnikov and Zaitsev, who carried on a life-long feud. Markovnikov, Butlerov's most devoted disciple, was a brilliant theoretician who did not appreciate the way that Butlerov's Structural Theory of Organic Chemistry was disregarded by many of the chemists in Western Europe, who tended to promote the views of Kekulé and Kolbe. In Markovnikov's view, Zaitsev was an apostate who had denigrated

Butlerov's theory in his ill-conceived kandidat dissertation from Marburg. Given Markovnikov's fanatical devotion to Butlerov and Russia (especially Russian chemistry), it is not unreasonable to suggest that this dissertation may have played a major role in the feud, as has been suggested before.[98]

However, Zaitsev may also have played a part, himself, in allowing things to fester. His own strengths were in the laboratory, doing experiments—his organozinc syntheses marked him as a synthetic chemist of the first water. Markovnikov, on the other hand, saw theory as the highest calling of a modern organic chemist, and he discounted Zaitsev's synthetic work rather as lesser science. And Markovnikov was never one to mince words, so it is doubtful whether he would hold back his disdain of his fellow student. Given his status as Butlerov's clear favorite student, this left Zaitsev with what we would now call an inferiority complex, despite Butlerov's many actions promoting him. Perhaps Zaitsev felt so intimidated and demeaned by Markovnikov and Markovnikov's close relationship with Butlerov that Butlerov may not have been able to do anything to soothe his feelings about the "teacher's pet."

The accomplished chemist-historian, Prof. Dr. Aleksei Vladimirovich Zakharov of the Butlerov Institute of Organic Chemistry of Kazan Federal University, has suggested to this author that things may have been even more serious and that Zaitsev may actually have been afraid of Markovnikov and of any other chemist showing theoretical brilliance. Prof. Dr. Zakharov suggested that this fear may have been behind Zaitsev's support of Wagner's move to St. Petersburg.

Of course, we will never know for certain …

Alkene oxidation

In 1885, Zaitsev's research again took a sharp turn into a new field, this time into what is now known as the dihydroxylation of alkenes. This work followed the early work of Tanatar[99–101] on the oxidation of fumaric and maleic acids. Zaitsev's work extended Tanatar's work into the oxidation of unsaturated fatty acids to include long-chain acids, especially oleic acid and its derivatives. In his first reports,[102–104] Zaitsev disclosed the oxidation of oleic acid with permanganate under neutral and alkaline conditions. Under neutral conditions, Zaitsev reported that oleic acid (**88**) was oxidized to azelaic acid (**90**), but that under alkaline conditions, dihydroxystearic acid (**91**) was the product (Scheme 4.11). From the neutral oxidation, he was able to deduce the structure for oleic acid: the structure of elaidic acid (**89**), which was known, placed the alkene double bond at position 9 of the oleic acid backbone. Although Zaitsev did not know the stereochemistry of his glycol product, we now know that the addition is suprafacial, and that the erythro isomer is formed.

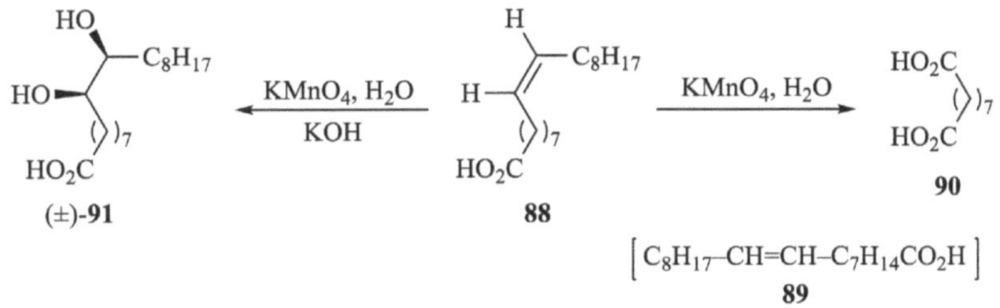

Scheme 4.11 Zaitsev's initial reports of the oxidation of oleic acid.

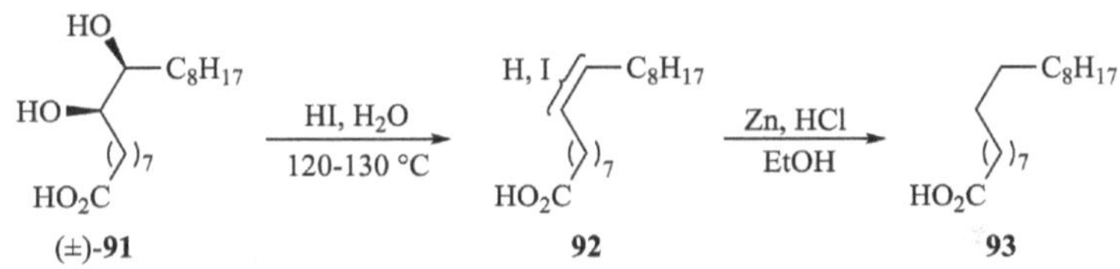

Scheme 4.12 Zaitsev's correlation of oleic acid with stearic acid.

In an early report, Zaitsev first correlated the structure of the diol **91** with the known stearic acid (**93**) (Scheme 4.12).

This he did by first reducing the diol **84** using the Berthelot conditions of heating in a sealed tube with concentrated hydriodic acid,[105,106] which gave the iodostearic acid derivative **92**. This was then further reduced with zinc dust in ethanol to give stearic acid. As part of doing this, he was also able to show that the work of Gröger,[107] whose analysis of the permanganate oxidation product of tallow acid that had led him to propose a molecular formula of $C_{26}H_{52}O_6$, was in error,[103] and that the correct molecular formula for the oxidation product was $C_{18}H_{36}O_4$ (i.e., the product was actually **93**). Zaitsev confirmed his formula by combustion and gravimetric analyses of the sodium, potassium, calcium, zinc, silver, and barium salts of the acid.

In addition to their studies on the chemistry of oleic acid, Zaitsev and his brothers also studied the chemistry elaidic acid (**94**), ricinoleic acid (**95**), and the elongated (C22) fatty acids erucic acid (**96**) and its *trans* isomer, brassidic acid (**97**)[108–111] (Fig. 4.10).

Following this initial foray, Zaitsev and his research group, including his brothers Konstantin and Mikhail, began a systematic study of the transformations of oleic acid under a variety of conditions. The results of their studies are summarized in Scheme 4.13.

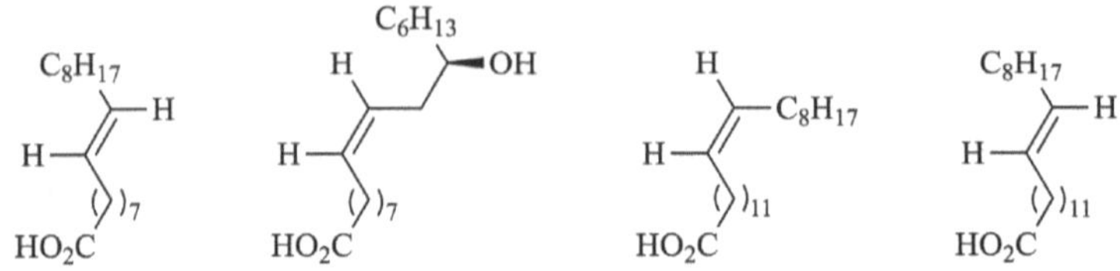

94 **95** **96** **97**

Figure 4.10 Unsaturated fatty acids that Zaitsev and his brothers worked with.

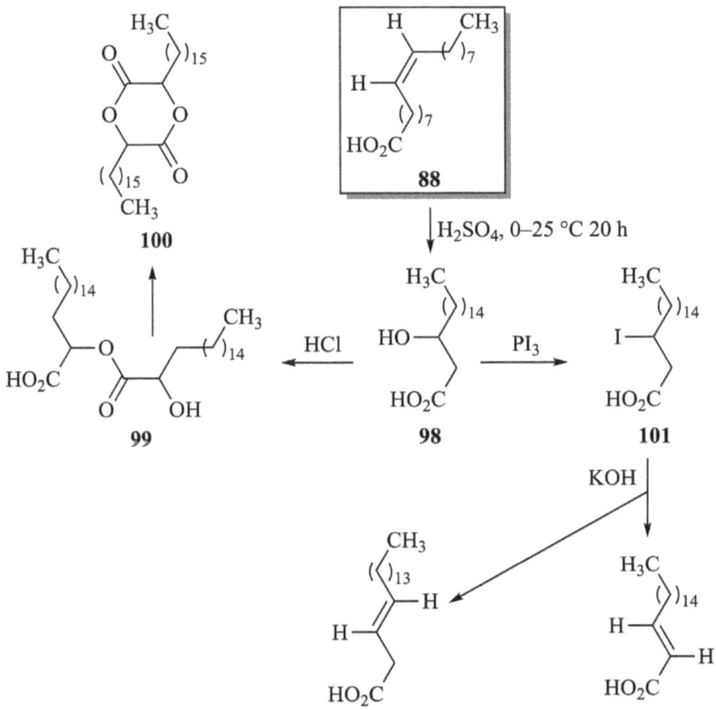

103 **102**

Scheme 4.13 Transformations of oleic acid with stearic acid.

Zaitsev's later career and legacy

Awards and honors

Although he had not been a founding member of the Russian Chemical Society, he served three terms as Chair and President of the Society (1905, 1909, and 1910). In 1903, he was elected a Corresponding Member of the Russian Academy of Sciences, where he served on the Award Committee of the Greater and Lesser Butlerov Prizes. He won the Greater Butlerov Prize himself in 1909 for services to chemistry.

Zaitsev's legacy

Like his mentor Butlerov before him, Zaitsev left a legacy not only in terms of the work he did himself, but in the work of his students. Three of these students developed their own eponymous reactions.

Yegor Yegorovich Wagner (Vagner) (1849–1903)

Yegor Yegorovich Vagner (Georg Wagner)[112–117] was a brilliant chemist who is much better known in the West under the German spelling of his name. It is important to note, however, that both he and his father used the name Yegor (an emphatically Russian version of the name, Georgii) to underscore the fact that they were Russian, not German. His mother died when Wagner was just a year old, and his upbringing was largely relegated to his grandmother. At the age of 16, Wagner was sent to Leffler boarding school, was near the city of Wenden, in Livonia (modern Cēsis, in Latvia). He was a brilliant student but got into a great deal of trouble because he was bored at the school. He eventually ran away and returned the thousand miles to Kazan. He ran out of money about two-thirds of the way home, so he had to finish the journey on foot, eating food that he begged. His indulgent father did not punish him for running away, but Wagner underwent intensive home-schooling to prepare him for entry into Kazan University. In 1867, he entered the Juridical Faculty as a student in Law. Wagner did not find the curriculum burdensome, which allowed him to explore the local theater scene, as well as lectures in other disciplines. It was Zaitsev's lectures in chemistry that decided him on becoming a chemist. In 1869, he petitioned the University to allow him to move to the Mathematics–Physics Faculty as a student in chemistry. The petition was granted, but at the price of having to start over again: none of his previous coursework was permitted to carry over. He graduated in 1874 with the degree of kandidat.

Wagner was one of the most brilliant chemists of his era. Had he not died tragically young in 1903, he would almost certainly have been a strong candidate for the Nobel Prize in Chemistry. As it is, with a professional career of just 15 years, he still managed to: (1) extend his mentor's syntheses of alcohols to permit the synthesis of unsymmetrical secondary alcohols by the reaction of organozinc reagents and aldehydes;[73,74,76,77] (2) extend his

104 **105** **106** **107**

Figure 4.11 The first rearrangements in the bornane series reported and explained by Wagner.

mentor's work on the alkaline permanganate oxidation of alkenes to water-insoluble alkenes[118–125]; and (3) become the first to provide an explanation of the isomerization of pinenes and camphenes to bornane derivatives.[126–128] Wagner's prototypical example of these rearrangements was the conversion of α-pinene (**104**) to bornyl chloride (**105**), and camphene (**106**) to isobornyl chloride (**107**) with hydrogen chloride (Fig. 4.11).

The importance of the permanganate oxidation lay in the fact that it occurred without rearrangement of the double bond, so the location of the double bond could be fixed unambiguously; in Russia, the reaction carries his name as the Wagner oxidation. The rearrangement that he proposed now carries the joint names of Wagner himself and the German chemist Hans Leb(e)recht Meerwein (1875–1965), who proposed the first reasonable mechanism for the reaction.

Sergei Nikolaevich Reformatskii (1860–1934)

The Reformatskii brothers, Sergei Nikolaevich (1860–1934, Fig. 4.12) and Aleksandr Nikolaevich (1864–1937, Fig. 4.12), both studied under Zaitsev at Kazan, where Aleksandr Reformatskii became one of Zaitsev's favorites. He eventually became Professor of Inorganic Chemistry at the Moscow State University and served at various times as Dean of the Faculty. His older brother, Sergei, was much closer to Zaitsev scientifically.

Sergei Reformatskii[129,130] entered Zaitsev's laboratory as a student, and in 1882, he graduated with the degree of kandidat and a gold medal for his thesis, which reported the extension of Zaitsev's allylation reactions with zinc and allyl iodide by replacing the allyl iodide with an α-haloester or an α-haloketone.[131] Reformatskii successfully defended both M. Khim.[132] and Dr. Khim.[133] dissertations on his reaction at Warsaw, and he published four key papers in the *Berichte der deutschen chemischem Gesellschaft* between 1887 and 1895.[134–137] In structural terms as they were understood at that time, this was an obvious substitution: replacing a methylene group at the

Figure 4.12 The Reformatskii brothers: Aleksandr (L) and Sergei (R). *(N.d.). Images retrieved from (A. N. Reformatskii) https://www.wikidata.org/wiki/Q16691992#/media/File:Aleksandro_Reformatskij_(1864-1937).jpg https://ru.wikipedia.org/wikiРеформатский_Александр_Николаевич#/media/Файл:Aleksandro_Reformatskij_(1864–1937).jpg and (S. N. Reformatskii) Lewis, D. E.,* Early Russian Organic Chemists and Their Legacy*; Springer: Heidelberg, 2012; p. 96.*

end of a double bond with an oxygen atom. The reaction of these organozinc reagents with aldehydes and ketones has been a reliable method for the production of β-hydroxycarbonyl compounds for over a century and a quarter, and now bears his name.[138–141] One of its most important characteristics of the Reformatskii reaction is that it can be used for the synthesis of highly hindered β-hydroxycarbonyl derivatives. A good example is provided by Reformatskii's own synthesis of the hindered ester **109** from ethyl α-bromoisobutyrate and acetone[137] (Scheme 4.14). Even with modern preformed lithium enolates, this reaction is impossible to accomplish by the aldol addition.

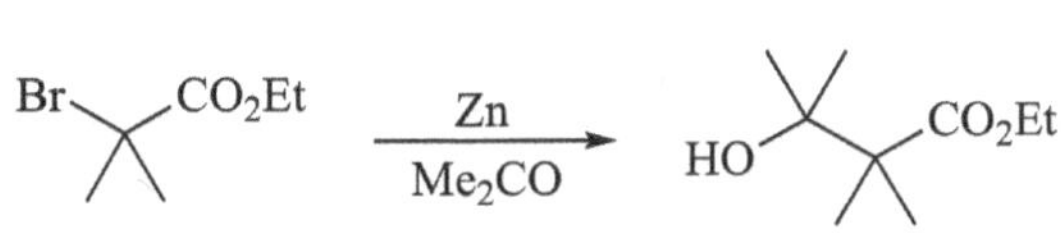

108 **109**

Scheme 4.14 An early Reformatskii reaction, showing its ability to form highly hindered β-hydroxyesters.

Aleksandr Yerminingel'dovich Arbuzov (1877–1968)

The last of Zaitsev's students to attain an international reputation and discover an eponymous reaction was Aleksandr Yerminingel'dovich Arbuzov 1877–1968, Fig. 4.13).[142–147] Arbuzov entered Kazan University, where he came under the influence of Zaitsev. The year that Arbuzov graduated at Kazan, Grignard published his first papers on his synthesis of alcohols using organomagnesium reagents. This led to a complete reassessment of Arbuzov's career path as a scientist. He did publish one article, in both Russian and German, on the synthesis of 2-phenyl-4-penten-1-ol by the reaction between allyl iodide, zinc metal, and acetophenone,[148,149] and a year later he made remarks, recorded in the minutes of the meeting of the Russian Physical–Chemical Society but never published as a formal paper, on the use of magnesium at Kazan for the Zaitsev synthesis of tertiary alcohols.[150]

The advent of the Grignard reaction obviously ended his graduate project on the synthesis of alcohols by Zaitsev's method, and Arbuzov quickly settled on a new project: the chemistry of phosphorous acid and its derivatives. In 1905, he published a series of three papers in the *Zhurnal Russkago Fiziko-Khimicheskago Obshchestva* containing four parts (glavy) that described his discovery of the rearrangement that bears his name.[151–153] He followed the initial disclosures in Russian with papers in German.[154,155] His work revealed that hydrogen halides, alkyl halides that would participate in

Figure 4.13 (L) Aleksandr Yerminingel'dovich Arbuzov (1914); (R) The International Arbuzovs Prize Diploma and Medal. *(N.d.) Images from (Arbuzov) https://commons.wikimedia.org/wiki/File:A._E._Arbuzov_1914.jpg and (Diploma and Medal) www.ksu.museum.ru/chmku/images/41b.jpg.*

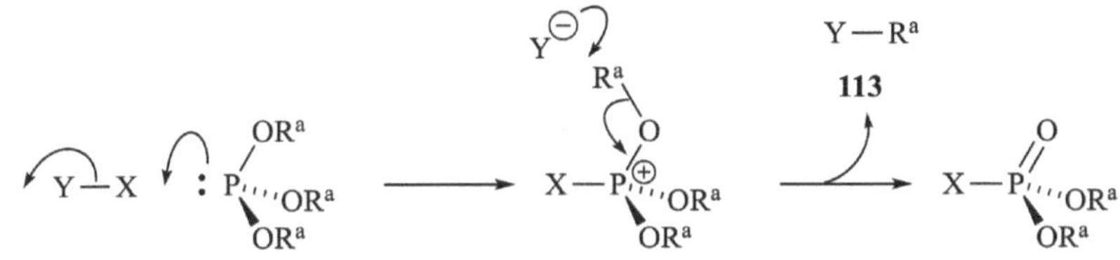

110 111 112 113

X=H; Me, Et, Pr, etc.; Br, I; Y=Br, I; R^a=Br, I

Scheme 4.15 The prototype eponymous rearrangements of phosphite esters reported by Arbuzov.

S_N2 substitution, and the halogens themselves would promote the formation of tetracoordinate phosphorous acid derivatives and phosphonic acid derivatives (Scheme 4.15).[156–158]

The Arbuzov rearrangement was critical in the development of reactions such as the Horner–Wadsworth–Emmons olefination.[159–162] After the Chair had been declined by Aleksandr Reformatskii and Vladimir Vasil'evich Chelintsev (1877–1947), both of whom decided to remain in Moscow, Arbuzov succeeded his mentor in the Chair of Chemistry at Kazan University, a position he held until 1930, when he became the Director of the Kazan Technical Institute of the USSR Academy of Sciences; in 1945, he was appointed permanent Director. This Institute is now named for him. In recognition of the contribution of the Arbuzovs, father and son, to science, the President and the Government of the Tatarstan Republic instituted the Arbuzovs International Prize (Fig. 4.12) for distinguished achievements in the chemistry of phosphorus compounds.

References

1. Lewis, D. E. Introduction to an English Translation of Markovnikov's First Paper Describing "Markovnikov's Rule," and Primary Documents. "On the Question of the Mutual Influence of Atoms in Chemical Compounds." V. *Bull. Hist. Chem.* **1869,** *1,* 242–247. Morkovnikov. Zhurnal Russkago Fiziko-Khimicheskago Obshchestva.
2. Markovnikoff, V. Sur Les Lois Qui Régissent Les Réactions de l'addition Directe. *C. R. Hebd. Séances Acad. Sci.* **1875,** *81,* 668–671.
3. Markovnikoff, V. Sur Les Lois Qui Régissent Les Réactions de l'addition Directe. *C. R. Hebd. Séances Acad. Sci.* **1875,** *81,* 728–730.
4. Markovnikoff, V. Sur Les Lois Qui Régissent Les Réactions de l'addition Directe. *C. R. Hebd. Séances Acad. Sci.* **1875,** *81,* 776–779.
5. Markovnikov, V. O Zakonakh Obrazovaniya Pryamikh Soedinenii Nepredel'nymi Organicheskimi Chastitsami [On the Laws of the Direct Formation of Compounds

with Unsaturated Organic Groups]. *Zh. Russ. Khim. O-va. I Fiz. O-va* **1876,** *8,* 16–29.
6. Markownikoff, W. Ueber Die Abhängigkeit Der Verschiedenen Vertretbarkeit Des Radikalwasserstoffs in Den Isomieren Buttersäuren. *Ann. Chem. Pharm* **1870,** *153,* 228–259. https://doi.org/10.1002/jlac.18701530204.
7. Morkovnikov, V. Prevrashenie Izobutil'nago Alkogolya v Tetichnyi Psevdobutil'nyi Alkogol [On the question of the mutual influence of atoms in chemical compounds. the conversion of isobutyl alcohol into tertiary pseudobutyl alcohol. *Zh. Russ. Khim. O-va* **1869,** *1,* 242–247.
8. Klyuchevich, A. S.; Bykov, G. V. *Aleksandr Mikhailovich Zaitsev. 1841–1910.; Izd-Vo. "Nauka": Moscow,* 1980.
9. Reformatskii, A. Biografiya Professora Aleksandra Mikhailovicha Zaitseva [A biography of Professor Aleksandr Mikhailovich Zaitsev]. *Zh. Russ. Fiz.-Khim. O-va* **1911,** *43,* 857–876.
10. Reformatskii, S. Spisok Trydov A. M. Zaitseva i Ego Uchenikov [List of works of A. M. Zaitsev and his students]. *Zh. Russ. Fiz.-Khim. O-va* **1911,** *43,* 933–942.
11. Reformatskii, S.; Al'bitskii, A. Ucheniya Deyatel'nost Prof. A. M. Zaitseva [Scientific works of Prof. A. M. Zaitsev]. *Zh. Russ. Fiz.-Khim. O-va* **1911,** *43,* 876–932.
12. Lewis, D. E.; Zaitsev, A. M. Lasting contributions of a synthetic virtuoso a century after his death. *Angew. Chem. Int. Ed.* **2011,** *50* (29), 6452–6458. https://doi.org/10.1002/anie.201100215.
13. Lewis, D. E. Aleksandr Mikhailovich Zaitsev (1841–1910). Markovnikov's Conservative Contemporary. *Bull. Hist. Chem.* **1995,** *17/18,* 21–30.
14. Lewis, D. E.; Saytzeff, A. M. Bleibendes Vermächtnis Eines Virtuosen Der Synthesechemie. *Angew. Chem.* **2011,** *123,* 6580–6586. https://doi.org/10.1002/ange.201100215.
15. Malov (Archpriest). *E. Istoricheskoe Opisanie Tserkvei Goroda Kazani [Historical Description of the Churches of the City of Kazan, Issue 1.; Tipografiya Imp; Tip. Imp. Un-ta.: Kazan,* 1884.
16. Morkovnikov, V. *Organicheskaya Khimiya, Sistavlennaya Po Lektsiyam Ord. Prof. A. M. Butlerova Stud. Vl. Morkovnikovym 1859–1860 [Organic Chemistry Compiled from the Lectures of Ord. Prof. A. M. Butlerov by Student Vl. Morkovnikov 1859-1860].; Kazan Litografiya: Kazan,* 1860.
17. Morkovnikov, V. *O al'degidakh i Ikh Otnosheniyakh k Alkogolyam i Ketonam [On Aldehydes and Their Relationship to Alcohols and Ketones]. Kazan,* 1860.
18. Bukhvostov, I. *Organicheskaya Khimiya, Sostavlennaya Po Lektsiyam Ord. Prof. A. M. Butlerova Stud. Iv. Bukhvostovym 1862–1863 [Organic Chemistry Compiled from the Lectures of Ord. Prof. A. M. Butlerov by Student Iv. Bukhvostov 1862-1863].; Kazan Litografiya: Kazan,* 1863.
19. Kolbe, H. Ueber den natürlichen Zusammenhang der organischen mit den unorganischen Verbindungen, die wissenschaftliche Grundlage zu einer naturgemässen Classification der organischen chemischen Körper. *Justus Liebigs Ann. Chem.* **1860,** *113* (3), 293–332. https://doi.org/10.1002/jlac.18601130305.
20. Zaitsev, A. *Teoreticheskie Vzglyady Kol'be Na Ratsional'nuyu Konstitutsiyu Organicheskikh Soedinenii i Ikh Svyaz s Neorganicheskimi [The Theoretical Views of Kolbe on the Rational Constitution of Organic Compounds and Their Relationship with Inorganic Ones;* (N.A.): Kazan, 1863.
21. Zakharov, A. V. *Khronologiya Stanovleniya Khimicheskoi Laboratorii i Kazanskoi Khimicheskoi Shkoly [The Chronology of the Formation of the Chemical Laboratory and the Kazan Chemical School]* Vol. 1; Kazan Federal University: Kazan, 2011; p 692.
22. Butlerow, A. Einiges über die chemische Structur der Körper. *Z. Chem. Pharm* **1861,** *4,* 549–560.

23. Butlerov, A. *O Khimicheskom Stroenii Veshchestv [On the Chemical Structure of Substances]. Uch. Zap. Kazan. Un-ta. (Otd. fiz.-mat. i med. nauk)* **1862,** *Vol. 1; pp 1–11.*
24. Zelinskii, N. D. Aleksandr Mikhailovich Zaitsev. *Russ. Chem. Bull.* **1910,** *4,* 512–513.
25. Gordin, M. D. Beilstein Unbound: The Pedagogical Unraveling of a Man and His Handbuch. In *Pedagogy and the Practice of Science;* Historical, Perspectives, C., Kaiser, D., Eds.; MIT Press: Cambridge, MA, 2005; pp 11–39.
26. Saytzeff, A. Sur l'acide Diamidosalicylique et Sur Ses Combinaisons Avec Les Acides. *Bull. Soc. Chim.* **1865,** *3,* 244–250. Paris, [2].
27. Saytzeff, A. XXXVIII. Ueber Diamidosalicylsäure. *Justus Liebigs Ann. Chem.* **1865,** *133* (3), 321–329. https://doi.org/10.1002/jlac.18651330308.
28. Zaitsev, A. M. O. *Diamidosalitsilovoi Kislote [On Diaminosalcylic Acid].; Kazan,* 1864.
29. Saytzeff, A. Action de Cyanate de Potasses Sur l'éther Monochloroacétique. *Bull Soc. Chim.* **1865,** *3,* 350–356. Paris, [2].
30. Saytzeff, A. Action de Cyanate de Potasses Sur l'éther Monochloroacétique. *C. R. Hebd. Séances Acad. Sci.* **1865,** *60,* 671–673.
31. Saytzeff, A. Ueber Die Einwirkung von Cyansaurem Kali Auf Monochlor-Essigsäther. *Ann. Chem. Pharm.* **1865,** *133,* 329–336. https://doi.org/10.1002/jlac.18651350215.
32. Saytzeff, A. Ueber Die Einwirkung von Cyansaurem Kali Auf Monochloressigsäther. *Ann. Chem. Pharm.* **1865,** *135,* 229–236. https://doi.org/10.1002/jlac.18651350215.
33. Saytzeff, A. XLIII. Ueber eine neue Reihe organischer Schwefelverbindungen. *Justus Liebigs Ann. Chem.* **1866,** *139* (3), 354–364. https://doi.org/10.1002/jlac.18661390310.
34. Saytzeff, A. Ueber Die Einwirkung von Saltpetersäure Auf Schwefelmethyl Und Schwefeläthyl. *Justus Liebigs Ann. Chem.* **1867,** *144,* 148–156. https://doi.org/10.1002/jlac.18671440205.
35. v. Oefele, A. Vorläufige Notiz über die Oxydationsproducte des Einfach-Schwefeläthyls. *Justus Liebigs Ann. Chem.* **1863,** *127* (3), 370–371. https://doi.org/10.1002/jlac.18631270315.
36. von Oefele, A. XXXIV. Ueber Diäthylsulfon. *Ann. Chem. Pharm* **1864,** *132,* 86–90. https://doi.org/10.1002/jlac.18641320112.
37. Zaitsev, M. O Srodstvie Sery v Sernistom Amil-Efile [On the affinity of the sulfur of amyl ethyl sulfide]. *Zh. Russ. Khim. O-va* **1870,** *2,* 47–49.
38. von Oefele, A. Untersuchungen aus dem academischen Laboratorium in Marburg: XXXIII. Ueber eine neue Klasse organischer Schwefelverbindungen. *Justus Liebigs Ann. Chem.* **1864,** *132* (1), 82–86. https://doi.org/10.1002/jlac.18641320111.
39. Saytzeff, A. Ueber die Einwirkung von Jodmethyl auf Schwefelamyläthyl. *Justus Liebigs Ann. Chem.* **1867,** *144* (2), 145–148. https://doi.org/10.1002/jlac.18671440204.
40. Saytzeff, A. *Ueber Eine Neue Reihe Organischer Verbindungen; Leipzig,* 1865.
41. Zaitsev, A. *O Deistvii Azotnoi Kisloty Na Nekotorye Organicheskie Soedineniya Dvuatomnoi Sery i o Novom Ryade Organicheskikh Sernistykh Soedinenii Poluchennom Pri Etoi Reaktsii Prolozhenie [On the Reactions of Nitric Acid with Divalent Sulfur Compounds and on a New Class of Organic Sulfur Compounds Obtained from This Reaction; Kazan,* 1867.
42. Dobrotin, R. B.; Karpilo, N. G.; Kerova, L. S.; Trofimov, D. N. *Letopis Zhizni i Deyatel'nosti D. I. Mendeleeva [A chronicle of the life of D. I. Mendeleev]; Nauka;* Leningrad Branch: Leningrad, 1984; pp 108.
43. Figurovskii, N. A. *Dmitrii Ivanovich Mendeleev. 1834–1907.; Izd-vo. Akad;* Nauk SSSR: Moscow, 1961; pp 295.
44. Mendeleev, D. I. *Sochineniya [Works]; Izv-do* Vol. 15; Nauk SSSR: Leningrad-Moscow, 1949; p 295.

45. Popov, A. *Po Povodu Srodstva Yglerondago Paya [On the Affinity of the Carbon Atom]; Kazan,* 1865.
46. Popov, A. *O Zakonnosti Okisleniya Ketonov i o Primenenii Ee k Opredeleniyu Stroeniya Alkogolei i Kislot [On the Law of Oxidation of Ketones and its Application to the Determination of the Structure of Alcohols and Acids]; Dr. Khim. Diss.: Warsaw,* 1872.
47. Zaitsev, A. *Novyi Sposob Prevrashcheniya Zhirnykh Kislot v Sootvestvuyushchie Im Alkogoli. Normal'nyi Butil'nyi Alkogol (Propil-Karbinol) i Ego Prevrashchenie vo Vtorichnyi Butil'nyi Alkogol (Mefil-Efil-Karbinol) [A New Way to Convert Fatty Acids to Their Corresponding Alcohols. Normal Butyl Alcohol (Propyl Carbinol) and its Conversion to Secondary Butyl Alcohol,* 1870.
48. Saytzeff, A. Ein Neues Verfahren, Die Fettsäuren in Die Entsprechenden Alkohole Umzuwandeln. *J. Prakt. Chem.* **1871,** *3,* 76–88. https://doi.org/10.1002/prac.18710030107.
49. Saytzeff, A. Neues Verfahren Zur Umwandlung Der Fettsäure in Deren Alkohole. Synthese Des Normalen Butylalkohols. *Z. Chem. N. F.* **1870,** *6,* 105–108.
50. Saytzeff, A. Eine Neue Methode Fettsäuren in Entsprechende Alkohole Umzuwandeln. *Z. Chem. N. F.* **1869,** *5,* 551–552.
51. Saytzeff, A. Ueber Das Produkt Der Reduction von Succinylchlorid Und Über Die Normale γ-Oxybuttersäure. *J. Prakt. Chem.* **1882,** *25,* 61–72. https://doi.org/10.1002/prac.18810250106.
52. Saytzeff, A. 4. Ueber die Reduction des Succinylchlorids. *Justus Liebigs Ann. Chem.* **1874,** *171* (2), 258–290. https://doi.org/10.1002/jlac.18741710216.
53. Boutlerow, A. *Sur Quelques Composés Organiques Simples. Bull. Soc. Chim. Paris* **1863,** *Vol. 1; pp 582–594.*
54. Butlerow, A. Studien über die einfachsten Verbindungen der organischen Chemie. *Z. Chem. Pharm* **1863,** *6,* 484–497.
55. Butlerow, A. *Will's Jahresberichte,* 1863; pp 474–476.
56. Boutlerow, A. Sur l'alcool Pseudobutylique Tertiaire Ou l'alcool Méthylique Triméthylé. *Bull. Soc. Chim.* **1864,** *Vol. 1, 106–116. Paris.*
57. Butlerow, A. Ueber den tertiären Pseudobutylalkohol (den trimethylirten methylalkohol). *Z. Chem. Pharm* **1864,** *7,* 385–402.
58. Lewis, D. E. The beginnings of synthetic organic chemistry: Zinc alkyls and the Kazan' school. *Bull. Hist. Chem.* **2002,** *27,* 37–42.
59. Grignard, V. Sur Quelques Nouvelles Combinaisons Organometalliques Du Magnésium et Leur Application à Des Synthèses d'alcools et d'hydrocarbures. *C. R. Hebd. Séances Acad. Sci.* **1900,** *130,* 1322–1324. Serraika Chron.
60. Grignard, V. Action Des Éthers d'acides Gras Monobasiques Sur Les Combinaisons Organomanésiennes Mixtes. *C. R. Hebd. Séances Acad. Sci.* **1901,** *132,* 336–338. Serraika Chron.
61. Grignard, V. Sur Les Combinaisons Organomanésiennes Mixtes et Leur Application a Des Synthèses d'acides, d'alcohols et d'hydrocarbures. *Ann. Chim.* **1901,** *24,* 433–490.
62. Grignard, V. Sur Les Combinaisons Organomanésiennes Mixtes. *C. R. Hebd. Séances Acad. Sci.* **1901,** 558–561.
63. Grignard, V. Über Gemischte Organomagnesiumverbindungen Und Ihre Anwendung Zu Synthesen von Säuren. *Chem. Zentralbl* **1901,** 622–625. pt. II.
64. Grignard, V. Action Des Combinaisons Organomanésiennes Sur Les Éthers A-Cétoniques. *C. R. Hebd. Séances Acad. Sci.* **1902,** *134,* 849–851. Serraika Chron.
65. Grignard, V.; Tissier, L. Action Des Combinaisons Organomanésiennes Mixtes Sur Le Trioxyméthylène: Synthèses D'alcools Primaires. *C. R. Hebd. Séances Acad. Sci.* **1902,** *134,* 107–108. Serraika Chron.

66. Tissier, L.; Grignard, V. Action Des Chlorures d'acides et Des Anhydrides d'acides Sur Les Composés Organo-Metalliques Du Magnésium. *C. R. Hebd. Séances Acad. Sci.* **1901,** *132,* 683–685. Serraika Chron.
67. Kanonnikoff, J.; Saytzeff, A. Neue Synthese des secundären Butylalkohols. *Justus Liebigs Ann. Chem.* **1875,** *175* (3), 374–378. https://doi.org/10.1002/jlac.18751750310.
68. Kanonnikoff, J.; Saytzeff, A. Versuch zur Darstellung von secundärem Alkohol mit den Radicalen Aethyl und Allyl. *Justus Liebigs Ann. Chem.* **1875,** *179* (3). https://doi.org/10.1002/jlac.18751790308, 338–338.
69. Kanonnikov, I.; Zaitsev, A. Novyi Sintez Vtorichnago Butil'nago Alkogolya [A New Synthesis of Secondary Butyl Alcohol. *Zh. Russ. Khim. O-va. Fiz. O-va* **1874,** *6,* 308–312.
70. Kanonnikov, I.; Zaitsev, A. O Deistvii Smesi Iodistago Allila i Iodistym Etilom i Tsinka Na Muraveinyi Etil'nyi Efir [On the treatment of a mixture of allyl and ethyl iodides and zinc on ethyl formate]. *Zh. Russ. Khim. O-va. Fiz. O-va* **1876,** *8,* 359–363.
71. Vagner, Y.; Zaitsev, A. Sintez Dietilkarbinola, Novago Izomera Amil'nago Alkogolya [The synthesis of diethyl carbinol]. *Zh. Russ. Khim. O-va. Fiz. O-va* **1874**, *6,* 290–308.
72. Wagner, G.; Saytzeff, A. Synthese des Diäthylcarbinols, eines neuen Isomeren des Amylalkohols. *Justus Liebigs Ann. Chem.* **1875,** *175* (3), 351–374. https://doi.org/10.1002/jlac.18751750309.
73. Vagner, Y. Deistvie tsinketila na uksusnyi al'degid [The action of ethylzinc on acetalde]. *Zh. Russ. Fiz.-Khim. O-va* **1876,** *8,* 37–40.
74. Wagner, G. Ueber die Einwirkung von Zinkäthyl auf Acetaldehyd. *Justus Liebigs Ann. Chem.* **1876,** *181* (2), 261–264. https://doi.org/10.1002/jlac.18761810211.
75. Zaitsev, M.; Zaitsev, A. Sintez i Svoistva Allildimetilkarbinola. [The synthesis and properties of allyl dimethyl carbinol]. *Zh. Russ. Khim. O-va. Fiz. O-va* **1876,** *8,* 363–380.
76. Vagner, Y. Sintez nepredel'nykh spirtov [The synthesis of unsaturated alcohols]. *Zh. Russ. Fiz.-Khim. O-va* **1884,** *16,* 283–353.
77. Wagner, G. Synthese ungesättigter Alkohole. *Ber. Dtsch. Chem. Ges.* **1894,** *27* (2), 2434–2439. https://doi.org/10.1002/cber.189402702242.
78. Kitamura, M.; Okada, S.; Suga, S.; Noyori, R. Enantioselective addition of dialkylzincs to aldehydes promoted by chiral amino alcohols. Mechanism and nonlinear effect. *J. Am. Chem. Soc.* **1989,** *111* (11), 4028–4036. https://doi.org/10.1021/ja00193a040.
79. Kitamura, M.; Suga, S.; Kawai, K.; Noyori, R. Catalytic asymmetric induction. highly enantioselective addition of dialkylzincs to aldehydes. *J. Am. Chem. Soc.* **1986,** *108* (19), 6071–6072. https://doi.org/10.1021/ja00279a083.
80. Noyori, R. Chiral metal complexes as discriminating molecular catalysts. *Science* **1990,** *248* (4960), 1194–1199. https://doi.org/10.1126/science.248.4960.1194.
81. Noyori, R.; Suga, S.; Kawai, K.; Okada, S.; Kitamura, M. Enantioselective alkylation of carbonyl compounds. From stoichiometric to catalytic asymmetric induction. *Pure Appl. Chem.* **1988,** *60* (11), 1597–1606. https://doi.org/10.1351/pac198860111597.
82. Yamakawa, M.; Noyori, R. An ab initio molecular orbital study on the amino alcohol-promoted reaction of dialkylzincs and aldehydes. *J. Am. Chem. Soc.* **1995,** *117* (23), 6327–6335. https://doi.org/10.1021/ja00128a023.
83. Saytzeff, A. Zur Kenntniss der Reihenfolge der Analgerung und Ausscheidung der Jodwasserstoffelemente in organischen Verbindungen. *Justus Liebigs Ann. Chem.* **1875,** *179* (3), 296–301. https://doi.org/10.1002/jlac.18751790304.
84. Zaitsev, A. K Voprosu o Poryadke Prisoedineniya i Vydeleniya Elementov Iodistago Vodoroda v Organicheskikh Soedineniyakh [On the question of the order of the

addition and elimination of the elements of hydrogen iodide in organic compounds]. *Zh. Russ. Khim. O-va. Fiz. O-va* **1875,** *7,* 289–293.
85. Popoff, A. Ueber die Isomerie der Ketone. *Justus Liebigs Ann. Chem.* **1868,** *145* (3), 283–292. https://doi.org/10.1002/jlac.18681450304.
86. Popoff, A. Ueber Oxydation Der Ketone. Zweite Abhandlung. *Ann. Chem. Pharm.* **1871,** *161,* 285–304.
87. Popoff, A. Zur Oxydation der Ketone. *Ber. Dtsch. Chem. Ges.* **1871,** *4* (2), 720–722. https://doi.org/10.1002/cber.18710040264.
88. Popoff, A. Die Oxydation der Ketone als Mittel zur Bestimmung der Constitution der fetten Säuren und der Alkohole. *Justus Liebigs Ann. Chem.* **1872,** *162* (1), 151–160. https://doi.org/10.1002/jlac.18721620112.
89. Popoff, A. Die Oxydation der Ketone, ein Mittel zur Bestimmung der Constitution der Säuren und der Alkohole. *Ber. Dtsch. Chem. Ges.* **1872,** *5* (1), 38–41. https://doi.org/10.1002/cber.18720050117.
90. Popoff, A. Oxydationsprodukte der Benzylketone. *Ber. Dtsch. Chem. Ges.* **1872,** *5* (1), 500–502. https://doi.org/10.1002/cber.187200501151.
91. Popov, A. O Khode Okisleniya Ketonov [On the course of the oxidation of ketones]. *Zh. Russ. Khim. O-va. Fiz. O-va* **1872,** *4,* 61–70.
92. Popov, A. Ob Okislenii Metil-Fenil'nago i Etil-Fenil'nago Ketonov [On the oxidation of methyl phenyl and ethyl phenyl ketones. *Zh. Russ. Khim. O-va. Fiz. O-va* **1872,** *4,* 61–64.
93. Popov, A. Okislenie Ketonov Kak Sredstvo Dlya Opredeleniya Stroeniya Alkogolei i Kislot [The oxidation of ketones as a method for determining the structures of alcohols and acids]. *Zh. Russ. Khim. O-va. Fiz. O-va* **1872,** *4,* 70–79.
94. Grabovskii, N.; Zaitsev, A. O Bromistom Butilene i Butilovom Glikole Iz Normal'-nago Butil'nago Alkogolya [On butylene bromide and butyl glycol from normal butyl alcohol]. *Zh. Russ. Khim. O-va. Fiz. O-va* **1875,** *7,* 317–328.
95. Grabowsky, N.; Saytzeff, A. Ueber Butylenbromür und Butylglycol aus dem normalen Butylalkohol. *Justus Liebigs Ann. Chem.* **1875,** *179* (3), 325–338. https://doi.org/10.1002/jlac.18751790307.
96. Vagner, Ye.; Zaitsev, A. Prevrashenie Dietilkarbinola v Meytilpropilkarbinol [The conversion of diethyl carbinol to methyl propyl carbinol]. *Zh. Russ. Khim. O-va. Fiz. O-va* **1875,** *7,* 302–312.
97. Wagner, G.; Saytzeff, A. Ueber amylenbromür und Amylglycol aus Diäthylcarbinol. *Justus Liebigs Ann. Chem.* **1875,** *179* (3), 302–313. https://doi.org/10.1002/jlac.18751790305.
98. Lewis, D. E. Feuding Rule Makers: Aleksandr Mikhailovich Zaitsev (1841–1910) and Vladimir Vasil'evich Markovnikov (1838–1904). A commentary on the origins of Zaitsev's rule. *Bull. Hist. Chem.* **2010,** *35* (2), 115–124.
99. Tanatar, S. *Ueber Bioxyfumarsäure. Berichte der deutschen chemischen Gesellschaft,* 1879; pp 2293–2298. https://doi.org/10.1002/cber.187901202285.
100. Tanatar, S. Reindarstellung der Bioxyfumarsäure. *Ber. Dtsch. Chem. Ges.* **1880,** *13* (1). https://doi.org/10.1002/cber.18800130145, 159–159.
101. Tanatar, S. Trioxymaleïnsäure. *Ber. Dtsch. Chem. Ges.* **1880,** *13* (2), 1383–1388. https://doi.org/10.1002/cber.18800130231.
102. Saytzeff, A. Ueber Die Oxydation Der Oelsäure Mit Kaliumpermanganat in Alkalischer Lösung. *J. Prakt. Chem.* **1885,** 541–542. https://doi.org/10.1002/prac.18850310143.
103. Saytzeff, A. Ueber die Oxydation der Oel- und Elaïdinsäure mit Kaliumpermanganat in alkalischer Lösung. *J. Prakt. Chem.* **1886,** *33,* 300–318. https://doi.org/10.1002/prac.18860330130.

104. Zaitsev, A. O Reaktsii Okisleniya Oleinovoi i Elaidinovoi Kislot Margantsovo-Kislym Kaliem v Shchelochnom Rastvore [On the Oxidation Reaction of Oleic and Elaidic Acids with Potassium Permanganate in Basic Solution. *Zh. Russ. Fiz.-Khim. O-va* **1885,** *17,* 417–435.
105. Berthelot, M. Nouvelles applications des méthodes de réduction en chimie organique. *Bull. Soc. Chim.* **1867,** *7,* 53–65. Paris [2].
106. Berthelot, M. Méthode Universelle Pour Réduire et Saturer d'hydrogène Les Composés Organiques. *Bull. Soc. Chim.* **1868,** *9,* 8–31. Paris [2].
107. Gröger, M. Vorläufige Mittheilung Über Die Oxydation von Talgfettsäuren Mit Kaliumpermanganat in Alkalischer Lösung. *Ber. Dtsch. Chem. Ges.* **1885,** *18,* 1268–1270. https://doi.org/10.1002/cber.188501801271.
108. Saytzeff, M.; Saytzeff, C.; Saytzeff, A. Zur Geschichte Der Oxystearinsäuren Verschiedenen Ursprungs. *J. Prakt. Chem.* **1887,** *35,* 369–390. https://doi.org/10.1002/prac.18870350141.
109. Saytzeff, M.; Saytzeff, C.; Saytzeff, A. Ueber Eine Neue Isoölsäre, Die Feste Oelsäure. *J. Prakt. Chem.* **1888,** *37,* 269–290.
110. Zaitsev, M.; Zaitsev, K.; Zaitsev, A. K Istorii Oksistearinovykh Kislot Razlichnago Proizkhozhdeniya [The history of the hydroxystearic acids of varied origin]. *Zh. Russ. Fiz.-Khim. O-va* **1886,** *18,* 328–346.
111. Zaitsev, M.; Zaitsev, K.; Zaitsev, A. Issledovanie Novago Izomera Oleinovoi Kisloty—Tverdoi Oleinovoi [The study of a new isomer of oleic acid—Solid oleic (acid)]. *Zh. Russ. Fiz.-Khim. O-va* **1888,** *20,* 388–407.
112. Goldsobel, A. G.; Ertschikowsky, G.; Lagorio, A.; Lawrow, W.; Slawinski, K.; Wagner, G., Jr. Georg Wagner. *Ber Dtsch. Chem. Ges* **1903,** *36,* 4591–4613. https://doi.org/10.1002/cber.190303604139.
113. Kuznetsov, V. I. Georg Wagner. In *Dictionary of Scientific Biography;* Gillisspie, C. C., Ed.; Charles Scribner & Sons: New York, 1976; p 549.
114. Lavrov, V. *Yegor Yegorovich Vagner, Ego Zhizn i Deyatel'nost [Yegor Yegorovich Vagner;* His Life and Activities: St. Petersburg, 1904.
115. Lewis, D. E. Yegor Yegorovich Vagner (1849–1903): A "Wondrously Sharpwitted" Chemist. In *Characters in Chemistry; A Celebration of the Humanity of Chemistry,* Vol. 1136, Symp, A. C. S., Ser, Patterson, G. D., Rasmussen, S. C., Eds.; American Chemical Society: Washington, DC, 2013; pp 143–1165.
116. Sementsov, A. Egor Egorovich Vagner and his role in terpene chemistry. *Chymia* **1966,** *11,* 151–155. https://doi.org/10.2307/27757265.
117. Starosel'skii, P. I.; Nikulina, E. P. *Yegor Yegorovich Vagner. 1849–1903; Izd-vo. "Nauka": Moscow,* 1977.
118. Vagner, Ye. Ob okislenii etilennykh uglevodorov i spirtov [On the oxidation of ethylenic hydrocarbons and alcohols]. *Zh. Russ. Fiz.-Khim. O-va* **1888,** *20,* 72–77.
119. Wagner, G. Ueber Die Oxydation Der Kohlenwasserstoffe, CnH2n−2. *Ber. Dtsch. Chem. Ges.* **1888,** *21,* 3343–3346. https://doi.org/10.1002/cber.188802102218.
120. Wagner, G. Ueber die Oxydation der Olefine und der Alkohole der Allyalalkoholreihe. *Ber. Dtsch. Chem. Ges.* **1888,** *21* (1), 1230–1240. https://doi.org/10.1002/cber.188802101235.
121. Wagner, G. Zur Frage Über Die Betheiligung Des Wassers an Der Oxydation Ungesättigter Verbindungen. *Ber. Dtsch. Chem. Ges.* **1888,** *21,* 3356–3360. https://doi.org/10.1002/cber.188802102220.
122. Wagner, G. Zur Oxydation Ungesättigter Verbindungen. *Ber. Dtsch. Chem. Ges.* **1888,** *21,* 3347–3355. https://doi.org/10.1002/cber.188802102219.
123. Wagner, G. Ueber Camphenglycol Und Den Vieratomigen Alkohol Aus Limonen. *Ber. Dtsch. Chem. Ges.* **1890,** *23,* 2307–2318. https://doi.org/10.1002/cber.18900230294.

124. Wagner, G. Zur Oxydation cyclischer Verbindungen. *Ber. Dtsch. Chem. Ges.* **1894,** *27* (2), 1636–1654. https://doi.org/10.1002/cber.189402702102.
125. Vagner, Y. *K Reaktsii Okisleniya Nepredel'nykh Uglerodistykh Soedinenii [Oxidation Reactions of Unsaturated Carbon Compounds]. Dr. Khim. Diss.; Warsaw,* 1888.
126. Vagner, Y. Ye. O Stroenii Kamfena [On the structure of camphene]. *Zh. Russ. Fiz.-Khim. O-va* **1899,** *31,* 680–684.
127. Wagner, G.; Brickner, W. Ueber die Beziehung der Pinenhaloïdhydrate zu den Haloïdanhydriden des Borneols. *Ber. Dtsch. Chem. Ges.* **1899,** *32* (2), 2302–2325. https://doi.org/10.1002/cber.189903202168.
128. Wagner, G.; Brykner, W. Bornylen, Ein Neues Terpen. *Ber. Dtsch. Chem. Ges.* **1900,** *33,* 2121–2125. https://doi.org/10.1002/cber.190003302117.
129. Lewis, D. E. Into a new century: Chemists advancing the legacies of Kazan, St. Petersburg, and Moscow. In *Early Russian Organic Chemists and Their Legacy;* Rasmussen, S. C., Ed.; Springer: Heidelberg, 2012; pp 96–97. Series.
130. Semenzow, A. Sergius Reformatsky. *Ber. Dtsch. Chem. Ges* **1935,** 68. https://doi.org/10.1002/cber.19350680558. A61–A61.
131. Reformatskii, S. *Issledovanie Uglevodoroda C10H18, Poluchaemogo Iz Allildipropilkarbinola [A Study of the Hydrocarbon C10H18 Produced from Allyl Dipropyl Carbinol; Kazan,* 1888.
132. Reformatskii, S. *Predel'nye Mnogoatomnye Alkogoli [Saturated Polyhydric Alcohols].; Warsaw,* 1889.
133. Reformatskii, S. N. *Deistvie Smesi Tsinka i Monoxloruksusnogo Efira Na Ketony i al'degidy [The Action of a Mixture of Zinc and Monochloroacetic Ester on Ketones and Aldehydes]; Warsaw,* 1890.
134. Reformatsky, S. Neue Synthese zweiatomiger einbasischer Säuren aus den Ketonen. *Ber. Dtsch. Chem. Ges.* **1887,** *20* (1), 1210–1211. https://doi.org/10.1002/cber.188702001268.
135. Reformatsky, S. Die Einwirkung eines Gemenges von Zink und Bromisobuttersäureester auf Isobutyraldehyd. Synthese der secundären β-Oxysäuren. *Ber. Dtsch. Chem. Ges.* **1895,** *28* (3), 2842–2847. https://doi.org/10.1002/cber.18950280399.
136. Reformatsky, S. Neue Darstellungsmethode der αα-Dimethylglutarsäure aus der entsprechenden Oxysäure. *Ber. Dtsch. Chem. Ges.* **1895,** *28* (3), 3262–3265. https://doi.org/10.1002/cber.189502803178.
137. Reformatsky, S.; Plesconossoff, B. Die Einwirkung eines Gemenges von Zink und Bromisobuttersäureester auf Aceton. Synthese der Tetramethyläthylenmilchsäure. *Ber. Dtsch. Chem. Ges.* **1895,** *28* (3), 2838–2841. https://doi.org/10.1002/cber.18950280398.
138. Diaper, D. G. M.; Kukis, A. Synthesis of alkylated alkanedioic acids. *Chem. Rev.* **1959,** *59* (1), 89–178. https://doi.org/10.1021/cr50025a003.
139. Ocampo, R.; Dolbier, W. R. The reformatsky reaction in organic synthesis. Recent advances. *Tetrahedron* **2004,** *60* (42), 9325–9374. https://doi.org/10.1016/j.tet.2004.07.018.
140. Rathke, M. W. The reformatsky reaction. *Org. React.* **1975,** *22,* 423–460. https://doi.org/10.1002/0471264180.or022.04.
141. Shriner, R. L. The reformatsky reaction. *Org. React.* **1942,** *1,* 1–37. https://doi.org/10.1002/0471264180.or001.01.
142. Kamai, G. Kh. Aleksandr Yerminingel'dovich Arbuzov. *Russ. Chem. Bull.* **1962,** *11,* 1625–1626.
143. Lewis, D. E.; Arbuzov, A. Ye. Father of Organophosphorus Chemistry in Russia. *Bull. Hist. Chem.* **2017,** *42,* 117–125.
144. Grechkin, N. P.; Kuznetsov, V. I. *Aleksandr Yerminingel'dovich Arbuzov. 1877–1968;* Nauka: Moscow, 1977.

145. Lozovoi, A. S. *Akademik Aleksandr Yerminingel'dovich Arbuzov: Mirovozzrenie, Nauka, Zhizn [Academician Aleksandr Yerminingel'dovich Arbuzov: Worldview, Science, Life].*; Tatarstan Book Publishing House: Kazan, 1985.
146. Arbuzov, B. A. Aleksandr Erminingel'dovich Arbuzov. In *Osnovopolozhnik Khimii Fosfororganicheskikh Soedinenii. Ocherki, Vospominaniya, Materialy (Aleksandr Erminingel'-Dovich Arbuzov. Founder of Organophosphorus Chemistry)*; Arbuzov, B. A., Ed.; Nauka: Moscow, 1989.
147. Figurovskii, N. A. *Aleksandr Yermningel'dovich Arbuzov. 1877–1968;* Nauka: Moscow, 1977.
148. Arbusoff, A. Ueber Das Allylmethylphenylcarbinol. *J. Prakt. Chem.* **1901,** *64,* 546–554. https://doi.org/10.1002/prac.19010640138.
149. Arbuzov, A. Ye. Allilmetilfenilkarbinoli, O. On allyl methyl phenyl carbinol. *Zh. Russ. Fiz.-Khim. O-va* **1901,** *33,* 38–45.
150. Arbuzov, A. Ye. Primenenie v Kazanskoi Laboratorii Magniya Dlya Polucheniya Tretichnykh Spirtov Po Sposobu Zaitseva [The use of magnesium in the laboratory of kazan for obtaining tertiary alcohols by the method of zaitsev]. *Zh. Russ. Fiz. -Khim. O-va* **1902,** *34,* 8–8.
151. Arbuzov, A. Ye. Glavy 1-2 [On the structure of phoshorous acid and its derivatives]. *Zh. Russ. Fiz.-Khim. O-va* **1906,** *38,* 187–228.
152. Arbuzov, A. Ye. O Stroenii Fosforistoi Kisloty i Ee Proizvodnik Glava 4 [On the structure of phoshorous acid and its derivatives]. *Zh. Russ. Fiz.-Khim. O-va* **1906,** *38,* 687–721.
153. Arbuzov, A. Ye. O Stroenii Fosforistoi Kisloty i Ee Proizvodnik. Glava 3 [On the structure of phoshorous acid and its derivatives]. *Zh. Russ. Fiz.-Khim. O-va* **1906,** *38,* 293–319.
154. Arbusoff, A. Zur Kenntniss der Phosphorigsäureester. *Ber. Dtsch. Chem. Ges.* **1905,** *38* (1), 1171–1173. https://doi.org/10.1002/cber.190503801212.
155. Arbusow, A. Über Die Struktur Der Phosphorigen Säure Und Ihrer Derivate. IV. Isomerisation Und Übergang Der Verbingungen Des Dreiwertigen Phosphors in Solche Des Fünfwertigen. *Chem. Zentralbl* **1906,** *II,* 1639–1641.
156. Arbusow, A. Über Isomerisations-Vorgänge Bei Einigen Phosphorverbindungen. *Chem. Zentralbl* **1910,** *II,* 453–454.
157. Arbuzov, A. Ye. O Protsessakh Izomerizatsii V Oblasti Nekotorykh Soedinenii Fosfora. Stat'ya 1 [On the process of isomerization of certain compounds of phosphorus]. *Zh. Russ. Fiz.-Khim. O-va* **1910,** *42,* 395–420.
158. Arbuzov, A. Ye. O Protsessakh Izomerizatsii v Oblasti Nekotorykh Soedinenii Fosfora. Stat'ya 2 [On the process of isomerization of certain compounds of phosphorus. Article 2]. *Zh. Russ. Fiz.-Khim. O-va* **1910,** *42,* 549–561.
159. Horner, L.; Hoffmann, H.; Wippel, H. G. Phosphororganische Verbindungen, XII. Phosphinoxyde als Olefinierungsreagenzien. *Chem. Ber.* **1958,** *91* (1), 61–63. https://doi.org/10.1002/cber.19580910113.
160. Horner, L.; Hoffmann, H.; Wippel, H. G.; Klahre, G. Phosphororganische Verbindungen, XX. Phosphinoxyde als Olefinierungsreagenzien. *Chem. Ber.* **1959,** *92* (10), 2499–2505. https://doi.org/10.1002/cber.19590921017.
161. Wadsworth, W. S., Jr. Synthetic applications of phosphoryl-stabilized anions. *Org. React.* **1977,** *25,* 73–253.
162. Wadsworth, W. S., Jr.; Emmons, W. D. The utility of phosphonate carbanions in olefin synthesis. *J. Am. Chem. Soc.* **1961,** *83* (7), 1733–1738. https://doi.org/10.1021/ja01468a042.

CHAPTER FIVE

August Wilhelm von Hofmann and Hofmann's rule for elimination

August Wilhelm von Hofmann (Fig. 5.1) was one of the most influential and productive organic chemists of the 19th century.[1–7] In a career lasting over half a century after he graduated into the Ph.D., he discovered three eponymous reactions: the Hofmann elimination,[8,9] which is the subject of this chapter, the Hofmann rearrangement,[10–15] and the Hofmann–Löffler–Freytag reaction.[16–20] His research into the chemistry of amines, and more especially anilines, paved the way for the founding of the British[21] and German[22] dye and pharmaceutical industries, and many eminent dye chemists worked in his research group.

August Wilhelm von Hofmann (1818–92)[4,5]

Family connections

Hofmann came from a well-connected and educated family. His father, Johann Philipp Hofmann (1776–1842, Fig. 5.2), began his career as an Officer in the Danish Service and later became an architect in Frieberg in Hesse. From here, he became a Privy Councilor at Court and University Architect in Giessen. One of his architectural projects would eventually provide a real family connection for his son: In his function as University architect, Johann oversaw the design and construction of the new chemistry laboratory for Justus Liebig (1803–73) at the University.[4]

The connections of Hofmann's family with the educational hierarchy at Giessen began with the 1833 marriage of his sister, Johannette Sophie Hofmann (1810–48, Fig. 5.2), to Johann Heinrich Buff (1805–78, Fig. 5.2), the docent of Physics at the Polytechnicum at Kassell. In 1848, Buff moved to Giessen as Professor of Physics. A family portrait of Hofmann's family taken in late summer after his Ph.D. graduation is given in Fig. 5.2.

Hofmann's marriages also cemented his ties to the German chemical establishment. He was married four times, although his first three marriages were cut short by the deaths of his wives. His first wife was Helene Ernestine Henriette Moldenhauer (1824–52), Liebig's niece by marriage, whom he married in 1836. He married her cousin, another Liebig niece, Elise

Addition, Elimination and Substitution: Markovnikov, Hofmann, Zaitsev and Walden
ISBN: 978-0-12-821027-7
https://doi.org/10.1016/B978-0-12-821027-7.00006-2

Figure 5.1 August Wilhelm von Hofmann as a Professor of Chemistry at the London Government School of Mines. Woodcut engraving by John Willard Cook based on a photograph from 1853 to 1861. *Portrait of August Wilhelm von Hofmann N.d. Engraving by Conrad Cook, 19th C., published by William Mackenzie, 1860. Smithsonian Libraries, image SIL-SIL14-h005-07.*

Moldenhauer (1815–71), in 1866. He married his second wife, Rosamond Wilson (1838–60) in 1856. His first three marriages were cut short by the early deaths of his wives. His last wife, Bertha Tiemann (1854–1922), was the younger sister of organic chemist and Hofmann student, Johann Karl Wilhelm Ferdinand Tiemann (1848–99), the codiscoverer of the Reimer–Tiemann reaction.[23–25]

Figure 5.2 From left: Adolph Buff, (1838–1901), Johannette Sophie (Hofmann) Buff (1810–48), Johann Heinrich Buff (1805–78), Johann Philipp Hofmann (1776–1842), August Wilhelm Hofmann (1818–92), Wilhelmine (Badenius) Hofmann, (1780–1854), Hermann Buff (1835–44). Taken in late summer 1841 (or possibly 1840) after Hofmann's Ph.D. graduation. *Adapted from an image provided by Dr. Olaf Skibbe, Director, Center for Advanced Materials, Universität Heidelberg, N.d. http://www.rzuser.uni-heidelberg.de/~gj7/HBuff/bilder.html.*

Student years: Giessen

While still a student at the Gymnasium, Hofmann traveled with his father to Switzerland and Italy; in Italy, they visited Milan, Genoa, Venice, and Naples. Hofmann had a close relationship with his older brother Friedrich (Fritz) and his mother, Wilhelmine (née Badenius) (Fig. 5.2), and he wrote to them regularly. Many of these letters, including one of his first (to Fritz, written in Latin), are included in his *Lebensbild*.[1,2]

In 1836, Hofmann entered the University of Giessen, intending to take a degree in philology (the study of literary texts as well as oral and written records, the establishment of their authenticity and their original form, and the determination of their meaning). This was very much over the objections of his father, who saw the discipline as one that had no future. His father would much rather that his younger son had chosen a career like his own, in architecture, where he could earn a solid living, but this August Wilhelm could not countenance. Eventually, he and his father compromised, and he entered the University to study law. Johann Hofmann could accept law as a course of study that would lead to a sustainable occupation.

While his son was studying at Giessen, Herr Hofmann was involved in the building of the new laboratory for Justus von Liebig (1803–73, Fig, 5.3), which led to frequent and cordial communication between the two men. During this period, the cameral movement was in force in Germany, so all students in law and economics (the preferred courses of study for students planning on entering government service as bureaucrats) were required to take 2 years of a science subject, typically physics, chemistry, or biology.

After 2 years of rudderless and mediocre performance in the study of law, Hofmann appeared to be adrift, as a concerned Herr Hofmann confided to his good friend Liebig. Liebig then suggested that perhaps the young man had not yet found his true calling,[4,26] and that he be sent to Liebig's laboratory to study chemistry as part of his preparation for a career in law.

Figure 5.3 Justus von Liebig. *Data from Justus Freiherr von Liebig (May 12, 1803–April 18, 1873). http://ihm.nlm.nih.gov/images/B17480 (Accessed January 1, 2021).*

Hofmann did so and was quickly captivated by the subject, which was vibrant and alive compared to the settled (stagnant) field of law. No longer did he seek to become a philologist or a lawyer.

Liebig's laboratory in Giessen (Fig. 5.4) had quickly become the model for a university chemistry laboratory throughout Europe, and it attracted many students. His approach to chemical education for his strongest students was not based on a lecture curriculum but, instead, on original experimental work carried out by the student—today we would describe it as "undergraduate research." On completing the work to Liebig's satisfaction, the student would pass an oral examination and be awarded the degree of Ph.D. directly. Outstanding performance on the oral examination would further qualify the student for the *habilitation*, thus allowing the graduate to begin an independent academic career immediately. This was what happened with Hofmann, who took his Ph.D. summa cum laude in April 1841, but because he placed a moratorium on his career for the next 2 years while he cared for his ailing father, he needed to complete his *habilitation* after he resumed his academic post. His father died in 1842, and Hofmann became an Assistant to Liebig at

Figure 5.4 Justus von Liebig and colleagues at work in the analytical laboratory in Giessen. Lithograph (?) after W. Trautschold, ca. 1840 (?). *Modified from Wellcome Library no. 568857i. Wellcome Collection. Attribution 4.0 International (CC BY 4.0).*

Giessen early in 1843. In April 1845, Hofmann formally qualified for the *habilitation*.

Hofmann's career

Bonn: docent and extraordinary professor, 1845

In the German higher education system, promotion was entrenched in seniority, so Hofmann could see that, as a very junior docent, his long-term prospects for advancement at Giessen were meager.[4] Accordingly, he began negotiations with Bonn, and while Liebig was in Britain, he transferred his laboratory to Bonn, where he became Docent in 1845. The move upset Liebig, who had his own plans for filling the position, but the fact that Hofmann was now his niece's fiancé may have stayed his hand from a more forceful expression of his disappointment. In any case, Hofmann's stay at Bonn proved to be very short thanks to events across the English Channel.

The Royal College of Chemistry: director, 1845–65

In 1842, Liebig made a tour of Britain following the publication of his textbook on the applications of organic chemistry in agriculture and physiology,[27] and this highly successful tour was followed the next year by the publication of his small, popular monograph, *Familiar Letters on Chemistry, and its Relation to Commerce, Physiology, and Agriculture*[28]; this monograph was instrumental in the rise of the popularity of chemistry in Britain in the 19th century. He followed his successful 1842 tour with another in 1845, the year after two of his former students, John Lloyd Bullock (1812–1905, Fig. 5.5)[29] and John Gardner (1804–80, Fig. 5.5), proposed a new institution for teaching chemistry as an experimental, rather than a theoretical science. They proposed that such an institution should be based on the Giessen model. These two chemists were able to gain the support of Sir James Clark (1788–1870, Fig. 5.6), Physician-in-Ordinary to Queen Victoria. Clark shared their ideas with Victoria's husband, Prince Albert of Saxe-Coburg and Gotha (1819–61, Fig. 5.6).

Although initially frustrated by the lack of substance and responsibility accorded his status as Prince Consort, Albert gradually became an influential supporter of progressive public causes, especially education reform and the abolition of slavery. Albert was a tireless promoter of science and was

Figure 5.5 Students in Liebig's Giessen laboratory ca. 1843, who were associated in the founding of the Royal College of Chemistry. *From Fresenius Will Bullock Gardener Hofmann in the Liebig Laboratory in Giessen, after a daguerreotype from the beginning of the 40th year, 1843. https://commons.wikimedia.org/wiki/Category:Heinrich_Will_(chemist)#/media/File:Liebig_Lab_Anf1. (Accessed December 29, 2020).*

committed to improving science and technology education in Britain.[30] He was happy to give Clark's proposal his enthusiastic support.

At the time in Britain, having the Ph.D. from a German university conferred considerable status on the degree holder, which meant that German candidates for the position of Director were desirable. Of course, the Directorship of the new College of Chemistry was offered to Liebig, but knowing that he was to be ennobled as *Freiherr* (baron) later that year, Liebig declined. Instead, he recommended three of his recent students as possible candidates for the Directorship: Heinrich Will (1812–90, Fig. 5.5), Remigius Fresenius (1818–97, Fig. 5.5), and Hofmann (Fig. 5.5). Will declined because he saw an improvement in his prospects for promotion at Giessen (he did, in face, succeed Liebig there); for similar reasons, Fresenius also declined.

This left Hofmann, whom Prince Albert had met while he and Queen Victoria were in Bonn for a Beethoven Festival. During that festival, Albert had dropped in to see the rooms where he had been a student and found them occupied by Hofmann, who had filled them with chemical glassware and assorted experiments. Albert's choice was inspired: the 27-year-old

Figure 5.6 Sir James Clark, 1867, and Prince Albert, 1840. *Public domain photograph of Sir James Clark by Ernest Edwards published in* Photographs of eminent medical men London; *John Churchill & Sons, 1867 v. 1, downloaded January 2, 2021 from https://commons.wikimedia.org/wiki/File:James_Clark_b1788.jpg Public domain painting of Prince Albert by John Partridge, 1840, downloaded January 2, 2021 from https://commons.wikimedia.org/wiki/File:Prince_Albert_-_Partridge_1840.jpg.*

Director was not only a charismatic young chemist who had established a strong record as a chemical researcher, but also an excellent and inspirational teacher.

Thanks to Prince Albert's personal invitation, Hofmann accepted the appointment as Director of the new Royal College of Chemistry in London. However, the uncertain outcome of the "English venture" meant that he did not officially accept until he had an insurance policy in the form of promotion to Extraordinary Professor and a 2-year leave of absence from Bonn. This would guarantee that he could resume his tenured position in Germany in the event that the London venture failed.

The Royal College of Chemistry (Fig. 5.7) opened in a building designed by the English architect James Lockyer in 1846, with the foundation stone being laid by Prince Albert. The college opened in 1846 with 26 students under Hofmann's directorship. Hofmann remained Director for the next two decades. The College quickly faced economic trouble, and Hofmann had to seriously cut back on expenses, including reducing his own salary and relinquishing the free accommodations for himself and the

Figure 5.7 The laboratories of the Royal College of Chemistry, 1846. *Library reference: ICV No 13875. Photo number: V0013588Full Bibliographic Record: http://catalogue.wellcomelibrary.org/record=b1196405. This file is licensed under the Creative Commons Attribution 4.0 International license.*

College secretary (who had to be physically evicted from his). By 1853, the College had become bankrupt. That year, it was absorbed into the Government School of Mines, although it did retain its name. The assimilation into the School of Mines did have the important positive effect of stabilizing the funding for the College by allowing it access to Government funds.

The University of Berlin: Professor, 1865–92

The death of Prince Albert in 1861 was a hard blow for Hofmann, both personally and professionally. In 1863, he wrote, writing in 1863, "[Albert's] early kindness exercised so powerful an influence upon the destinies of my existence. Year by year do I feel more deeply the debt of gratitude which I owe to him ... it is to him, I feel, that I owe my opportunities through life."[31] With his passing, British science lost its most powerful and visible supporter; science lost its status as a priority for the British government, leading to reduced funding for the College. In 1864, Hofmann received calls to the Chairs of Chemistry at Bonn and Berlin. He took his time, carefully weighing the two offers, and in 1865, he returned to Germany to the Chair at Berlin as the successor to Eilhard Mitscherlich (1794–1863). He held the position of Professor and Director of the Chemical Laboratories until his death in 1892. In 1888, Hofmann was ennobled as Freiherr (baron) August Wilhelm von Hofmann.

Hofmann's chemistry

Coal-tar chemistry: aniline and aniline dyes

Hofmann's doctoral project at Giessen concerned the subject that would consume him for his entire career: the study of amine bases. Liebig set him to the task of studying the compounds in coal tar, so he sent his young student to the asphalt factory of a former student, Ernst Sell. At Sell's factory, coal tar was distilled in iron retorts to give three fractions: a lower-boiling fraction—the "light" coal tar oil, which is less dense than water and consists of lower aromatic hydrocarbons and phenols—a higher-boiling fraction—the "heavy" coal tar oil, which is denser than water and consists of aromatic bases (anilines and quinolines)—and the pot residue, a black material that solidified to a resinous asphalt.

At Sell's factory, Hofmann eventually obtained approximately a kilogram of basic oils from 500 to 600 kg of coal tar oil by a method involving extraction of the crude oil with crude concentrated hydrochloric acid, siphoning off the acid layer, and adding fresh oil to the acid. The process was continued until a fairly concentrated solution of the hydrochlorides was produced. The mixture of free bases (mainly quinoline and aniline) was liberated from the salts with limewater, and the bases were isolated by steam distillation. Subsequent fractionation gave aniline (b. 184°C) as the first fraction, and quinoline (b. 237°C) as the second. The work was not published until 1843, after a preliminary notice[32] reporting his confirmation of the identity of Runge's *Kyanol*,[33] with aniline. His first publication described the isolation Runge's *Kyanol [aniline]* and *Leukol [quinoline]* by extraction from coal tar oil[32,34–38] with hydrochloric acid and confirmed the identity of his *Kyanol* with Unverdorben's *Krystallin*,[39] Zinin's *Benzidam*[40–42] and Fritsche's *Anilin*[43–46] (Fritsche's name for the compound, based on the Portuguese, *anil*, from the Arabic, *alnayliu*, indigo, eventually became the choice of the chemical community at large). Hofmann published the same work in English in 1844.[47–49] Hofmann's next German publication was a review, published in two parts, of the work of Dumas, Erdmann, and Laurent into indigo and its transformations.

Hofmann began his research studies at a time when the Substitution Theory being developed by Dumas.[50–53] Gerhardt[54,55] and Laurent[56–60] was at its zenith, making substitution of hydrogen by halogen a popular reaction to carry out. In an exhaustive study,[61,62] Otto Linné Erdmann had shown that although indigo was resistant to the action of dry chlorine, it did react with chlorine in the presence of water. The exhaustive chlorination of an aqueous paste of indigo gave chlorinated isatin derivatives (e.g., 5-chloroisatin, **115**) as intermediate compounds, and chloranil (**116**) as the final product, and that Fritzsche's *Anilin* gave chloranil as the exhaustive chlorination product in aqueous solution (Scheme 5.1). By following Erdmann's procedure, Hofmann was able to show that his lower-boiling fraction was identical with aniline, which Runge had shown[33] to give an intense blue color on treatment with "chloride of lime" (ClCaOCl); Perkin later showed this coloration to be due not to mauveine, but to a blue dye that could be not be isolated pure, but decomposed to mauveine salts.[63]

With the advent of Zinin's method for the reduction of nitroarenes to anilines,[40–42] pure aniline became a much more readily available commodity. However, the use of hydrogen sulfide as the reducing agent did have its drawbacks in terms of its toxicity and malodorous nature, which limited its wider use. The solution to these problems was provided 12 years later by Antoine Béchamp (1816–1908),[64] who used ferrous salts (ferrous acetate,

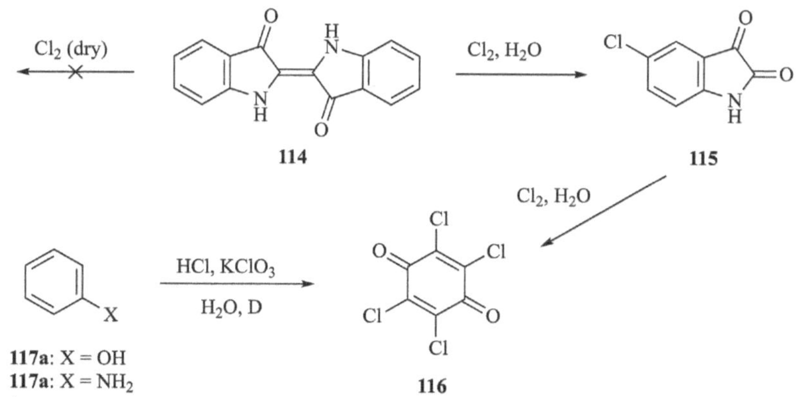

Scheme 5.1 Chlorination of indigo and aniline to chloranil.

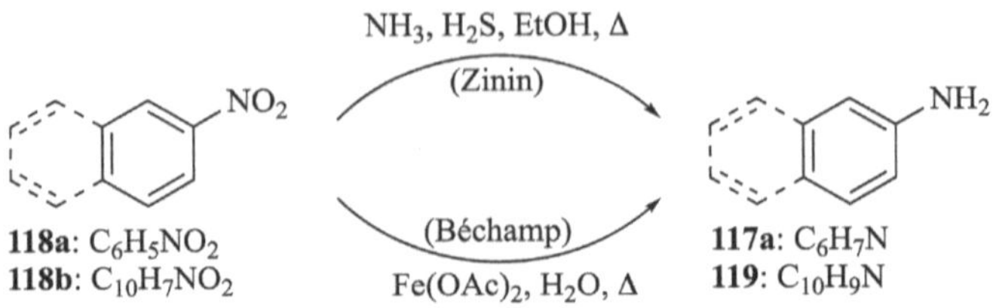

118a: $C_6H_5NO_2$
118b: $C_{10}H_7NO_2$
117a: C_6H_7N
119: $C_{10}H_9N$

Scheme 5.2 The Zinin and Béchamp reductions of nitrobenzene and nitronaphthalene.

Scheme 5.2) as the reducing agent; Béchamp's method now allowed the manufacture of aniline "by the ton."[65,66]

During the course of his work developing methods for the synthesis of aniline, in which he collaborated with Auguste Laurent, he came to accept Laurent's view that phenol and aniline were related such that the oxygen atom in phenol was replaced by an NH group in aniline. Among the first methods that Laurent reported for the synthesis of aniline was the dry distillation of alkaline earth salts of anthranilic acid[67] (Scheme 5.3). This new method provided a method to complement the Zinin reduction for obtaining aniline from nitroarenes. With pure aminoarenes now readily available in quantity from the easily prepared nitroarenes, Hofmann and his students could proceed apace with their research into aniline chemistry.

The work of Hofmann and Muspratt[68] extended the work of Laurent to show that pyrolysis of derivatives of anthranilic acid (**120a**) and salicylic acid (**120b**) gave aniline and phenol, respectively. In addition, they showed the direct replacement of the nitro group of nitrobenzene to give phenol (Scheme 5.4), although the reaction almost certainly proceeds by a benzyne mechanism.

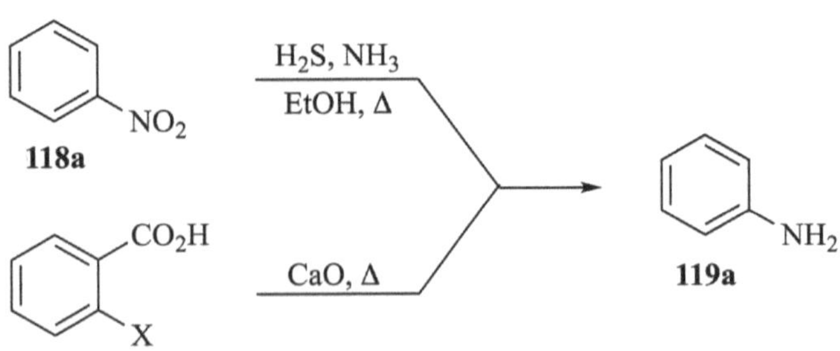

118a
119a
120a: X = NH_2
120b: X = OH

Scheme 5.3 The Zinin reduction of nitrobenzene and Laurent's pyrolysis of the calcium salt of anthranilic acid.

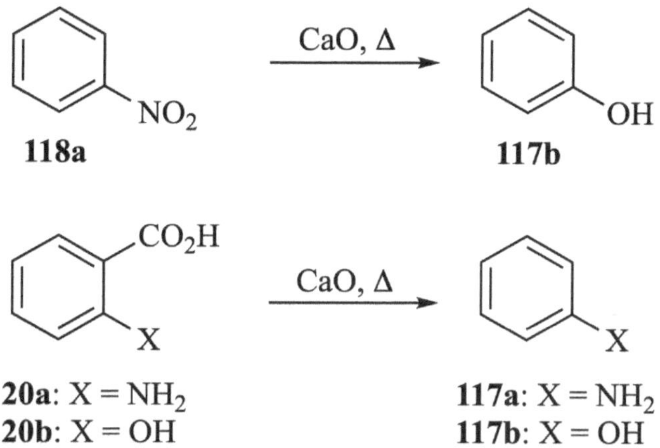

118a **117b**

120a: X = NH_2
120b: X = OH

117a: X = NH_2
117b: X = OH

Scheme 5.4 Comparison of pyrolytic syntheses of phenol and aniline.

Hofmann followed up these early works with a systematic study of the reaction of aniline with halogens and pseudohalogens. Hofmann noted that although iodine reacted only to give *p*-iodoaniline (**121**), chlorine and bromine both reacted to give the 2,4,6-trihaloanilines (**122a** and **122b**).[69,70–72] He also showed that iodoaniline would react with excess chlorine to give chloranil (**116**) (Scheme 5.5).

Cyanogen was identified fairly early on as a pseudohalogen that reacted with (particularly alkali) metals, in particular, to give metal cyanides, just as halogens gave metal halides. Consequently, Hofmann studied the reaction of cyanogen with aniline, expecting to obtain the cyanoanilines. Instead, he obtained the formal adduct of aniline across the C—N π (Scheme 5.6)[73,74].

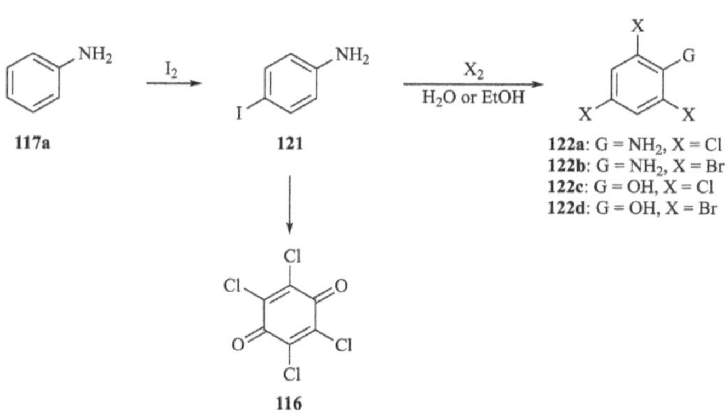

117a **121** **116**

122a: G = NH_2, X = Cl
122b: G = NH_2, X = Br
122c: G = OH, X = Cl
122d: G = OH, X = Br

Scheme 5.5 Comparison of pyrolytic syntheses of phenol and aniline.

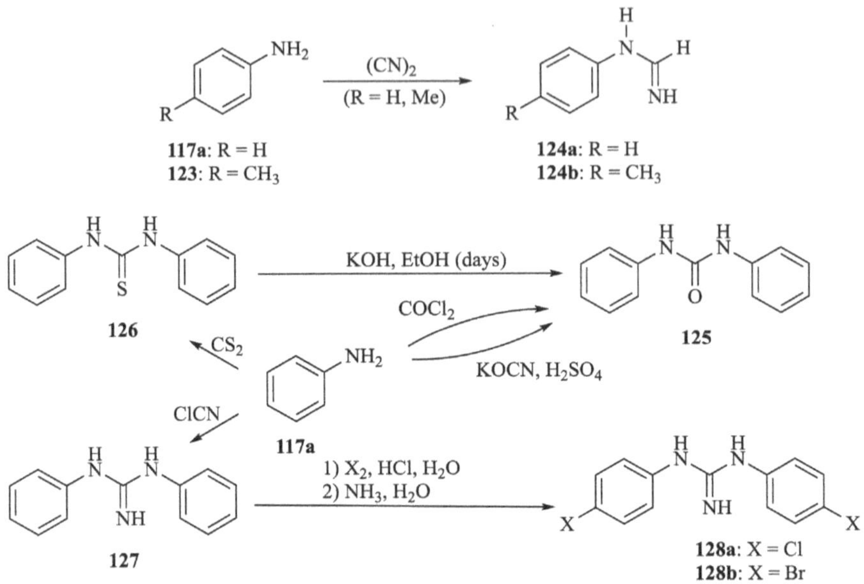

Scheme 5.6 The reaction of anilines with cyanogen and cyanogen halides to give amidines.

Perkin and aniline dyes (mauveine)

Perhaps the best known example of serendipitous aniline chemistry was provided by Hofmann's student, (later Sir) Willian Henry Perkin (1838–1907, Fig. 5.8). At age 15, Perkin had entered the Royal College of Chemistry and begun his studies under Hofmann. It is worth remembering here that organic chemistry was still relatively primitive, with the concept of organic structure still over a decade in the future and not even the atomic weights of the elements settled. Hofmann had hypothesized that quinine ($C_{20}H_{24}N_2O_2$) might be prepared by the oxidation of aniline derivatives.

During the Easter holidays of 1856, while he was at home, the 18-year-old Perkin tried the synthesis by oxidizing *N*-allyltoluidine ($C_{10}H_{13}N$) with chromic acid, expecting that an oxidative dimerization would take place: $2\ C_{10}H_{13}N + 3\ O \rightarrow C_{20}H_{24}N_2O_2 + H_2O$. Of course, the experiment failed, giving an orange-brown precipitate. As a means of simplifying the outcome, Perkin repeated the experiment using aniline and a mixture of toluidines, perhaps envisaging a transformation such as the following: $C_6H_7N + 2\ C_7H_9N + 4\ O \rightarrow C_{20}H_{24}N_2O_2 + HNO_2$ (Fig. 5.9).

Alkylation of amines: the "Ammonia Type"

By 1840, the efforts of Dumas,[52,53,75–81] Gerhardt,[54,55] and Laurent[56–60] had established the "Hydrogen Type" and the "Hydrogen Chloride

Figure 5.8 Sir William Henry Perkin, 1906. *William Henry Perkin (* 12. März 1838 in East End, London; † 14. Juli 1907 in Sudbury), n.d. https://commons.wikimedia.org/wiki/File: William_Henry_Perkin.jpg.*

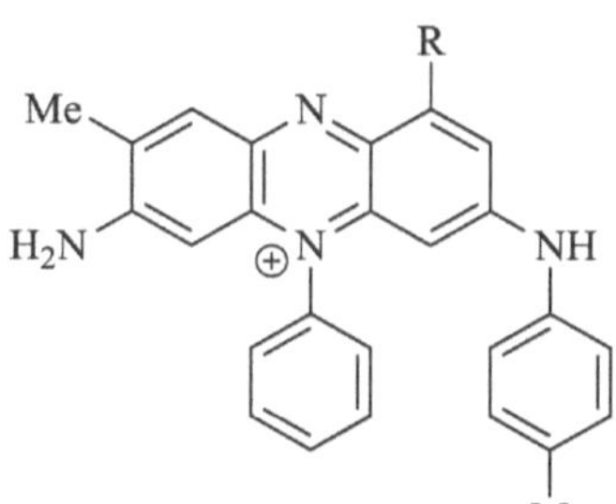

129a: R = H (mauveine A)
129b: R = Me (mauveine B)

Figure 5.9 Mauveines A and B, the first synthetic aniline dyes.

Type" and the associated Type Theory. To this, Hofmann added the "Ammonia Type" (Fig. 5.10),[82] which he followed up a year later with his descriptions of quaternary ammonium ions.[9]

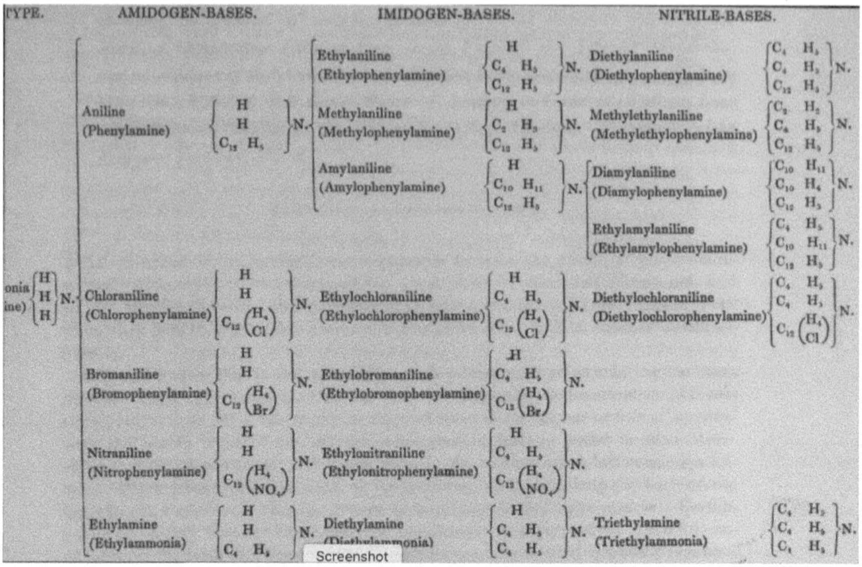

TYPE.	AMIDOGEN-BASES.		IMIDOGEN-BASES.		NITRILE-BASES.	
	Aniline (Phenylamine)	$\left\{ \begin{matrix} H \\ H \\ C_{12}\ H_5 \end{matrix} \right\}$ N.	Ethylaniline (Ethylophenylamine)	$\left\{ \begin{matrix} H \\ C_4\ H_5 \\ C_{12}\ H_5 \end{matrix} \right\}$ N.	Diethylaniline (Diethylophenylamine)	$\left\{ \begin{matrix} C_4\ H_5 \\ C_4\ H_5 \\ C_{12}\ H_5 \end{matrix} \right\}$ N.
			Methylaniline (Methylophenylamine)	$\left\{ \begin{matrix} H \\ C_2\ H_3 \\ C_{12}\ H_5 \end{matrix} \right\}$ N.	Methylethylaniline (Methylethylophenylamine)	$\left\{ \begin{matrix} C_2\ H_3 \\ C_4\ H_5 \\ C_{12}\ H_5 \end{matrix} \right\}$ N.
			Amylaniline (Amylophenylamine)	$\left\{ \begin{matrix} H \\ C_{10}\ H_{11} \\ C_{12}\ H_5 \end{matrix} \right\}$ N.	Diamylaniline (Diamylophenylamine)	$\left\{ \begin{matrix} C_{10}\ H_{11} \\ C_{10}\ H_4 \\ C_{12}\ H_5 \end{matrix} \right\}$ N.
					Ethylamylaniline (Ethylamylophenylamine)	$\left\{ \begin{matrix} C_4\ H_5 \\ C_{10}\ H_{11} \\ C_{12}\ H_5 \end{matrix} \right\}$ N.
onia ine) $\left\{ \begin{matrix} H \\ H \\ H \end{matrix} \right\}$ N.	Chloraniline (Chlorophenylamine)	$\left\{ \begin{matrix} H \\ H \\ C_{12} \binom{H_4}{Cl} \end{matrix} \right\}$ N.	Ethylochloraniline (Ethylochlorophenylamine)	$\left\{ \begin{matrix} H \\ C_4\ H_5 \\ C_{12} \binom{H_4}{Cl} \end{matrix} \right\}$ N.	Diethylochloraniline (Diethylochlorophenylamine)	$\left\{ \begin{matrix} C_4\ H_5 \\ C_4\ H_5 \\ C_{12} \binom{H_4}{Cl} \end{matrix} \right\}$ N.
	Bromaniline (Bromophenylamine)	$\left\{ \begin{matrix} H \\ H \\ C_{12} \binom{H_4}{Br} \end{matrix} \right\}$ N.	Ethylobromaniline (Ethylobromophenylamine)	$\left\{ \begin{matrix} H \\ C_4\ H_5 \\ C_{12} \binom{H_4}{Br} \end{matrix} \right\}$ N.		
	Nitraniline (Nitrophenylamine)	$\left\{ \begin{matrix} H \\ H \\ C_{12} \binom{H_4}{NO_4} \end{matrix} \right\}$ N.	Ethylonitraniline (Ethylonitrophenylamine)	$\left\{ \begin{matrix} H \\ C_4\ H_5 \\ C_{12} \binom{H_4}{NO_4} \end{matrix} \right\}$ N.		
	Ethylamine (Ethylammonia)	$\left\{ \begin{matrix} H \\ H \\ C_4\ H_5 \end{matrix} \right\}$ N.	Diethylamine (Diethylammonia)	$\left\{ \begin{matrix} H \\ C_4\ H_5 \\ C_4\ H_5 \end{matrix} \right\}$ N.	Triethylamine (Triethylammonia)	$\left\{ \begin{matrix} C_4\ H_5 \\ C_4\ H_5 \\ C_4\ H_5 \end{matrix} \right\}$ N.

Figure 5.10 Hofmann's 1850 Table illustrating his "Ammonia Type". *Hofmann, A. W. Research Studies Regarding the Molecular Constitution of the Volatile Organic Bases.* Philos. Trans. Royal Soc. ***1850**, 140, 93–131.*

Halogenation of amines and amides

In addition to the alkylation of amines with alkyl halides and the reactions of aniline derivatives with halogens, Hofmann also studied the reactions of aliphatic amines and amides with halogens in the presence of base. The reactions of amines with bromine in the presence of base lead to substitution of the N—H bonds with N—Br bonds.[14] The reaction is not easily controlled: substitution of every N—H bond by an N—Br bond occurs (Scheme 5.7). In the course of his systematic examination of the chemistry of these substances, Hofmann studied the synthesis of these *N*-halo compounds with both acids and bases.

The Hofmann—Löffler—Freytag reaction

Having synthesized *N*-bromoamines, Hofmann next studied their reactions with both base and acid. Of the two types of reactions, the reactions of the *N*-haloamines in concentrated sulfuric acid gave the most interesting results. In 1885, Hofmann extended his studies of coniine (**132b**) by converting the alkaloid to its *N*-bromo derivative (**133b**). When Hofmann heated the *N*-bromoamine with concentrated sulfuric acid, the product that he

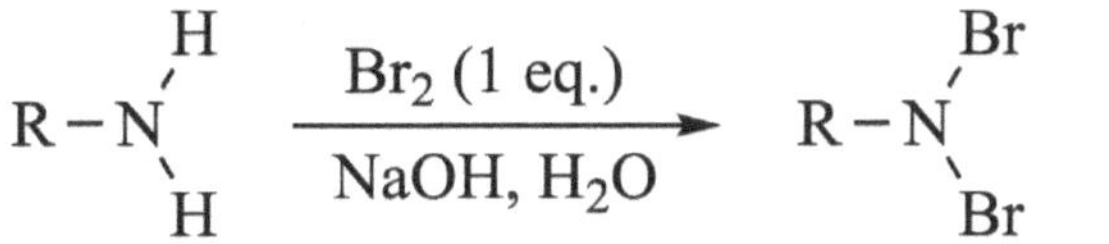

130a: R = Me
130b: R = Et

131a: R = Me
131b: R = Et

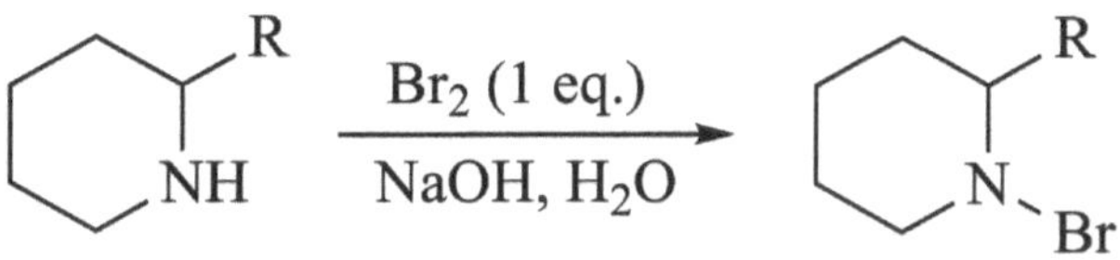

132a: R = H
132b: R = n-Pr

133a: R = H
133b: R = n-Pr

Scheme 5.7 Synthesis of N-bromoamines.

isolated after neutralization of the sulfate salts with base was Δ-coniceine (**135**) (Scheme 5.8).

In 1909, Karl Löffler, a newly habilitated Privatdozent at Breslau, extended Hofmann's work by publishing two papers, one with his student, Curt Freytag,[19] based on Freytag's Ph.D. Dissertation,[83] and one with another student, Samy Kober,[20] based on Kober's Ph.D. Dissertation.[84] He followed these papers the following year with another that acknowledged contributions from three students (Freytag, Marian Lukowsky, and Waldemar Bobiloff).[18] In these studies, the highly regioselective nature of the reaction, which gives pyrrolidine derivatives (**137**) instead of isomeric piperidine derivatives, was demonstrated (Scheme 5.9). This culminated in the synthesis of nicotine (**140**) from metanicotine (**138**) by reduction of the conjugated double bond using the two-reaction sequence of addition of hydrogen iodide and reduction of the iodide with zinc dust to hive the dihydro compound **139**, which was then converted to the pyrrolidine (Scheme 5.9).

(Hofmann, 1885)

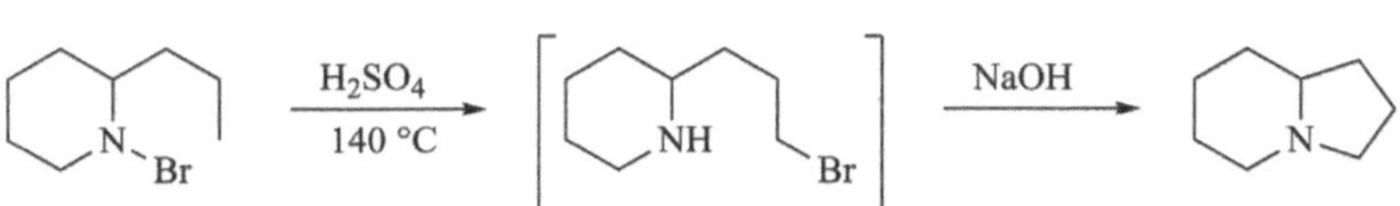

132b **134** **135**

Scheme 5.8 Hofmann's conversion of coniine to Δ-coniceine.

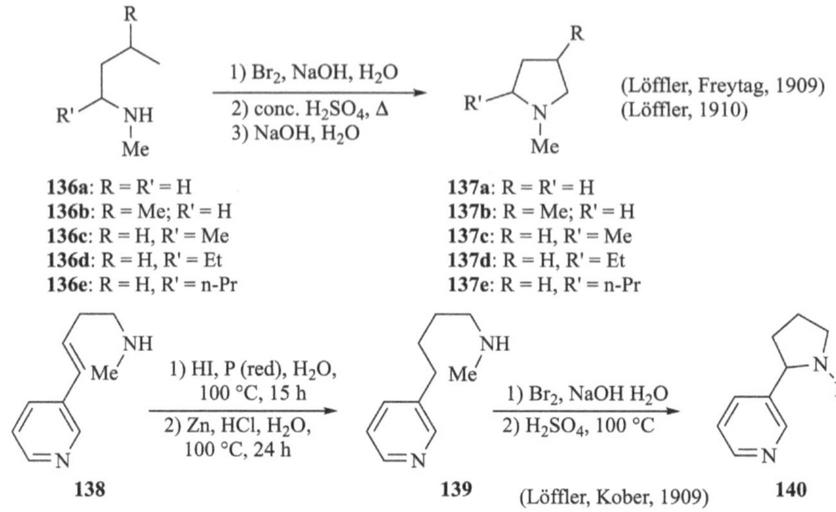

Scheme 5.9 The synthesis of pyrrolidines reported by Löffler and his students.

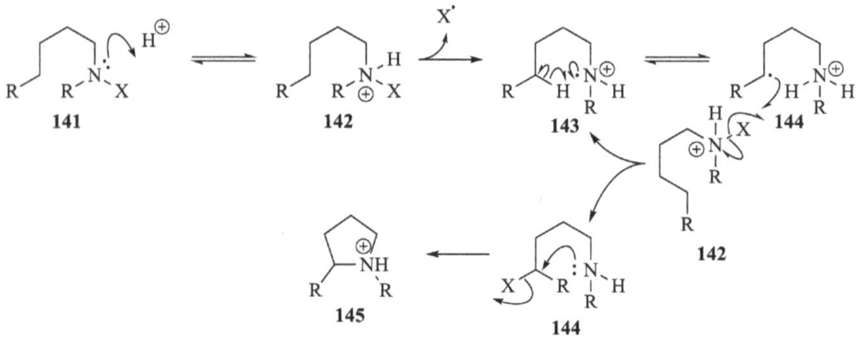

Scheme 5.10 The free radical mechanism of the Hofmann–Löffler–Freytag reaction.

The accepted mechanism for the Hofmann–Löffler–Freytag reaction (Scheme 5.10) proceeds through free radical intermediates, starting from the conjugate acid of the *N*-haloamide (**142**). The homolysis of the N—X bond generates the halogen atom and the radical cation **143**, which undergoes an intramolecular hydrogen atom transfer through a six-membered transition state to give the radical cation **144**. This radical cation abstracts the halogen atom from **142** to generate the aminohalide product **145** and regenerate the chain-carrier radical cation **143**. Intramolecular displacement of the halide then gives the pyrrolidine **146**.[85–87]

The Hofmann rearrangement

The *N*-haloamides react analogously to the amines under acid conditions, but it was the reaction of *N*-haloamides with base that gave Hofmann his third eponymous reaction: the Hofmann rearrangement. In his first paper,[10] Hofmann reported the formation of the *sym*-dialkylurea **147** from the amide **146**, and the subsequent hydrolysis of the urea to the next-lower homologous primary amine **148**. Hofmann continued with a systematic study of the reaction between amides and bromine in base. In the next two papers,[11,12] Hofmann reported that the treatment of the *N*-bromoacetamide (**149**, R=Me) with silver carbonate gave methyl isocyanate (**150**, R=Me), and that when one half equivalent of the base is used, the product is the *N*-monoacylurea **151**, and with one equivalent of bromine in dilute base, the primary amine was isolated directly. Finally,[15] with three equivalents of bromine, the product is the next-lower homologous nitrile **153**, presumably formed by elimination from the *N,N*-dibromoamine **152**. These transformations are collected in Scheme 5.11.

The Hofmann rearrangement is accepted to proceed by the mechanism shown in Scheme 5.12. In this mechanism, the initial deprotonation of the bromoamide **146** to give the conjugate base **154** is followed by a concerted rearrangement of alkyl group with concomitant displacement of the halide leaving group. Because this reaction proceeds by a concerted mechanism, the alkyl group remains bound to carbon and nitrogen in the transition state (**155**), and the reaction occurs with retention of configuration.

The original version of the Hofmann rearrangement often gave relatively poor yields due to overoxidation or the poor solubility of some amides in aqueous base. Developments over the last two decades, in particular, have focused on refining both these two factors affecting the reaction. Scheme 5.13

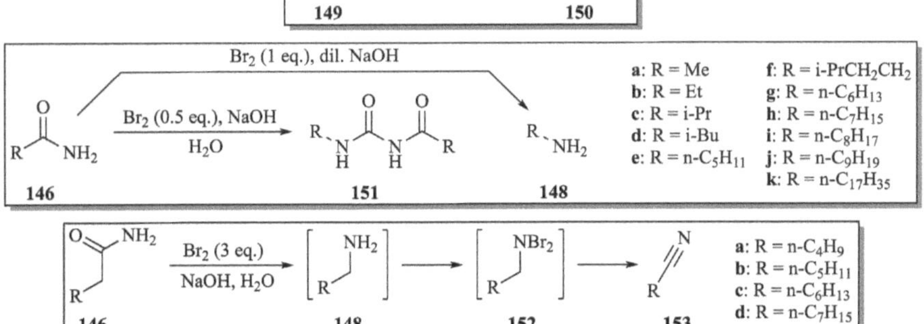

Scheme 5.11 Variations of the Hofmann rearrangement reported by Hofmann.

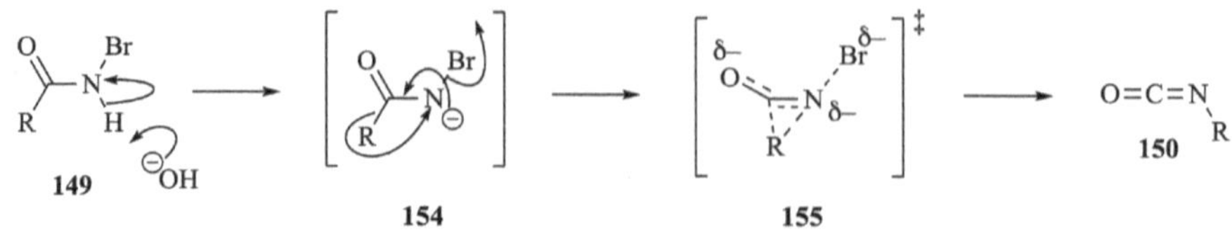

Scheme 5.12 Variations of the Hofmann rearrangement reported by Hofmann.

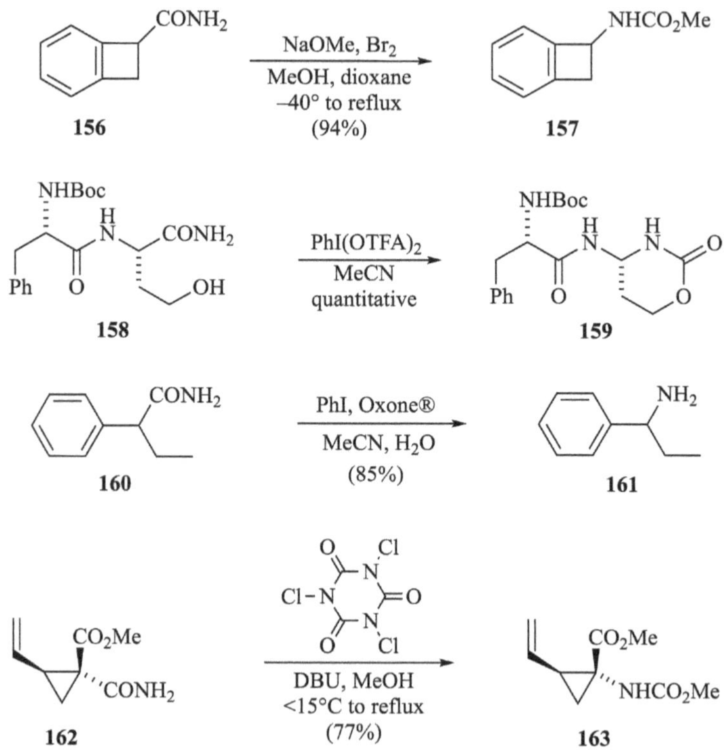

Scheme 5.13 Recent examples of Hofmann rearrangements that mitigate problems with the reaction.

shows four recent examples of the Hofmann rearrangement: Radlick and Brown[88] showed that methyl hypobromite in methanol made a useful modification of the reaction, giving the urethane, which is relatively resistant to further oxidation, rather than the free amine. During the first decade of the 21st century, hypervalent iodine compounds have emerged as reagents that avoid both the strong base and the free halogens in the reaction.[89,90] The final entry in Scheme 5.13 contains another alternative source of the halogen oxidant: trichlorocyanuric acid.[91]

The Hofmann elimination[92]

When Hofmann described the synthesis of quaternary ammonium halides, he also reported their reaction with moist silver oxide to give the quaternary ammonium hydroxides that were, in contrast to the tertiary amines from which they were formed, very strong bases. He reported that when an ethyl group was present and could be eliminated as ethylene, this was the only reaction observed (Scheme 5.14).

Hofmann's original work was followed fairly quickly by a report by British chemist J. Norman Collie (1859–42) and his student, S. B. Schryver, in which the effects of the quaternary ammonium ion structure on the outcome of the elimination were studied.[93] Their findings, which are summarized in Table 5.1, confirmed Hofmann's own conclusions about ethylammonium salts with a range of different quaternary salts, although they did also observe that benzyltrimethylammonium ion gave benzyl alcohol.

Beginning in 1911, German chemist Julius von Braun undertook a two-decade program of research into the pyrolysis of quaternary ammonium hydroxides that was one of the most extensive of the era.[94–100] Von Braun especially focused on the "anomalous" elimination reaction of alkyltrimethylammonium hydroxides, which gave methanol by the competing S_N2 displacement at the trimethylammonium group. Von Braun found that as the length of the alkyl chain increases, the relative amount of the substitution product rises from 0% to 75%. Von Braun's results are summarized in Table 5.2.

Another of von Braun's studies involved the Hofmann elimination of cyclic quaternary ammonium ions.[101] The six representative spirocyclic quaternary hydroxides pyrolyzed under vacuum by von Braun are given in Scheme 5.15. In general, he found that the piperidine rings in these

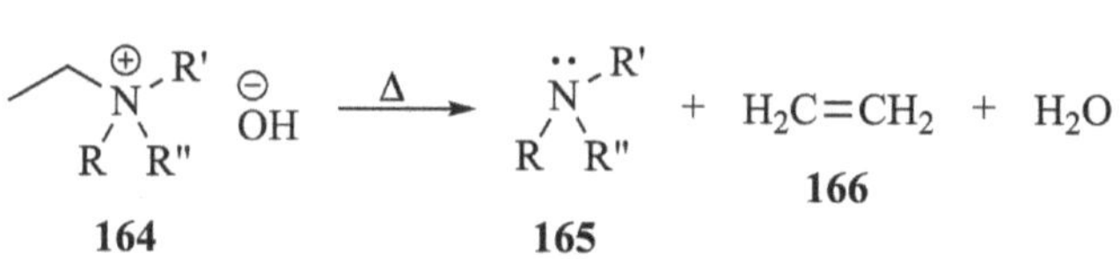

R = Me, Et, n-C_5H_{11}, Ph, $PhCH_2$
R' = Me, Et, n-C_5H_{11}
R'' = Me, Et, n-C_5H_{11}

Scheme 5.14 Hofmann's original elimination reactions.

Table 5.1 The observations of Collie and Schryver on the Hofmann elimination from quaternary ammonium salts.

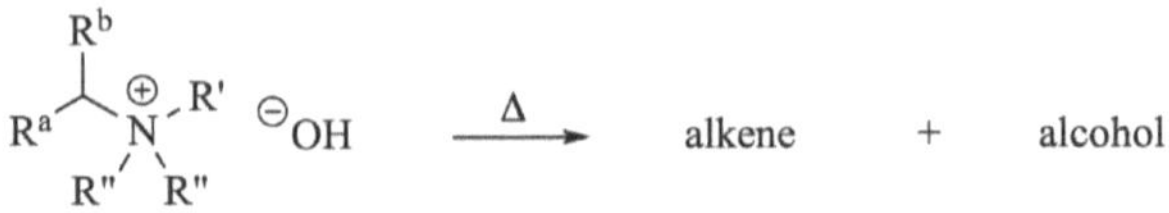

	R^a	R^b	R'	R"	alkene	alcohol
167	Me	H	Me	Me	$H_2C{=}CH_2$	none
168	Et	H	Me	Me	$MeCH{=}CH_2$	MeOH (trace)
169	Me	Me	Me	Me	$MeCH{=}CH_2$	MeOH (trace)
170	i-Pr	H	Me	Me	trace	MeOH (trace)
171	i-Bu	H	Me	Me	*i*-$PrCH{=}CH_2$	MeOH ("small")
172	$H_2C{=}CH_2$	H	Me	Me	$MeCH{=}CH_2$	MeOH (trace)
173	Et	H	Me	Me	$Me_2CHCH{=}CH_2$	
174	H	H	Ph	Me		MeOH only
175	Ph	H	Me	Me		$PhCH_2OH$
176	Me	H	Me	Et	$H_2C{=}CH_2$	
176	Me	H	Me	Me	$H_2C{=}CH_2$	

Table 5.2 The percent of substitution product formed in the pyrolysis of alkyltrimethylammonium hydroxides.

$CH_3(CH_2)_nNMe_3OH$	CH_3OH	$RX(CH_2)_nNMe_3OH$	CH_3OH
$CH_3CH_2NMe_3OH$	0	$C_6H_5O(CH_2)_3NMe_3OH$	10
$CH_3(CH_2)_2NMe_3OH$	5-10	$C_6H_5O(CH_2)_4NMe_3OH$	55
$CH_3(CH_2)_3NMe_3OH$	50	$CH_3O(CH_2)_4NMe_3OH$	60
$CH_3(CH_2)_4NMe_3OH$	60	$C_6H_5O(CH_2)_5NMe_3OH$	60
$CH_3(CH_2)_5NMe_3OH$	73	$NH_2(CH_2)_5NMe_3OH$	60
$CH_3(CH_2)_6NMe_3OH$	75	$C_6H_5CONH(CH_2)_5NMe_3OH$	60
$CH_3(CH_2)_7NMe_3OH$	75		
$CH_3(CH_2)_{15}NMe_3OH$	75		

spirocyclic ions underwent ring opening by the Hofmann eliminations. Thus, cations **183, 189,** and **191** all undergo elimination to give the *o*-vinylbenzylamines **184, 190,** and **192** as the major (or sole) products of the reaction.

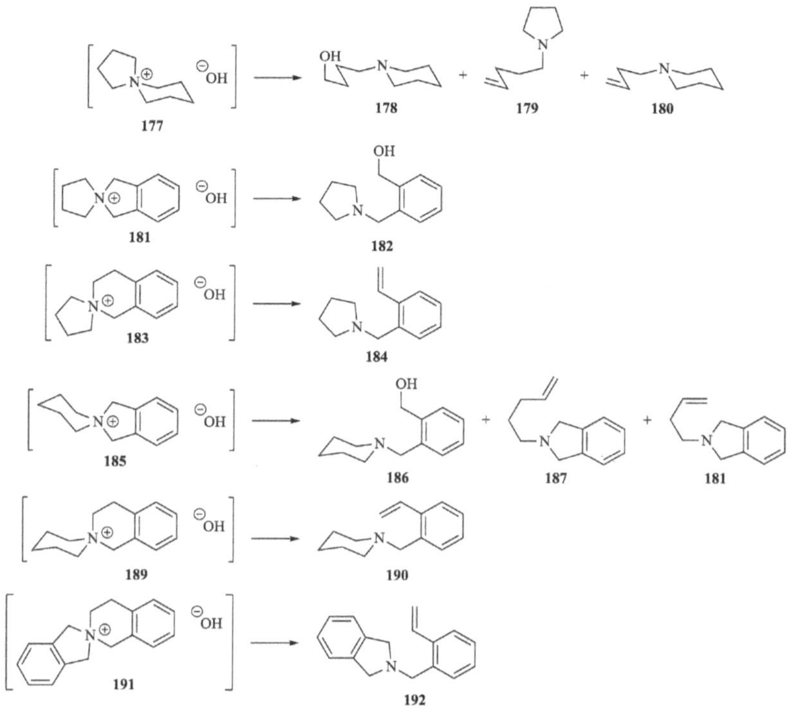

Scheme 5.15 Representative cyclic quaternary ammonium hydroxides whose pyrolysis was studied by von Braun.

In reactions where the major product is the ring-opened alcohol formed by S_N2 substitution by hydroxide anion, the pyrrolidine ring undergoes the more ready cleavage, as was confirmed by Jewers and McKenna for cation **177**.[102] This is illustrated by the reactions of the cations **177**, **181**, and **185**. In each of these reactions, the major product was formed by the S_N2 cleavage of the pyrrolidine ring; *tert.*-aminoalkenes were formed by cleavage of both heterocyclic rings in the Hofmann elimination.

A more detailed discussion of the Hofmann elimination with respect to regiochemistry, stereochemistry and mechanism will be found in Chapter 7.

References

1. Volhard, J.; Fischer, E. *August Wilhelm von Hofmann, Ein Lebensbild*; R. Friedländer u. Sohn: Berlin, 1902.
2. Volhard, J.; Fischer, E. August Wilhelm von Hofmann, Ein Lebensbild. A Complete Copy of the 1902 Book Published in Berlin, with Original Pagination. *Ber. Dtsch. Chem. Ges.* **1902,** *35* (4), 1–284.

3. Ronge, G. Hofmann, August Wilhelm v. In *Neue Deutsche Biographie*; Vol. 9; Duncker & Humblot: Berlin, 1972, pp 446–450.
4. Meinel, C. August Wilhelm Hofmann—"Reigning Chemist-in-Chief.". *Angew Chem. Int. Ed. Engl.* **1992,** *31* (10), 1265–1282 https://doi.org/10.1002/anie.199212653.
5. Partington, J. R. A. *History of Chemistry* IV; MacMillan and Company: London, 1964, pp 432–444.
6. Travis, A. S. August Wilhelm Hofmann (1818–1892). *Endeavour* **1992,** *16* (2), 59–65. https://doi.org/10.1016/0160-9327(92)90003-8.
7. Jackson, C. M. Synthetical Experiments and Alkaloid Analogues: Liebig, Hofmann, and the Origins of Organic Synthesis. *Hist. Stud. Nat. Sci.* **2014,** *44* (4), 319–363. https://doi.org/10.1525/hsns.2014.44.4.319.
8. Hofmann, A. W. Beiträge zur Kenntniss der flüchtigen Basen. X. Uebergang der flüchtigen Basen in eine Reihe nichtflüchtiger Alkaloïde. *Ann. Chem. Pharm.* **1851,** *78,* 253–286.
9. Hofmann, A. W. Researches into the Molecular Constitution of the Molecular Bases. *Phil. Trans. Roy. Soc. Lond.* **1851,** *141,* 357–398.
10. Hofmann, A. W. Ueber die Einwirkung des Broms in alkalischer Lösung auf Amide. *Ber. Dtsch. Chem. Ges.* **1881,** *14* (2), 2725–2736. https://doi.org/10.1002/cber.188101402242.
11. Hofmann, A. W. Ueber die Einwirkung des Broms in alkalischer Lösung auf Amide. *Ber. Dtsch. Chem. Ges.* **1882,** *15* (1), 407–416. https://doi.org/10.1002/cber.18820150184.
12. Hofmann, A. W. Ueber die Einwirkung des Broms in alkalischer Lösung auf Amide. *Ber. Dtsch. Chem. Ges.* **1882,** *15* (1), 752–762. https://doi.org/10.1002/cber.188201501165.
13. Hofmann, A. W. Ueber die Einwirkung des Broms in alkalischer Lösung auf Amide. *Ber. Dtsch. Chem. Ges.* **1882,** *15* (1), 762–775. https://doi.org/10.1002/cber.188201501166.
14. Hofmann, A. W. Ueber die Einwirkung des Broms in alkalischer Lösung auf die Amine. *Ber. Dtsch. Chem. Ges.* **1883,** *16* (1), 558–560. https://doi.org/10.1002/cber.188301601120.
15. Hofmann, A. W. Ueber die Einwirkung des Broms in alkalischer Lösung auf Amide. *Ber. Dtsch. Chem. Ges.* **1884,** *17,* 1406–1412. https://doi.org/10.1002/cber.188401701360.
16. Hofmann, A. W. Zur Kenntniss der Coniin-Gruppe. *Ber. Dtsch. Chem. Ges.* **1885,** *17* (1), 109–131.
17. Hofmann, A. W. Zur Kenntniss der Coniin-Gruppe. *Ber. Dtsch. Chem. Ges.* **1885,** *18* (1), 5–23. https://doi.org/10.1002/cber.18850180103.
18. Löffler, K.; with Freytag, C.; Lukowsky, M.; et al. Über eine neue Bildungsweise *N*-alkylierter Pyrrolidine. *Ber. Dtsch. Chem. Ges.* **1910,** *43* (2), 2035–2048. https://doi.org/10.1002/cber.191004302146.
19. Löffler, K.; Freytag, C. Über eine neue Bildungsweise von *N*-alkylierten Pyrrolidinen. *Ber. Dtsch. Chem. Ges.* **1909,** *42* (3), 3427–3431. https://doi.org/10.1002/cber.19090420377.
20. Löffler, K.; Kober, S. Über die Bildung des *i*-Nicotins aus *N*-Methyl-*p*-pyridyl-butylamin (Dihydrometanicotin). *Ber. Dtsch. Chem. Ges.* **1909,** *42* (3), 3431–3438. https://doi.org/10.1002/cber.19090420378.
21. Perkin, W. H. The Origin of the Coal-Tar Colour Industry, and the Contribution of Hofmann and His Pupils. *J. Chem. Soc.* **1896,** *63,* 596–637. https://doi.org/10.1039/CT8966900596.

22. Beer, J. J. *The Emergence of the German Dye Industry* Vol. XLIV; University of Illinois Press: Urbana, 1950.
23. Dauben, W. G. The Reimer-Tiemann Reaction. *Org. React.* **1982,** *28,* 1–36. https://doi.org/10.1002/0471264180.or028.01. ISBN 978–0471861416.
24. Wynberg, H. The Reimer-Tiemann Reaction. *Chem. Rev.* **1960,** *60* (2), 169–184. https://doi.org/10.1021/cr60204a003.
25. Reimer, K.; Tiemann, F. Ueber die Einwirkung von Chloroform auf Phenole und besonders aromatische Oxysäuren in alkalischer Lösung. *Ber. Dtsch. Chem. Ges.* **1876,** *9,* 1268–1278. https://doi.org/10.1002/cber.18760090270.
26. Will, W. Festrede bei der Enthüllung des Denkmals für A. W. v. Hofmann in Gießen am 8. Juli 1918. *Ber. Dtsch. Chem. Ges.* **1918,** *52* (2), 1693–1704. https://doi.org/10.1002/cber.19180510256.
27. Liebig, J. *Die organische Chemie in ihrer Anwendung auf Agricultur und Physiologie;* Vlg. Friedrich Vieweg u. Sohn: Braunschweig, 1840.
28. Liebig, J. In *Familiar Letters on Chemistry, and its Relation to Commerce, Physiology, and Agriculture; Gardner, J;* James, M., Ed.; Campbell & Co.: Philadelphia, 1843.
29. Brock, W. H. The Curious Story of the Chemical Society's Missing Obituary of John Lloyd Bullock. *Ambix* **2020,** *67* (4), 389–399. https://doi.org/10.1080/00026980.2020.1826822.
30. Crowther, J. G. The Prince Consort and Science. *New Sci.* **1961,** *265,* 689–691.
31. Brock, W. H. *Justus von Liebig: The Chemical Gatekeeper;* Cambridge University Press: Cambridge, 1997.
32. Wöhler, F.; Liebig, J. Vorläufige Notizen über Kyanol und Pseuderythrin. *Justus Liebigs Ann. Chem.* **1843,** *45* (2), 250. https://doi.org/10.1002/jlac.18430450213, 250.
33. Runge, F. F. Ueber einige Produkte der Steinkohlendestillation. *Annalen der Physik und Chemie* **1834,** *31* (5), 65–78. https://doi.org/10.1002/andp.18341070502.
34. Hofmann, A. W. Chemische Untersuchung der organischen Basen im Steinkohlen-Theeröl. *Justus Liebigs Ann. Chem.* **1843,** *47* (1), 37–87. https://doi.org/10.1002/jlac.18430470106.
35. Hofmann, A. W. Action Des Bases Alcalines Sur Différents Produits de Décomposition de l'indigo. *J. Pharm. Chim. Sér. 3* **1845,** *7,* 241–267.
36. J.L. [Liebig]. Verläufige Notiz über die Identität des Leukols und Chinolins. *Ann. Chem. Pharm.* **1845,** *53,* 427–427.
37. Liebig, J. Notice préliminaire sur l'identité du leukole et de la chinoline. *J. Pharm. Chim.* **1845,** *8,* 122–122.
38. Wilhelm Hofmann, A. Action des bases alcalines sur différents produits de décomposition de l'indigo. *J. Pharm. Chim. Sér 3* **1845,** *7,* 193–214.
39. Unverdorben, O. Ueber das Verhalten der organischen Körper in höheren Temperaturen. *Ann. Phys.* **1826,** *84* (11), 397–410. https://doi.org/10.1002/andp.18260841109.
40. Sinin, N. Beschreibung einiger neuer organischer Basen, dargestellt durch die Einwirkung des Schwefelwasserstoffes auf Verbindungen der Kohlen-Wasserstoffe mit Untersalpetersäure. *Bull. Classe Math. Acad. Imp. Sci. St.-Pétersbourg* **1842,** *10,* 273–285.
41. Zinin, N. Beschreibung einiger neuer organischer Basen, dargestellt durch die Einwirkung des Schwefelwasserstoffes auf Verbindungen der Kohlenwasserstoffe mit Untersalpetersäure. *J. Prakt. Chem.* **1842,** 140–153. https://doi.org/10.1002/prac.18420270125.
42. Zinin. Organische Salzbasen, aus Nitronaphtalose und Nitrobenzid mittelst Schwefelwasserstoff entstehend. *Ann. Chem. Pharm.* **1842,** *44,* 283–287. https://doi.org/10.1002/jlac.18420440328.

43. Fritsche, J. Ueber das Anilin, ein neues Zersetzungsproduct des Indigo. *Bull. Classe Math. Acad. Imp. Sci. St.-Pétersbourg* **1840,** 7 (12), 161–165.
44. Fritsche, J. Ueber das Anilin, ein neues Zersetzungsprodukt des Indigo. *Annalen der Chemie und Pharmacie* **1840,** *36* (1), 84–90. https://doi.org/10.1002/jlac.18400360108.
45. Fritsche, J. Ueber das Anilin, ein neues Zersetzungsprodukt des Indigo. *J. Prakt. Chem.* **1840,** *20* (1), 453–459. https://doi.org/10.1002/prac.18400200161.
46. Fritsche, J. Bemerkung zu der Abhandlung des Hrn. Sinin in No. 234 des Bulletin. *Bull. Classe Math. Acad. Imp. Sci. St.-Pétersbourg* **1842,** *10.* col 352–352.
47. Hofmann, A. W. A Chemical Investigation of the Organic Bases Contained in Coal-Gas Naphtha. *Phil. Mag.* **1844,** *24,* 261–267.
48. Hofman, A. W. A. Chemical Investigation of the Organic Bases Contained in Coal-Gas Naphtha. *Phil. Mag.* **1844,** *24,* 115–128.
49. Hofman, A. W. A Chemical Investigation of the Organic Bases Contained in Coal-Gas Naphtha. *Phil. Mag.* **1844,** *24,* 193–205.
50. Dumas, J. Ueber die Chloressigsäure, die Constitution einiger organischer Körper und über die Substitutionstheorie. *Annalen der Pharmacie* **1839,** *32* (1), 101–119. https://doi.org/10.1002/jlac.18390320109.
51. Dumas, J. Mémoire sur la constitution de quelques corps organiques et sur la théorie des substitutions. *Compt. rend. hebd. Séances Acad. Sci.* **1839,** *8,* 609–622.
52. Dumas, J. Recherches sur les Combinaisons de l'Hydrogène et du Carbone. *Ann. Chem. Phys.* **1832,** *50,* 182–197.
53. Dumas, J. Mémoire sur la loi des substitutions et la théorie des types. *Compt. rend. hebd. Séances Acad. Sci.* **1840,** *10,* 149–178.
54. Gerhardt, C. Sur la Constitution des Sels organiques à Acides complexes, et leurs Rapports avec les Sels ammoniacaux. *Ann. Chem. Phys.* **1839,** *72,* 184–214.
55. Gerhardt, C. Sur une nouvelle classe des composés organiques. *Compt. Rend. Hebd. Séances Acad. Sci* **1839,** *20,* 1031–1038.
56. Laurent, A. *Thèse de Docteur; Paris,* 1837; pp 11, 88, 102.
57. Laurent, A. *Méthode de Chimie (avec un avis au lecteur par J. B. Biot);* Mallet-Bachelier: Paris, 1854.
58. Laurent, A. Essai sur l'Action du Chlore sur la Liqueur des Hollandais et sur quelques Éthers. *Ann. Chem. Phys.* **1836,** *63,* 377–389.
59. Laurent, A. Suite des Recherches diverses de Chimie organique. *Ann. Chem. Phys.* **1837,** *66,* 314–335.
60. Laurent, A. Théorie des Combinaisons Organiques. *Ann. Chem. Phys.* **1836,** *61,* 125–146.
61. Erdmann, O. L. Untersuchungen über den Indigo. Erste Abhandlung. *J. Prakt. Chem.* **1840,** *19,* 321–362. https://doi.org/10.1002/prac.18400190161.
62. Erdmann, O. L. Untersuchungen über den Indigo. Zweite Abhandlung. *J. Prakt. Chem.* **1841,** 257–299. https://doi.org/10.1002/prac.18410220148.
63. Perkin, W. H. Note on the Action of Chloride of Lime on Aniline. *J. Chem. Soc.* **1869,** *22,* 25–27. https://doi.org/10.1039/JS8692200025.
64. Béchamp, A. De l'action des protosels de fer sur la nitronaphtaline et la nitrobenzine. Nouvelle méthode de formation des bases organiques artificielles de Zinin. *Ann. Chim. Phys.* **1854,** *42,* 186–196.
65. Perkin, Mr. [W. H.]. On the Colouring Matters derived from Coal-Tar. *Chem. News* **June 1861,** *8,* 346–354.
66. Perkin, W. H. On Colouring Matters derived from Coal-Tar. *Quart. J. Chem. Soc.* **1862,** *14,* 230–255. https://doi.org/10.1039/QJ8621400230.
67. Laurent, A. Sur un nouveau mode de formation de l'aniline. *Compt. rend. hebd. Séances Acad. Sci.* **1843,** *17,* 1366–1368.

68. Muspratt, J. S.; Hofmann, A. W. On certain Processes in which Anilene Is formed. *Mem. Proc. Chem. Soc.* **1843,** *2,* 249–254. https://doi.org/10.1039/MP8430200249.
69. Hofmann, A. W. Researches on the Volatile Organic Bases. II. On the action of iodine on aniline. *Quart. J. Chem. Soc.* **1849,** *1,* 269–281. https://doi.org/10.1039/QJ8490100269.
70. Hofmann, A. W. Beiträge zur Kenntniss der flüchtigen Basen. I. Ueber die Einwirkung des Cyans auf Anilin, Toluidin und Cumidin. *Justus Liebigs Ann. Chem.* **1848,** *66* (2), 129–149. https://doi.org/10.1002/jlac.18480660202.
71. Hofmann, A. W. Beiträge zur Kenntniss der flüchtigen organischen Basen. III. Einwirkung des Chlorcyans, Bromcyans und Jodcyans auf Anilin. Melanilin, eine neue gepaarte Base. *Justus Liebigs Ann. Chem.* **1848,** *67* (2), 129–173. https://doi.org/10.1002/jlac.18480670202.
72. Hofmann, A. W. Researches on the Volatile Bases. III. Action of Chloride, Bromide and Iodide of Cyanogen on Aniline. *Quart. J. Chem. Soc.* **1849,** *1,* 285–317. https://doi.org/10.1039/QJ8490100285.
73. Perkin, W. H. *Producing a New Coloring Matter for Dyeing with a Lilac or Purple Color Stuffs of Silk, Cotton, Wool, or Other Materials.* Great Britain Patent No. 1984, 1856.
74. Travis, A. S. Perkin's Mauve: Ancestor of the Organic Chemical Industry. *Technol. Cult.* **1990,** *31* (1), 51–82. https://doi.org/10.2307/3105760.
75. Dumas, J. Note relative aux réclamations de M. Laurent. *Compt. rend. hebd. Séances Acad. Sci.* **1840,** *10,* 511–524.
76. Dumas, J. Ueber das Gesetz der Substitutionen und die Theorie der Typen. *Justus Liebigs Ann. Chem.* **1840,** *33* (3), 259–300. https://doi.org/10.1002/jlac.18400330302.
77. Dumas, J. Ueber die Constitution der Essigsäure und Chloressigsäure. *Justus Liebigs Ann. Chem.* **1840,** *33* (2), 179–182. https://doi.org/10.1002/jlac.18400330203.
78. Dumas, J. Ueber die Einwirkung des Chlors auf den aus essigsauren Salzen entstehenden Kohlenwasserstoff. *Justus Liebigs Ann. Chem.* **1840,** *33* (2), 187–189. https://doi.org/10.1002/jlac.18400330205.
79. Dumas, J.; Peligot, E. Ueber die chemischen Typen. *Justus Liebigs Ann. Chem.* **1840,** *35* (3), 281–292. https://doi.org/10.1002/jlac.18400350304.
80. Dumas, J.; Piria, R. Ueber die chemischen Typen. *Justus Liebigs Ann. Chem.* **1842,** *44* (1), 66–100. https://doi.org/10.1002/jlac.18420440107.
81. Dumas, J.; Stass, J. S. Ueber die chemischen Typen. *Justus Liebigs Ann. Chem.* **1840,** *35* (2), 129–173. https://doi.org/10.1002/jlac.18400350202.
82. Hofmann, A. W. Researches regarding the Molecular Constitution of the Volatile Organic Bases. *Philos. Trans. Royal Soc.* **1850,** *140,* 93–131. https://doi.org/10.1098/rstl.1850.0006.
83. Freytag, C. Neue Synthese *N*-Alkylierter Pyrrolidine. *Inaug.-Diss.; H. Fleischmann: Breslau* **1910**.
84. Kober, S. Ueber die Bildung des *i*-Nikotins aus Dihydrometanikotin (*N*-Methyl-β-Pyridyl-Butylamin). *Inaug.-Diss. H. Fleischmann: Breslau* **1910**.
85. Wawzonek, S.; Culbertson, T. P. The Formation of 4-Chlorodibutylamine from N-Chlorodibutylamine. *J. Am. Chem. Soc.* **1959,** *81* (13), 3367–3369. https://doi.org/10.1021/ja01522a053.
86. Wawzonek, S.; Thelan, P. J. Preparation of N-Methylgranatanine. *J. Am. Chem. Soc.* **1950,** *72* (5), 2118–2120. https://doi.org/10.1021/ja01161a068.
87. Wawzonek, S.; Thelan, M. F., Jr.; Thelan, P. J. Preparation of Quinuclidines. *J. Am. Chem. Soc.* **1951,** *73* (6), 2806–2808. https://doi.org/10.1021/ja01150a111.
88. Radlick, P.; Brown, L. R. A Versatile Modification of the Hofmann Rearrangement. *Synthesis* **1974**, (4), 290–292. https://doi.org/10.1055/s-1974-23298.

89. Yu, C.; Jiang, Y.; Liu, B.; et al. A facile synthesis of 2-oxazolidinones via Hofmann rearrangement mediated by bis(trifluoroacetoxy)iodobenzene. *Tetrahedron Lett.* **2001,** *42* (8), 1449–1452. https://doi.org/10.1016/S0040-4039(00)02297-8.
90. Zagulyaeva, A. A.; Banek, C. T.; Yusubov, M. S.; Zhdankin, V. V. Hofmann Rearrangement of Carboxamides Mediated by Hypervalent Iodine Species Generated In Situ from Iodobenzene and Oxone: Reaction Scope and Limitations. *Org. Lett.* **2010,** *12* (20), 4644–4647. https://doi.org/10.1021/ol101993q.
91. Crane, Z. D.; Nichols, P. J.; Sammakia, T.; et al. Synthesis of Methyl-1-(*tert*-butoxycarbonylamino)-2-vinylcyclopropanecarboxylate via a Hofmann Rearrangement Utilizing Trichloroisocyanuric Acid as an Oxidant. *J. Org. Chem.* **2011,** *76,* 277–280. https://doi.org/10.1021/jo101504e.
92. Cope, A. C.; Trumbull, E. R. Olefins from Amines: The Hofmann Elimination Reaction and Amine Oxide Pyrolysis. *Org. React.* **1960,** *11,* 317–493. https://doi.org/10.1002/0471264180.or011.05.
93. Collie, N.; Schryver, S. B. LIII.—The Action of Heat on the Chlorides and Hydroxides of Mixed Quaternary Ammonium Compounds. *J. Chem. Soc. Trans.* **1890,** 767–782. https://doi.org/10.1039/CT8905700767.
94. von Braun, J. Über den Zerfall quartärer Ammoniumhydroxyde. *Justus Liebigs Ann. Chem.* **1912,** *386* (3), 273–303. https://doi.org/10.1002/jlac.19123860302.
95. Braun, J.; Anton, E. Über den Zerfall quartärer Ammoniumhydroxyde (VI. Mitteil.). *Ber. Dtsch. Chem. Ges.* **1931,** 2865–2869. https://doi.org/10.1002/cber.19310641122.
96. Braun, J.; Hamann, K. Über den Zerfall quartärer Ammoniumhydroxyde (VII. Mitteil.). *Ber. Dtsch. Chem. Ges.* **1932,** 1580–1586. https://doi.org/10.1002/cber.19320650907.
97. Braun, J.; Teuffert, W.; Weißbach, K. Über den Zerfall quartärer Ammonium- und Sulfoniumhydroxyde. IV. *Justus Liebigs Ann. Chem.* **1929,** *472* (1), 121–142. https://doi.org/10.1002/jlac.19294720104.
98. von Braun, J.; Buchman, E. R. Über den Zerfall Quartärer Ammoniumhydroxyde (V. Mitteil.). *Ber. Dtsch. Chem. Ges.* **1931,** 2610–2617. https://doi.org/10.1002/cber.19310641007.
99. von Braun, J.; Cahn, R. S. Über den Zerfall Quartärer Ammoniumhydroxyde. *Justus Liebigs Ann. Chem.* **1924,** *436* (1), 262–273. https://doi.org/10.1002/jlac.19244360115.
100. von Braun, J. Über den Zerfall Quartärer Ammoniumhydroxyde. *Justus Liebigs Ann. Chem.* **1911,** *382* (1), 1–49. https://doi.org/10.1002/jlac.19113820102.
101. von Braun, J. Die relative Ringfestigkeit cyclischer Basen bei der Hofmannschen Aufspaltung. *Ber. dtsch. Chem. Ges.* **1916,** *49* (2), 2629–2642. https://doi.org/10.1002/cber.191604902131.
102. Jewers, K.; McKenna, J. Stereochemical Investigations of Cyclic Bases. Part II. Hofmann Degradation of Some Cyclic Quaternary Ammonium Salts. *J. Chem. Soc.* **1958,** 2209–2217. https://doi.org/10.1039/JR9580002209.

CHAPTER SIX

Paul Walden and the Walden inversion

Introduction

And now we come to Paul Walden (1863–957)[1–6] (Fig. 6.1), whose eponymous reaction, the Walden inversion, is the only nucleophilic substitution that will be discussed. For Walden to be able to accomplish what he did, the development two major theories was required: the Structural Theory of Organic Chemistry,[7] first proposed by Kekulé[8,9] and Couper,[10–13] which was modified into a more manageable and intuitive form by Butlerov in 1861,[14,15] and the proposal of the tetrahedral carbon atom by van't Hoff[16–19] and Le Bel,[20] which introduced the third dimension into organic chemistry.[21]

By the time that Walden began his research work, stereochemistry was both well established and a major focus of research in organic chemistry. It should be pointed out that shortly before the time that Walden was carrying out his own studies in the 1890s, Emil Fischer was carrying out his remarkable research culminating in the structure and relative stereochemistry of glucose.[22]

Paul Walden (Pauls Valdens, Pavel Ivanovich Val'den)

Walden was a Latvian chemist who was the 13th child born into the large peasant farming family of Jānis and Anna Valdens on the family farm in Rozulas in the Livonia Governate of the Russian Empire (now Pipēni in Stalbe Parish, Pāraguaja Municipality, Latvia); only five of the children survived to adulthood. When he was 4 years old, his father died, and his mother died soon after, leaving the children orphaned. After the death of his mother, the family farm went bankrupt and had to be sold. His older brothers in Riga, Johann, a merchant and Boris, a military officer, gave him the financial support needed for his education.

Addition, Elimination and Substitution: Markovnikov, Hofmann, Zaitsev and Walden
ISBN: 978-0-12-821027-7
https://doi.org/10.1016/B978-0-12-821027-7.00003-7

Figure 6.1 Paul Walden (186357). *Paul Walden ca. 1910. n.d. Public domain image downloaded from https://commons.wikimedia.org/wiki/File:Paul_Walden.jpg.*

Education

He began his education in the district school in Cēsis, a German-language school where he graduated with honors in 1876; at the age of 10, he was handed over to the care of teacher, Erdmanis Rubens. While in that school, he learned both German and Russian and remained fluent in both for the rest of his life. In 1882, he graduated from the Riga Technical High School. By the time of his graduation, his brothers' financial situation had improved enough to allow him to continue his education. He entered the Riga Polytechnic Institute (Fig. 6.2), the oldest Polytechnic in the Russian Empire, in December the same year. At that time, approximately 5%–6% of the population of the Baltic Provinces (including modern Latvia and Estonia) were Baltic Germans, an ethnic group who enjoyed considerable autonomy as well as both political and economic power in the region. The Institute retained the privilege of delivering instruction in German,

Figure 6.2 Riga Polytechnic Institute. *The image of Riga Polytechnic Institute courtesy of RTU Research Center for Engineering History, N.d.*

rather than Russian (a rare situation in the Russian Empire at the end of the 19th century) until 1896, when it was nationalized as part of Nicholas II's program of Russification and became the State Riga Polytechnic Institute with instruction in Russian.

Student years at the Polytechnic Institute (the Polytechnicum)

In December 1882, Walden entered the Riga Polytechnicum as an undergraduate student. Almost simultaneously with Walden's admission to the Institute, Wilhelm Ostwald (1853–932, Fig. 6.3), who would win the Nobel Prize in Chemistry in 1909, was appointed to the Polytechnicum as Professor of Chemistry.[23] Ostwald had returned to his home city from Dorpat (now Tartu, in Estonia), where he had been a student and professor. This was a singularly fortuitous coincidence for the young Walden, who found in Ostwald both a mentor and a friend.[24] As an undergraduate at the Polytechnicum, Walden quickly turned to the study of chemistry and began working with Ostwald. In April 1887, while still an undergraduate student, Walden was approved for membership in the Russian Physical–Chemical Society on the recommendation of Nikolai Kurnakov, Nikolai Menshutkin and I. Schroeder—an unusual move by the Society. By the time he had graduated (with honors) from the Department of Chemical

Figure 6.3 (Friedrich) Wilhelm Ostwald. *Public domain image of Ostwald from https://commons.wikimedia.org/wiki/File:Wilhelm_Ostwald_by_Nicola_Perscheid.jpg, n.d.*

Technology as a certified chemical engineer in 1888, he had already been working with Ostwald, and then, after Ostwald's departure for Leipzig, with Carl Adam Bischoff (1855–908, Fig. 6.3), Ostwald's successor. In 1888, he published three papers: two in the *Zeitschrift für physikalisch Chemie*,[25,26] on the use of electrical conductivity in determining the molecular weights of solutes in aqueous solutions, and one concerning the detection of nitric acid, in the *Zh. Russ. Fiz-Khim, O-va.*[27]

Graduate studies and first academic appointments

Ostwald's successor, Bischoff, was an organic chemist whose focus was organic stereochemistry. Walden accepted Bischoff's invitation to remain at Riga as an Assistant in the Department of Chemistry; in 1890, he was promoted to Senior Assistant, and in 1892, he was promoted to Extraordinary (Associate) Professor.[28] His scientific career was strongly influenced by his work with both Ostwald and Bischoff: Walden made important contributions in the physical chemistry and electrochemistry of nonaqueous solutions and in organic stereochemistry.

During the summers of 1890 and 1891, Walden worked in Ostwald's laboratory in Leipzig, where Ostwald persuaded him to write and defend a dissertation for the degree of Ph.D. Walden submitted his dissertation for the degree of Ph.D., "Über die Affinitätsgrößen einiger organischer Säuren und ihre Beziehung zur Konstitution derselben [About the magnitude of the affinities of some organic acids and their relation to the constitution of the same]," in September, 1891.[29] For this dissertation, which was also published in several papers in the *Zeitschrift für physikalische Chemie*,[30–32] he received his Ph.D. degree with the highest level (summa cum laude). The same year, he published[33,34] studies of the tetronic acids that had been prepared and described by Eugene Demarçay.[35] Following Walden's successful defense of his Ph.D., Ostwald offered him the position of Privat-Docent at Leipzig, but Walden chose, instead, to return to Riga, where he assessed the opportunities for advancement as being better.

Having his Ph.D. from Leipzig meant that Walden's position in 1891 was similar to that of Zaitsev in 25 years earlier: he held a doctoral degree that had required defense of a dissertation based on original research, from a Western university. However, during that quarter of a century, nothing had changed in terms of the requirements for advancement in an academic career in the Russian Empire. His advancement still required that he hold an advanced degree from a Russian University. Without the *Magistr Khimii* degree, he could not be appointed to the permanent rank of Extraordinary (Associate) Professor. In Spring 1893, Walden passed the examinations for the *M. Khim.*, and he successfully defended his dissertation, "Experiments in the Study of Osmotic Phenomena with Semi-Permeable Membranes," at Novorossiisk University, in Odessa (Ukraine) the same year.[36]

Leader at the Riga Polytechnic Institute Professor and Rector

In 1894, Walden was appointed as Professor of Analytical and Physical Chemistry at the Riga Polytechnicum. In one way, his promotion was unusual for the time, because the *Doktor Khimii* was normally the required degree for promotion to a Professorial Chair. Whatever the reason, Walden was appointed professor. He did eventually obtain his *Dr. Khim.* degree—five years later, in 1899—from St. Petersburg University for a dissertation on optical isomerism.[37]

In 1902, Walden was elected to the first of two terms as Director (the position became Rector after 1917) of the Riga Polytechnic Institute. During the second year of his first term, the Institute introduced the title of Honorary Member. The first person to be elected Honorary Member of the Riga Polytechnic Institute was Walden's friend and former mentor, Wilhelm Ostwald. Walden's first term of office as Director of the Institute also included the tumultuous year of 1905, which began with Russia's defeat in the Russo-Japanese war of 1904–05,[38,39] and ended with the Revolution of 1905.[40,41] This was a time when mass political and social unrest, whose roots can be traced to the social and economic effects of that defeat, spread across the Empire, leading to the establishment of the Duma (an elected parliamentary body designed to limit the power of the autocrat) and an official curtailing of the power of the central government. In reality, however, the reforms ended up being effectively *pro forma*—the Tsar retained the power of veto and could dissolve the Duma in the event of an impasse, which he did for the first time after less than 3 months.

During this period of unrest, Walden, as Director of the Institute, tried to strike a compromise position between the revolutionaries and the more conservative members of the administration. The website of Riga Technical University reports that he later wrote, "The young people were right, that a lot had to be reviewed at university. The teaching staff had to initiate changing the direction of reforms for freer and more effective cooperation, which was what the young revolutionaries were trying to accomplish." At the same time, he proposed a number of initiatives to improve the functioning of the Institute, but almost all of them failed to clear the bureaucratic hurdles. There is a story that Walden was forced to resign his position on January 1, 1906 as a result of his refusal to report the students who had participated in protests and riots, to the Tsar's secret police.[42] His generally prostudent stance does lend credibility to this story. Following his dismissal, Walden returned to his position as Professor of Chemistry. When Bischhoff retired in 1898, the responsibility for inorganic, organic, and physical chemistry devolved upon Walden.

In 1896, the Polytechnicum was nationalized, and German was replaced as the language of instruction: henceforth, all teaching was to be delivered in

Russian. Prior to that, Walden had been unusual because he already taught his Physical Chemistry classes in Russian rather than German. During the process of nationalization, the Riga Polytechnic Institute underwent widespread reforms. Walden was responsible for reorganizing the Chemistry Department, which he accomplished with advice from Ostwald, who sent suggestions from Leipzig. The change in status of the institution brought about by its nationalization was important for both the Polytechnic and its students because it opened up opportunities for both. It allowed the Institute to receive subsidies from the Russian government and also lifted restrictions on the rights of its graduates in the civil service in the Russian Empire.

By the last years of the 19th century, the laboratory facilities at the Institute had become demonstrably inadequate in meeting the expanding needs of chemical research at the Institute. Thanks to the joint efforts of Bischhoff and Walden, a new, modern chemical laboratory building was constructed between 1899 and 1900[43]; the Chemistry Department moved into the new facilities, which they shared with the laboratory of chemical technology.

Consulting within the empire

Walden's stature as Director at Riga made him an attractive consultant to a number of institutions of higher learning. During his first term, Walden served with Nikolai Aleksandrovich Menshutkin on the Commission in the planning of the metallurgical department and the electrochemical subdivision of the newly founded Saint Petersburg Polytechnic Institute (now St. Petersburg State Polytechnic University). In 1902, Menshutkin was appointed Professor there, jointly with his Professorship at St. Petersburg University (now St. Petersburg State University). That same year, Kaiser Wilhelm II gave assent to the establishment of the Royal School of Technology in Wrocław (Königliche Technische Hochschule Breslau). Construction began in 1905, but before building began, Walden had been working on setting up the chemical department at the Hochschule. In a letter to Menshutkin's son, the chemist-historian Boris Nikolaevich Menshutkin, he wrote, "The Commission of Chemists was to submit sample curricula for the metallurgical department with an electrochemical subdivision. Similar work was done after a detailed discussion on the part of us, the three chemists—namely, Nikolai Alexandrovich, N.S. Kurnakov and me—and presented by me at the meeting of the entire commission. We defended our curricula at the meeting, and these were entered without change into the presentation to the Highest Name *[the Tsar]* about the establishment of St. Petersburg Polytechnic Institute."

The Imperial Academy of Sciences in St. Petersburg

In October 1906, Friedrich Konrad (Fyodor Fyodorovich) Beilstein, a full member of the St. Petersburg Academy of Sciences and holder of one of the two Academy Chairs, died in St. Petersburg. In March 1910, a group of influential Academicians proposed Walden as a candidate for Beilstein's Chair, which had remained unfilled since his death. In May 1910, Walden was elected to the Academy as a Full Member (Academician) in the Physics—Mathematics Division. This election greatly enhanced the prestige of the Polytechnic because one of their own had been elected to one of only two Chairs in the Academy.

Walden was granted a special consideration while he occupied his Chair—he was exempted from the normal rules so, instead of being required to live in St. Petersburg, he received the privilege of living and working in Riga, where he had more favorable opportunities for scientific research. In his memoirs, he wrote, "Every week or every 10 days, I just took a fast train to St. Petersburg, took part there in meetings of the physics and mathematics department of the Academy, gave instructions and instructions for research work in the laboratory and after a few days returned to Riga at night." He remained in this post until 1919.

World War I: evacuation of Riga

In 1915, the German forces pushed north into the Baltic states, occupying first Lithuania, then Latvia and Estonia. This invasion led to the evacuation of Riga Polytechnic to Ivanovo-Voznesensk, 250 km northeast of Moscow. Ahead of the invading forces, the Russians dismantled much of the industrial infrastructure in Riga and transported it east into Russia. The Polytechnic and its laboratories were moved to the Polytechnic Institute (Fig. 6.4) there, which, in 1918 became the first institution of higher education established by the Soviet Union.

Walden began his second term as Rector in 1917, overseeing the return of the Polytechnic to Riga in 1918. At the same time, the Ivanovo-Voznesensk Polytechnic Institute was formally established based on the

Figure 6.4 The Ivanovo-Voznesensk Polytechnic Institute in 1918, after its official founding by the Soviet Union.

Riga model, taking advantage of the expertise of the Riga scientists who preferred to remain in the Soviet Union. Latvia had fallen into disarray following the October Revolution of 1917, and the 5-year civil war that followed it simply exacerbated the situation.

One week after the armistice ending World War I was signed, Latvia declared its independence from the Russian Empire. Soviet Russia invaded less than 3 weeks later, leading to the Latvian War of Independence. This new war lasted until August 1920, prolonging the chaos begun by the German invasion of Latvia during World War I. Walden's second term as Rector was marked by events that were much more chaotic than the student protests of his first term.

Although Walden was sympathetic to the ideals of the October revolution, as was evident from his actions during and after the student protests of 1905, the chaos in the capital caused by this continued war was a major distraction from his scientific work. These difficulties and the political uncertainty that they engendered prompted Walden to send his wife and family to Germany at the beginning of summer, 1919. In August 1919, Walden was sent on a *komandirovka* (paid study leave) to Germany. He resigned his position as Rector and never returned to Riga. He did, however, retain his connections to Russia, and in 1927, he was elected an Honorary Foreign Member of the USSR Academy of Sciences.

Rostock

In 1919, Walden was elected Professor of Inorganic Chemistry at the University of Rostock. He remained in this post until his retirement at age 70, in 1934. In 1932, he was elected to membership of the Leopoldina (the German Academy of Sciences), and in 1936, he became Chair of the Section on the History of Chemistry at the Association of German Chemists (Verein Deutscher Chemiker, VDCh). He was invited to return to Riga in April 1924 to deliver guest lectures at the University of Latvia and was once again offered the position of Rector. This time, he declined. The same year, he also declined two offers from the Soviet Union: one to become Head of the Department of Chemistry at Leningrad University and the other to become Head of the Chemical Section at the Institute of Experimental Medicine in Leningrad. In September 1934, he led the German delegation to the seventh Mendeleev Congress in Leningrad, celebrating the centenary of Mendeleev's birth.

Walden's scientific career as an independent researcher

Walden's scientific work can be divided into three major periods: 1) the period up to 1892, where he worked with Ostwald in physical chemistry; 2) the period from 1892 to 1899 was focused on organic chemistry, with particular emphasis on stereochemistry and synthesis; and 3) the 20th century, when he returned to physical chemistry and electrochemistry.

Research in physical chemistry

Between 1911 and 1915, Walden published 14 papers on the electrochemistry of nonaqueous solutions, including the electrical conductivity of various substances dissolved in hydrocarbons, the association of binary salts in various media, etc., in the *Izvestia Imperatorskoi Akademii Nauk* (until 1894, the *Bulletin de l'Académie Impériale des Sciences de St.-Petersbourg*).[44–48] One of these papers contained the disclosure of the first ionic liquid, ethylammonium nitrate.[44]

Walden made an important contribution with the Walden Viscosity Rule. This empirical rule, which concerns ions in solutions, states that the product of the molar conductivity, Λ_m, and the viscosity, η, is approximately constant for the same ions in different solvents.[49–51] The relationship is not exact, however, as the data in Table 6.1 show: the average value of $\Lambda_m \cdot \eta$ is

Table 6.1 Walden's data from his 1906 paper describing his rule relating conductivity and viscosity of electrolyte solutions in nonaqueous solvents.

Bei 0°.

Ionisierungsmittel	Innere Reibung η^0_∞	Grenzwert Λ^0_∞	$\Lambda^0_\infty \cdot \eta^0_\infty$ = Konst.
1. Aceton	0·00397	177	0·707
Propionitril	0·00541	129	0·698
Nitromethan	0·00829	93·5	0·775
Methylalkohol	0·00846	90	0·761
Äthylsenföl	0·00846	82	0·694
Methylrhodanid	0·0101	70	0·707
Acetylaceton	0·0115	57	0·656
Sym. Schwefligsäureäthylester	0·0119	55	0·655
Essigsäureanhydrid	0·0130	52·5	0·683
10. Epichlorhydrin	0·0156	43·2	0·674
Äthylalkohol	0·0179	37	0·662
Benzonitril	0·0194	35·5	0·689
Furfurol	0·0248	30	0·743
Schwefelsäurediäthylester	0·0264	26	0·686
Schwefelsäuredimethylester	0·0270	28	0·756
Nitrobenzol	0·0307	25	0·768
Benzylcyanid	0·0338	19	0·642
Asym. Schwefligsäureäthylester	0·0440	14·4	0·634
19. Citrakonsäureanhydrid	0·0740	10·5	0·777
		im Mittel	0·700

From Walden, P. Über Organische Lösungs-Und Ionisierungsmittel. III. Teil: Innere Reibung Und Deren Zusammenhang Mit Dem Leitvermögen. *Z. Phys. Chem.* **1906**, 55, 207–246. https://doi.org/10.1515/zpch-1906-5408.

0.704 for all the solvents studied, with an RMS error of 6.4%. Nevertheless, the Rule has been widely used in the field of ionic polymers[52] and ionic liquids.[53]

Handbuch der Stereochemie

In 1892, following Walden's initial appointment as Extraordinary Professor, Bischoff suggested that he focus his research on organic chemistry, with an emphasis on stereochemistry. For the next 2 years, he worked with Bischoff, coauthoring their magnum opus, the *Handbuch der Stereochemie* (Fig. 6.5).[54] This work, which was encyclopedic in its coverage, details the optical and physical properties of a large number of the stereoisomers of substances known at the time; it was modeled on Beilstein's *Handbuch der organischen Chemie*.[55] The first 178 pages of the book are given to describing the history of stereochemistry, and this is followed by another 900 pages in which the properties of each compound—melting point, boiling point,

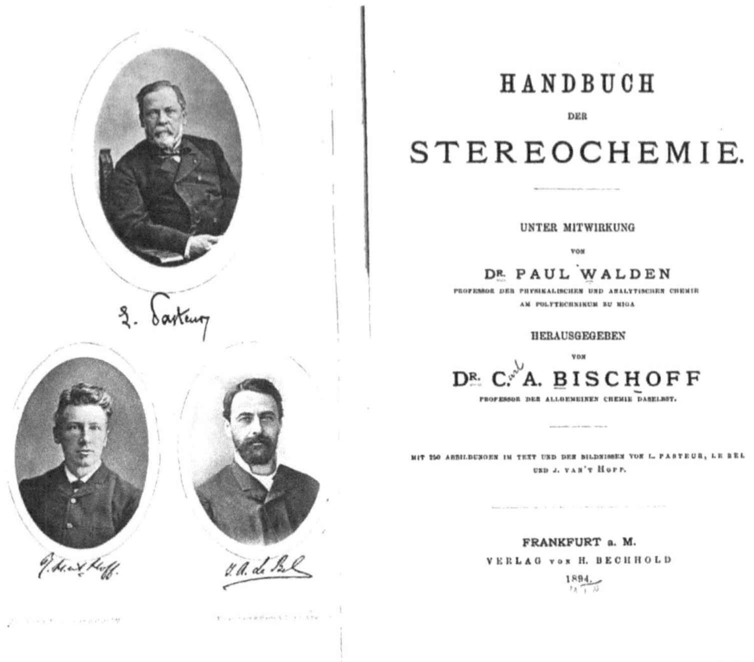

Figure 6.5 The title pages of the *Handbuch der Stereochemie. From Walden, P.; Bischoff, C. A. Handbuch Der Stereochemie; Verlag von H. Bechhold: Frankfurt am Main, 1894.*

density, refractive index, crystal form for solids, and, for chiral compounds, the specific rotation—are tabulated along with references to the original literature.

While he was working on the *Handbuch*, Walden continued his research in physical chemistry, thus carrying out work in two disparate fields. His work on the *Handbuch* often required him to make compounds and measure their physical constants.[56–65]

Research in organic chemistry

The Walden inversion and the Walden cycle

In 1896, Walden first reported his observation that the reaction of malic acid (**186**) with phosphorus pentachloride gave the chlorosuccinic acid with the opposite configuration (**187**), and that the reaction of the chlorosuccinic acid with silver oxide gave the malic acid with retention of configuration (**188**) (Scheme 6.1).[66]

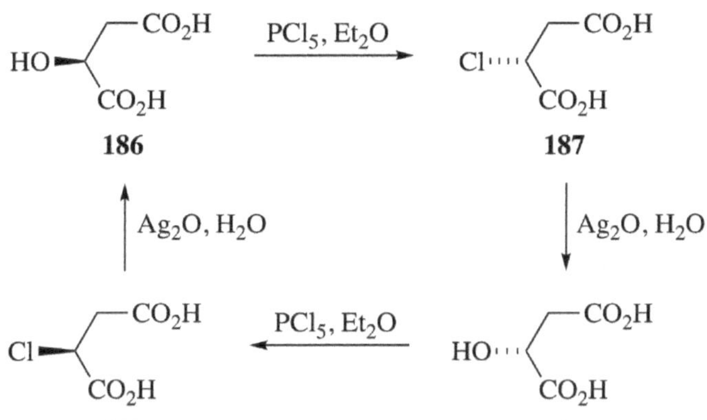

Scheme 6.1 The Walden cycle that was used to illustrate the Walden inversion.

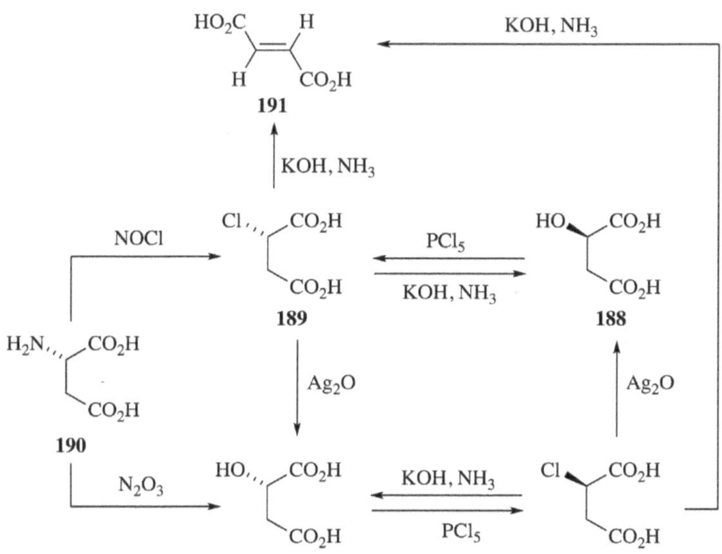

Scheme 6.2 Further elaborations by Walden on the Walden cycle.

This initial paper was followed by four more in the next 3 years,[67–70] all of them expanding on the original paper, as shown in Scheme 6.2 for a sequence beginning with aspartic acid (**190**) and ultimately terminating in fumaric acid (**191**). The results obtained are summarized in Table 6.2; the optical rotations reported are for the uranyl complexes of the malic acids, which Walden had shown to result in a remarkable enhancement of the rotation magnitude.[71]

Table 6.2 Optical rotations of the malic acids obtained by hydrolysis of bromo- and chlorosuccinic acids with various metal oxides and hydroxides as bases.

Active Halogenbernsteinsäure (Ausgangsmaterial)	Angewandtes Hydroxy-lirungsmittel	Isolirte Aepfelsäure und deren spec. Drehung	
I. *l*-Chlor-Bernsteinsäure	Ag_2O	*l*-Aepfelsäure	$[\alpha]_D$ = ca. − 460°
l-Brom- »	H_2O	*l*- »	$[\alpha]_D$ = − 212°
l-Brom- »	TlOH	*l*- »	$[\alpha]_D$ = − 203°
l-Brom- »	HgO	*l*- »	$[\alpha]_D$ = − 172°
l-Brom- »	Hg_2O	*l*- »	$[\alpha]_D$ = − 130°
l-Brom- »	PdO	*l*- »	$[\alpha]_D$ = − 125°
II. *l*-Brombernsteinsäure	Tl_2O_3	*i*-Aepfelsäure	$[\alpha]_D$ = ∓ 0°
III. *l*-Chlor-Bernsteinsäure	RbOH	*d*-Aepfelsäure	$[\alpha]_D$ = + 440°
l-Chlor- »	KOH	*d*- »	$[\alpha]_D$ = + 425°
l-Chlor- »	$NH_4.OH$	*d*- »	$[\alpha]_D$ = ca. + 460°
l-Chlor- »	CuO	*d*- »	$[\alpha]_D$ = + 330°
l-Brom- »	$Cd(OH)_2$	*d*- »	$[\alpha]_D$ = + 301°
l-Chlor- »	$Ba(OH)_2$	*d*- »	$[\alpha]_D$ = + 170°
l-Chlor- (Brom)- »	$Pb(OH)_2$	*d*- »	$[\alpha]_D$ = + 170°
l-Chlor- »	NaOH	*d*- »	$[\alpha]_D$ = + 150°
l-Chlor- »	$Sn(OH)_2$	*d*- »	$[\alpha]_D$ = + 133°
l-Chlor- »	LiOH	*d*- »	$[\alpha]_D$ = + 100°

From Walden, P. Ueber die gegenseitige Umwandlung optischer Antipoden. Berichte der deutschen chemischen Gesellschaft **1899**, 32 (2), 1833–55. https://doi.org/10.1002/cber.18990320276.

The results collected in Table 6.2 reveal that there is a competition between inversion and retention in this reaction. The data in Table 6.2 are divided into three groups. In the first group, water and the transition metal oxides (except Cu) are used as the hydrolyzing reagent. In this group, there is a net retention of configuration with e.e. ranging from 100% (Ag) to 27% (Pd). The second group contains only a single member, with thallium (III) oxide, a Lewis acid, being used as the hydrolyzing reagent; this reaction gives the racemic product. In the third group, main group metal oxides, along with copper oxide and cadmium hydroxide, are used as the hydrolyzing reagent. In all these cases, the reaction proceeds with a net inversion ranging from 100% e.e. with aqueous ammonia to 22% e.e. with lithium hydroxide. The rationalization of these results would have to await the work of Hughes and Ingold three decades later.

Walden's discovery was epoch-changing, because it took the van't Hoff-Le Bel theory of optical activity, which was based solely on structural considerations, and added a dynamic dimension to it, much the same as Markovnikov had done with the Structural Theory of Organic Chemistry in 1869.[72–74] At the same time, it produced the first experimental support for that theory and evidence for role of stereochemistry in organic reactions.

There was a problem, however: the mechanism of the inversion was not known, and nor were the reasons why the reagent used should have such a dramatic effect on the stereochemistry of the reaction. It would be 40 years before a reasonable rationalization of the Walden cycle would emerge. This question will be discussed in much more detail in the next chapter, where the efforts of physical organic chemists to provide a reasonable rationalization of the experimental observations will be discussed.

Following his retirement from his Professorship in 1934, Walden turned his attention full-time to the history of chemistry. He had begun this line of research because of the difficulties associated with carrying out experimental chemistry while under evacuation to Ivanovo-Voznesensk during World War I. During the evacuation, he wrote an account of the history of chemistry in Russia that was published as a 329-page addendum to the 1917 Russian translation of Ladenburg's *Vorlesungen über die Geschichte der Chemie von Lavoisier bis heute.*[75]

Walden in the Third Reich

Although it became clear after Hitler's rise to power that membership in the Nazi Party would be an essential requirement for advancement, Walden's decision to join the Nazi Party in 1934 did not have any suggestion of self-interest. He was sympathetic toward the Nazi ideology, and the same year he became a member of the SS, advocating for them and supporting them financially.[76] He did not protest at the Nazification of science and the expulsion of Jewish scientists from membership in German learned societies, nor did he protest the erasure of their accomplishments from the Nazi history of German science. Still, in his silence, he was not very different from many of his fellow German scientists.[77]

The push from certain factions of the regime to propagandize science was especially focused on the history of science. Chemistry was characterized as a "Northern Racial science; " this racial—political perspective was then used to as justification for the Nazi claim of German leadership in chemistry. The famed alchemist, Paracelsus (1493/1494–1541, Philippus Aureolus Theophrastus Bombastus von Hohenheim), was reinvented by the Nazis as a racist anti-Semitic fanatic, making him the forerunner of Nazi anti-Semitism. Paracelsus did hold the anti-Semitic views common among European Christians at the time, but in fact, the anti-Semitism based on race attributed to him by the Nazis was not part of his ideology. As a historian of chemistry, Walden must have been aware of the racial purges and the false representations they engendered, but his writings at the time showed that he actually supported these actions. His position of international eminence among scientists (he was on friendly terms with many of the leading scientists in Europe) meant that he was not as rigidly controlled by the Nazi government as many others. For example, he was never punished for refusing to write pieces hostile to the Soviet Union, and his trips abroad were never hindered by the government.

Walden in retirement

When Walden retired from his position in 1934, he received a regular pension from Rostock University. This allowed him to turn his attention full-time to the history of chemistry. He had begun this line of research while under evacuation to Ivanovo-Voznesensk during World War I because of the difficulties associated with carrying out experimental chemistry there. During the evacuation, he wrote an account of the history of chemistry in Russia that was published as a 329-page addendum to the 1917 Russian translation of Ladenburg's *Vorlesungen über die Geschichte der Chemie von Lavoisier bis heute*.[75]

During the bombing of Rostock in April 1942, Walden lost his home, his book collection, which contained over 10,000 rare volumes, and much of his correspondence with other scientists, including Arrhenius, Ramsay, Ostwald, and Mendeleev. He and his wife would also have been killed had they not been sheltering in the basement of neighbor's house. Following this personal catastrophe, Walden and his wife moved to an apartment for a visiting scientist in Berlin. Here, they were bombed once again, when the allies began the spring bombing of that city. Walden was seriously wounded by a glass shard in one of these attacks. His health rapidly deteriorated after this, and he suffered more than one heart attack. In March 1943, he and his wife moved again, to Frankfurt am Main, where he accepted an appointment as Visiting Professor in the History of Chemistry at the Goethe University.

Walden did not need to be reinstated to his academic position immediately after the war, so he was not immediately subject to the postwar blacklist of former Nazi Party members; most had fled to the British, American, and French zones ahead of the Soviet advance (Walden himself ended up in the French zone). Under the denazification policy of occupying powers, no former members of the Nazi Party could occupy any position of responsibility in Germany. That included teaching and research positions in German universities. However, Walden did suffer one unforeseen consequence at the end of the war—all his retirement savings and accounts were held in Rostock, which was in the Soviet zone; he could not access any of them.

Faced with the resultant financial difficulties, Walden sent a request to the Presidium of the USSR Academy of Sciences that he be reinstated to his full (as opposed to honorary) membership of the Academy and provided with financial assistance. His request went unanswered. Finally, Walden and his wife were placed in a hospital *pensionat* in Hammertingen, a small town 40 km south of Tübingen in the Baden-Württemberg district of Germany, with funds raised by other German chemists (many of whom were in distress, themselves). Walden lived there until his death in 1957.

It was not until 1947, when the denazification sanctions against former Nazi Party members were eased, that Walden was able to obtain an appointment as Visiting Professor in the History of Chemistry at the University of Tübingen; thereafter, he regularly visited the university to deliver lectures. His last lecture took place in Tübingen on July 23, 1953. The Waldens had only very limited means, which meant that they had to live frugally, with Walden supplementing his meager income by writing and reviewing articles in German and foreign journals and writing books and memoirs. The payment of his pension was not resumed until 1955. Walden died in 1957, at the age of 93, and was buried in the Bergfriedhof cemetery in Tübingen.

Walden, the historian of chemistry

Walden's career as a historian of chemistry began shortly after he became Professor of Analytical Chemistry at Riga Polytechnic. In 1896, he wrote his first major work in the history of chemistry, a biography of

Louis Pasteur.[78] After his retirement, and in the 20 years that followed, he established an international reputation as a preeminent scholar in that field. In all, Walden wrote 135 papers and book in the history of chemistry between 1895 and 1957. He continued his work writing biographies and obituaries of important chemists,[79–90] His major works in the History of Chemistry include his 1917 history of chemistry in Russia,[79–90] a book on the influence of physics on the development of chemistry,[91] on Goethe's science,[92] and general works on the history of chemistry.[93–98]

Afterword

Over the four decades from 1914 to 1954, Walden was nominated nine times for the Nobel Prize in Chemistry: 1914, 1920 (two nominations), 1927, 1934 (two nominations), 1947, and 1954. Like his younger contemporary, Walter Reppe (1892–969), it is possible that Walden's willing support of the Nazi Party and its ideology may have cost him his chance at the Prize.

References

1. Everts, S. Latvia Celebrates Paul Walden. *C&EN* **2013,** *91* (29), 28–30.
2. Morachevskii, A. G. Academician Pavel Ivanovich Walden (On 140th Anniversary of His Birthday). *Russ. J. Appl. Chem.* **2003,** *76* (7), 1186–1190. https://doi.org/10.1023/A:1026399420965.
3. Morachevskii, A. G.; Firsova, E. G. Akademik Pavel Ivanovich Val'den (k 150-Letno Co Dnya Rozhdeniya) Academician Pavel Ivanovich Walden (On the Occasion of His 150th Birthday). *Khronika, SPbSPU Sci. Tech. Statements Phys. Math. Sci.* **2013,** 245–249.
4. Stradinš, J. W.; Valden, P. I. In *Complete Dictionary of Scientific Biography;* Gillispie, C. C., Holmes, F. L., Koertge, N., Eds.; Charles Scribner's Sons: Detroit, 2008.
5. Stradinš, J. P.; Solov'ev, Y. I. *Pavel Ivanovich (Paul) Walden: 1863-1957;* Nauka: Moscow, 1988.
6. Hückel, W. P. Walden. 26 Juli 1863 - 22 Januar 1957. *Chem. Ber.* **1957,** *26* (7), XIX–LXV.
7. Lewis, D. E. 150 Years of Organic Structures. *ACS (Am. Chem. Soc.) Symp. Ser.* **2010,** *1044,* 35–57. https://doi.org/10.1021/bk-2010-1044.ch004.
8. Kekulé, A. Ueber die s. g. gepaarten Verbindungen und die Theorie der mehratomigen Radicale. *Justus Liebigs Ann. Chem.* **1857,** *104* (2), 129–150. https://doi.org/10.1002/jlac.18571040202.
9. Kekulé, A. Ueber die Constitution und die Metamorphosen der chemischen Verbindungen und über die chemische Natur des Kohlenstoffs. *Justus Liebigs Ann. Chem.* **1858,** *106* (2), 129–159. https://doi.org/10.1002/jlac.18581060202.
10. Couper, A.-S. Sur Une Nouvelle Théorie Chimique. *Compt. rend. hebd. séances Acad. Sci.* **1858,** *46,* 1157–1160.
11. Couper, A.-S. Sur une nouvelle théorie chimique. *Ann. Chim. Phys. Sér. 3* **1858,** *53* (3), 469–489.

12. Couper, A. S. Ueber eine neue chemische Theorie. *Justus Liebigs Ann. Chem.* **1859,** *110* (1), 46–51. https://doi.org/10.1002/jlac.18591100105.
13. Couper, A. S. On a New Chemical Theory. *Phil. Mag.* **1858,** *16,* 104–116. https://doi.org/10.1080/14786445808642541.
14. Butlerow, A. Einiges über die chemische Structur der Körper. *Z. Chem. Pharm.* **1861,** *4,* 549–560.
15. Butlerow, A. Ueber Die Verwandtschaft Der Mehraffinen Atome. *Z. Chem. Pharm.* **1862,** *5,* 297–304.
16. van't Hoff, J. H. *Voorstel tot Uitbreiding der tegenwoordig in die scheikunde gebruikte Struktuur-Formels in die ruimte; benevens een daarme samenhangende opmerkung omtrent het verband tusschen optisch actief Vermogen en Chemische Constitutie van Organische Verbindingen;* Greven: Utrecht, 1874.
17. van't Hoff, J. H. Sur Les Formules de Structure Dans l'espace. *Bull. Soc. Chim. Paris, N. S* **1875,** *23,* 295–301.
18. van't Hoff, J. H.; Hermann, F. *Translator. Die Lagerung der Atome im Raume;* Vieweg: Braunschweig, 1875.
19. van't Hoff, J. H. *La Chimie Dans l'espace;* Bazendijk: Rotterdam, 1875.
20. Le Bel, J.-A. Sur Les Relations Qui Existent Entre Les Formules Atomiques Des Corps Organiques et Le Pouvoir Rotatoire de Leurs Dissolutions. *Bull. Soc. chim. Paris* **1874,** *22,* 337–347.
21. Ramberg, P. J. *Chemical Structure, Spatial Arrangement: The Early History of Stereochemistry, 1874-1914;* Taylor and Francis: United States, 2017; pp 1–399.
22. Hudson, C. S. Emil Fischer's Discovery of the Configuration of Glucose: A Semicentennial Retrospect. *J. Chem. Educ.* **1941,** *18* (8), 353–357. https://doi.org/10.1021/ed018p353.
23. van't Hoff, J. H. Friedrich Wilhelm Ostwald. *Z. Phys. Chem.* **1903,** *46,* V–XV.
24. Walden, P. Schriften von W. Ostwald. *Z. Phys. Chem.* **1903,** *46,* XVI–XXVII.
25. Walden, P. Über Die Bestimmung Der Molekulargrössen von Salzen Aus Der Elektrischen Leitfähigkeit Ihrer Wässerigen Lösungen. I. *Z. Phys. Chem.* **1887,** *1,* 529–549. https://doi.org/10.1515/zpch-1887-0155.
26. Walden, P. Über die Bestimmung der Molekulargrössen von Salzen aus der elektrischen Leitfähigkeit ihrer wässerigen Lösungen. II. *Z. Phys. Chem.* **1888,** *2,* 49–77. https://doi.org/10.1515/zpch-1888-0211.
27. Val'den, P. Sravnitel'naya Otsenka Reaktsii, Predlozhennykh dlya otkrytiya azotnoi kisloty [Comparative Evaluation of the Reactions Proposed for the Detection of Nitric Acid]. *Zh. Russ. Fiz.-Khim. O-va.* **1887,** *19,* 274–295.
28. Stradyn, Ya. P. Khimiya v Rizhskom Politekhnicheskom Institute (1862 – 1918) [Chemistry at the Riga Polytechnic Institute (1862 – 1918)]. *Uch. zap. Latv. gos. un-ta. Khim. fiz.-tekh.-t* **1958,** *XXII* (No. 6), 307–325.
29. Walden, P. *Über die Affinitätsgrößen einiger organischer Säuren und Ihre Beziehung zur Konstitution derselben.* Ph.D. Diss.: Leipzig, 1891.
30. Walden, P. Über die Affinitätsgrössen organischer säuren und ihre Beziehungen zur Konstitution derselben. *Z. Phys. Chem.* **1891,** *8,* 433–503. https://doi.org/10.1515/zpch-1891-0837.
31. Walden, P. Über die Affinitätsgrössen organischer säuren und ihre Beziehungen zur Konstitution derselben. II. Teil: Tricarbonsäure. *Z. Phys. Chem.* **1892,** *10,* 563–579. https://doi.org/10.1515/zpch-1892-1040.
32. Walden, P. Über die Affinitätsgrössen organischer säuren und ihre Beziehungen zur Konstitution derselben. III. Teil: Monocarbonsäure. *Z. Phys. Chem.* **1892,** *10,* 638–664. https://doi.org/10.1515/zpch-1892-1044.

33. Walden, P. Zur Kenntniss der Tetrinsäure, Oxytetrinsäure und ihrer Homologen. *Ber. Dtsch. Chem. Ges.* **1891,** *24* (1), 2025–2039. https://doi.org/10.1002/cber.189102401343.
34. Val'den, P. I. K voprosu o stroenii kislot Demarsé [On the Question of the Structures of Demarçay's Acids]. *Zh. Russ. Fiz.-Khim. O-va.* **1891,** *23,* 632–652.
35. Demarçay, E. Sur Les Acides Tétrique et Oxytétrique et Leurs Homologues. *Ann. Chem. Phys.* **1880,** *20,* 433–503.
36. Walden, P. Ueber Diffusionserscheinungen an Niederschlagsmembranen. *Z. Phys. Chem.* **1892,** *10,* 699–732. https://doi.org/10.1515/zpch-1892-1047.
37. Val'den, P. *Materialy k izucheniyu opticheskoy izomerii [Materials for the Study of Optical Isomerism];* Dr. Khim. Diss.: St. Petersburg, 1898.
38. Steinberg, J. W. Was the Russo-Japanese War World War Zero? *Russ. Rev.* **2008,** *67,* 1–7 https://doi.org/10.1111/j.1467-9434.2007.00470.x.
39. *The Russo-Japanese War in Global Perspective. World War Zero, Volume* 1; Steinberg, J., Menning, B., Schimmelpenninck van der Oye, D., Wolfe, D., Yokote, S., Eds.; Brill: Leiden, 2005; Vol. 29.
40. Ascher, A. *The Revolution of 1905: Russia in Disarray;* Stanford University Press: Stanford, CA, 1994.
41. Harcave, S. *The Russian Revolution;* Collier Books: London, 1970.
42. See: "Our Rectors". https://www.rtu.lv/en/university/history/Our-Rectors (Accessed September 2020).
43. Leimanis, E. The Polytechnical Institute of Riga and its Role in the Development of Science. *J. Baltic Stud.* **1972,** *3* (2), 113–123. https://doi.org/10.1080/01629777200000021.
44. Walden, P. Ueber die Molekulargrösse und elektrische Leitfähigkeit einiger geschmolzenen Salze. *Izv. Imp. Akad. Nauk. Ser. 6* **1914,** *8,* 405–422.
45. Walden, P. Ueber die Dissoziationsgrad eines gelösten Elektrolyten beim Sättigungspunkt in verschiedenen Solventien. *Izv. Imp. Akad. Nauk Ser. 6* **1913,** *7,* 427–448.
46. Walden, P. Neue Materialien Über den Zusammenhang zwischen den Grenzwerten der Molarleitfähigkeit und der innerung Reibung in nichtwässrigen und wässrigen Lösungen. *Izv. Imp. Akad. Nauk Ser* **1913,** *6,* 559–582.
47. Walden, P. Ueber das elektrische Leitvermögen in Kohlenwasserstoffen und deren Halogenderivaten, sowie in Estern und Basen als Solventien. I Teil. I. *Izv. Imp. Akad. Nauk Ser. 6* **1913,** *7,* 907–936.
48. Walden, P. Ueber das elektrische Leitvermögen in Kohlenwasserstoffen und deren Halogenderivaten, sowie in Estern und Basen als Solventien. I Teil. II. Gruppe Basen. *Izv. Imp. Akad. Nauk, Ser. 6* **1913,** *7,* 987–996.
49. Walden, P. Über organische Lösungs- und Ionisierungsmittel. I. *Z. Phys. Chem.* **1903,** *46,* 103–188. https://doi.org/10.1515/zpch-1903-0115.
50. Walden, P. Über oorganische Lösungs- und Ionisierungsmittel. II Teil. Messungen der rlektrischen Leitfähigkeit. *Z. Phys. Chem.* **1905,** *54,* 129–230. https://doi.org/10.1515/zpch-1906-5408.
51. Walden, P. Über organische Lösungs- und Ionisierungsmittel. III. Teil: Innere Reibung und deren Zusammenhang mit dem Leitvermögen. *Z. Phys. Chem.* **1906,** *55,* 207–246. https://doi.org/10.1515/zpch-1906-5408.
52. Wang, Y.; Fan, F.; Agapov, A. L.; Yu, X.; Hong, K.; Mays, J.; Sokolov, A. P. Design of Superionic Polymers - New Insights from Walden Plot Analysis. *Solid State Ionics* **2014,** *262,* 782–784. https://doi.org/10.1016/j.ssi.2013.09.026.
53. Harris, K. R. On the Use of the Angell-Walden Equation to Determine the Ionicity of Molten Salts and Ionic Liquids. *J. Phys. Chem. B* **2019,** *123* (32), 7014–7023. https://doi.org/10.1021/acs.jpcb.9b04443.

54. Walden, P.; Bischoff, C. A. *Handbuch Der Stereochemie;* Verlag von H. Bechhold: Frankfurt am Main, 1894.
55. Beilstein, F. *Beilsteins Handbuch Der Organischen Chemie;* Hamburg & Leipzig: Leopold Voss, 1881.
56. Bischhoff, C. A.; Walden, P. Ueber die physikalischen Constanten der stiuirten Aethenyltricarbonsäureester. *Ber. Dtsch. Chem. Ges.* **1890,** 660–664. https://doi.org/10.1002/cber.189002301105.
57. Bischhoff, C. A.; Walden, P. I. Ueber Derivate der Glycolsäure. *Justus Liebigs Ann. Chem.* **1894,** *279,* 45–70. https://doi.org/10.1002/jlac.18942790106.
58. Bischoff, C. A.; Walden, P. II. Ueber Derivate der Milchsäure. *Justus Liebigs Ann. Chem.* **1894,** *279,* 71–99. https://doi.org/10.1002/jlac.18942790107.
59. Bischoff, C. A.; Walden, P. III. Ueber Derivate der beiden α-Oxybuttersäuren. *Justus Liebigs Ann. Chem.* **1894,** *279,* 100–118. https://doi.org/10.1002/jlac.18942790108.
60. Bischoff, C. A.; Walden, P. IV. Ueber Derivate der Mandelsäure. *Justus Liebigs Ann. Chem.* **1894,** *279,* 118–129. https://doi.org/10.1002/jlac.18942790109.
61. Bischoff, C. A.; Walden, P. V. Ueber Derivate der Aepfelsäure. *Justus Liebigs Ann. Chem.* **1894,** *279,* 130–137. https://doi.org/10.1002/jlac.18942790110.
62. Bischoff, C. A.; Walden, P. VI. Ueber Derivate der Weinsäure. *Justus Liebigs Ann. Chem.* **1894,** *279,* 138–152. https://doi.org/10.1002/jlac.18942790111.
63. Bischoff, C. A.; Walden, P. Ueber Anomalien bei Verkettungen. *Ber. Dtsch. Chem. Ges.* **1894,** 27 (2), 1491–1495. https://doi.org/10.1002/cber.18940270262.
64. Bischoff, C. A.; Walden, P. Ueber die Aethylenester der Glycol- und Oxalsäure und die Anhydride α-halogenisierter Fettsäuren. *Ber. Dtsch. Chem. Ges.* **1894,** *27* (3), 2939–2952. https://doi.org/10.1002/cber.18940270354.
65. Walden, P.; Kernbaum, A. Ueber die Isomerieverhältnisse in der Stilbengruppe. *Ber. Dtsch. Chem. Ges.* **1890,** *23,* 1958–1961. https://doi.org/10.1002/cber.18900230240.
66. Walden, P. Ueber die gegenseitige Umwandlung optischer Antipoden. *Ber. Dtsch. Chem. Ges.* **1896,** *29* (1), 133–138. https://doi.org/10.1002/cber.18960290127.
67. Walden, P. Ueber die gegenseitige Umwandlung optischer Antipoden. [III Mittheilung]. *Ber. Dtsch. Chem. Ges.* **1897,** *30* (3), 3146–3151. https://doi.org/10.1002/cber.189703003139.
68. Walden, P. Ueber die gegenseitige Umwandlung optischer Antipoden. *Ber. Dtsch. Chem. Ges.* **1899,** *32* (2), 1833–1855. https://doi.org/10.1002/cber.18990320276.
69. Walden, P. Ueber die gegenseitige Umwandlung optischer Antipoden. *Ber. Dtsch. Chem. Ges.* **1899,** *32* (2), 1855–1864. https://doi.org/10.1002/cber.18990320277.
70. Walden, P.; Lutz, O. Ueber die gegenseitige Umwandlung optischer Antipoden. [II Mittheilung]. *Ber. Dtsch. Chem. Ges.* **1897,** *30* (3), 2795–2798. https://doi.org/10.1002/cber.18970300369.
71. Walden, P. Ueber ein neues, die Drehungsgrösse steigerndes Mittel. *Ber. Dtsch. Chem. Ges.* **1897,** *30* (3), 2889–2895. https://doi.org/10.1002/cber.18970300391.
72. Lewis, D. E. Static to Dynamic: Kazan Chemists and the Transformation from Chemical Structure to Reaction Regiochemistry. *Angew. Chem. Int. Ed.* **2019,** *58* (12), 3694–3705. https://doi.org/10.1002/anie.201804036.
73. Lewis, D. E. 1860-1861: Magic Years in the Development of the Structural Theory of Organic Chemistry. *Bull. Hist. Chem.* **2019,** *44,* 77–91.
74. Lewis, D. E. The Logic Behind Markovnikov's Rule. Was It an Inspired Guess? …No!. *Angew. Chem. Int. Ed.* **2021,** *60* (9), 4412–4421. https://doi.org/10.1002/anie.202008228.
75. Walden, P. I. Ocherk istorii khimii v Rossii [An Account of the History of Chemistry in Russia] (Überblick über die Geschichte der chemie in Russland). In *Vorlesungen über die Geschichte der Chemie von Lavoisier bis heute.* Mathesis: Odessa, 1917; pp 361–690.

76. Maier, H. *Chemiker Im "Dritten Reich." Die Deutsche Chemische Gesellschaft und der Verein Deutscher Chemiker im NS-Herrschaftsapparat;* VCH: Weinheim, 2015; pp 434–436.
77. Deichmann, U. The Expulsion of Jewish Chemists and Biochemists from Academia in Nazi Germany. *Pers. Sci.* **1999,** 7, 1–86. https://doi.org/10.1162/posc.1999.7.1.1.
78. Val'den, P. *Lui Paster (Louis Pasteur); [in Russian];* n/a: Riga, 1896.
79. Walden, P. *Wilhelm Ostwald : Mit zwei Heliogravüren und einer Bibliographie;* W. Engelmann: Leipzig, 1904.
80. Walden, P. Erinnerungen an W. Ostwalds Rigaer Professorentätigkeit 1881-1887. *Angew. Chem.* **1953,** 511–513. https://doi.org/10.1002/ange.19530652005.
81. Walden, P. Wilhelm Ostwald. *Ber. Dtsch. Chem. Ges.* **1932,** *65* (8), A101–A141.
82. Walden, P. Nachruf für D. J. Mendelejeff. *Chem. Ztg.* **1907,** *31,* 167–172.
83. Walden, P. Dmitri Iwanowitsch Mendelejeff (27. Januar 1834—20. Januar 1907 a. St.). *Ber. Dtsch. Chem. Ges.* **1908,** *41* (3), 4719–4800. https://doi.org/10.1002/cber.190804103191.
84. Walden, P. Marcellin Berthelot †. *Chem. Ztg.* **1907,** *31,* 367–373.
85. Walden, P. Carl Adam Bischoff. *Chem. Ztg.* **1908,** *32,* 1053–1053.
86. Val'den, P. I. *Lomonosov kak khimik (Lomonosov as a Chemist);* Tip. Imp. Akad. Nauk: St. Petersburg, 1911.
87. Val'den, P. I. Ser Vil'yam Ramsai [Sir William Ramsay]. Nekrolog. *Izv. Imp. Akad. Nauk, Ser. 6* **1916,** *10,* 1233–1238.
88. Walden, P. Svante Arrhenius. *Naturwissenschaften* **1928,** *16* (19), 325–333. https://doi.org/10.1007/BF01505422.
89. Walden, P. Martin Heinrich Klaproth (1. XII. 1743 Bis 1. I. 1817). Ein Gedenkblatt zu seinem 200. Geburtstag. *Z. Anorg. Chem.* **1944,** *252,* 187–189. https://doi.org/10.1002/zaac.19442520502.
90. Val'den, P. Nikolai Nikolaevich Beketov. 1827-1911. Nekrolog. *Izv. Imp. Akad. Nauk* **1912,** *6,* 123–135.
91. Val'den, P. I. *Vliyanie fiziki na razvitie khimii [The Influence of Physics on the Development of Chemistry];* n.a.: Odessa, 1914.
92. Walden, P. *Goethe und die Naturwissenschaften;* G. A. v. Halem: Leipzig, 1933.
93. Walden, P. *Maß, Zahl und Gewicht in der Chemie der Vergangenheit;* Verlag F. Enke: Stuttgart, 1931.
94. Walden, P. *Geschichte der organischen Chemie seit 1880;* J. Springer-Verlag: Berlin, 1941.
95. Walden, P. *Geschichte der Chemie;* Universitäts-Verlag: Bonn, 1947.
96. Walden, P. *Geschichte Der Chemie,* 2 erg. Aufl.; Athenäum Verlag: Bonn, 1950.
97. Walden, P. *Chronologische Übersichtstabellen zur Geschichte der Chemie von den ältesten Zeiten bis zur Gegenwart;* Springer-Verlag: Berlin, Göttingen, Heidelberg, 1952.
98. Walden, P. Darmois, E., Translator *Histoire de La Chimie;* Lamarre: Paris, 1953.

CHAPTER SEVEN

Mechanistic studies

Early mechanistic and stereochemical investigations

The rules relating to reaction regiochemistry (the Markovnikov, Hofmann, and Zaitsev Rules), as well as the inversion of configuration during the Walden cycle, were all based on empirical observations for the given reactions. The Hofmann elimination, in particular, occupied much of the interest of early workers in this field for two reasons: (1) quaternary ammonium salts were crystalline, and most easily purified; and (2) the eliminations from quaternary ammonium ions gave the least substituted of the possible alkenes, in direct contrast to the observations of Zaitsev with alkyl halides and the observations of others in the acid-catalyzed dehydration of alcohols. Fairly early on, it was discovered that when alkyltrimethylammonium hydroxides with alkyl groups larger than ethyl were pyrolyzed, an alcohol was formed in addition to the alkene.

Elimination reactions

Hofmann elimination

Hofmann's original work had shown that the pyrolyses of quaternary ammonium hydroxides where the nitrogen carries an ethyl group (**192**) all result in elimination of ethylene.[1,2] Wilhelm Lossen confirmed Hofmann's observations 25 years later.[3] In 1890, Hofmann's observation was extended by Collie and Schryver, who carried out a systematic study of the pyrolysis of quaternary ammonium chlorides and hydroxides.[4] The results Hofmann's work are shown in Scheme 7.1, and those of Collie and Schryver are gathered in Scheme 7.2.

Beginning in 1911, German chemist Julius von Braun[5–10] undertook a two-decade program of research into the pyrolysis of quaternary ammonium hydroxides that was one of the most extensive of the era. The main focus of this research was the exhaustive methylation of amines, and the "anomalous" hydrolysis that often accompanied it. This anomalous reaction occurs by the S_N2 displacement of methyl from the trimethylammonium group to

Addition, Elimination and Substitution: Markovnikov, Hofmann, Zaitsev and Walden
ISBN: 978-0-12-821027-7
https://doi.org/10.1016/B978-0-12-821027-7.00002-5

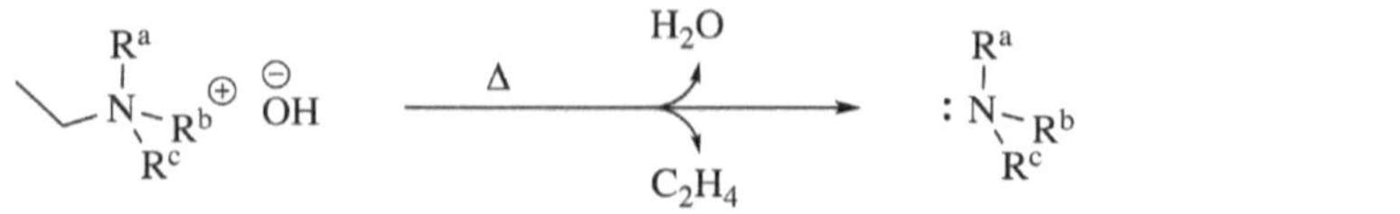

192a: $R^a = R^b = R^c = Et$
192b: $R^a = Me; R^b = R^c = Et$
192c: $R^a = n\text{-}C_5H_{11}; R^b = R^c = Et$
192d: $R^a = Ph; R^b = R^c = Et$
192e: $R^a = Ph; R^b = R^c = n\text{-}C_5H_{11}$
192f: $R^a = Ph; R^b = Me; R^c = Et$
192g: $R^a = Me; R^b = n\text{-}C_5H_{11}; R^c = Et$

193a: $R^a = R^b = R^c = Et$
193b: $R^a = Me; R^b = R^c = Et$
193c: $R^a = n\text{-}C_5H_{11}; R^b = R^c = Et$
193d: $R^a = Ph; R^b = R^c = Et$
193e: $R^a = Ph; R^b = R^c = n\text{-}C_5H_{11}$
193f: $R^a = Ph; R^b = Me; R^c = Et$
193g: $R^a = Me; R^b = n\text{-}C_5H_{11}; R^c = Et$

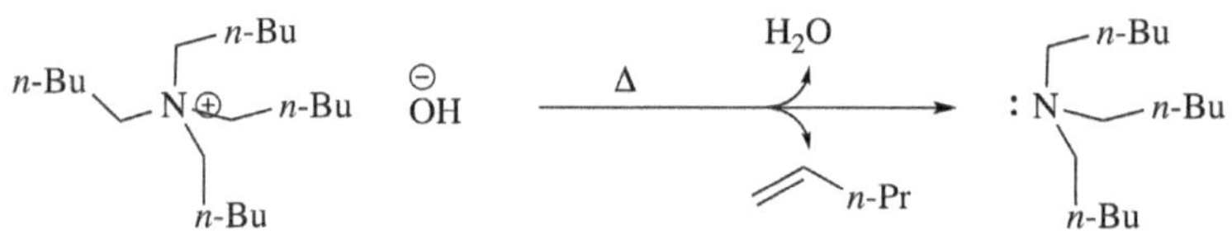

194 **195**

Scheme 7.1 Pyrolyses of quaternary ammonium hydroxides carried out by Hofmann.

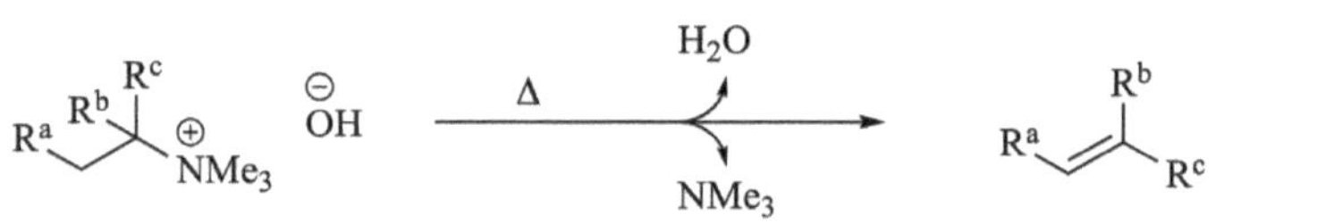

196a: $R^a = Me; R^b = R^c = H$
196b: $R^a = R^c = H; R^b = Me$
196c: $R^a = i\text{-}Bu; R^b = R^c = H$
196d: $R^a = Et; R^b = R^c = Me$

197a: $R^a = Me; R^b = R^c = H$
197b: $R^a = R^c = H; R^b = Me$
197c: $R^a = i\text{-}Bu; R^b = R^c = H$
197d: $R^a = Et; R^b = R^c = Me$

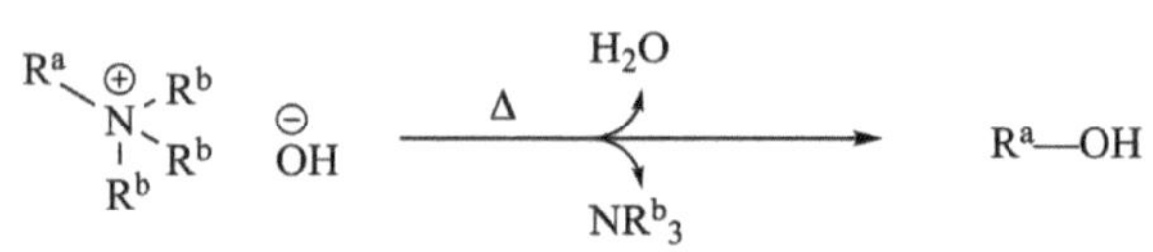

198a: $R^a = PhCH_2; R^b = Me$
198b: $R^a = PhCH_2; R^b = Et$
198c: $R^a = H_2CCH{=}CH_2; R^b = Me$

199a: $R^a = PhCH_2$
199b: $R^a = PhCH_2$
199c: $R^a = H_2CCH{=}CH_2$

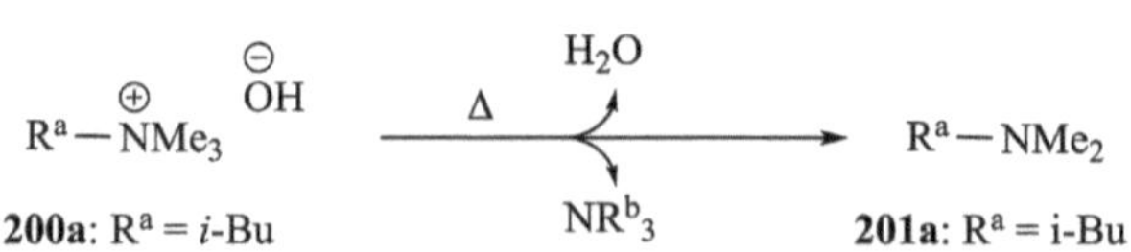

200a: $R^a = i\text{-}Bu$
200b: $R^a = Ph$

201a: $R^a = i\text{-}Bu$
201b: $R^a = Ph$

Scheme 7.2 Pyrolyses of quaternary ammonium hydroxides carried out by Collie and Schryver.

Table 7.1 von Braun's studies of the pyrolysis of quaternary ammonium hydroxides.

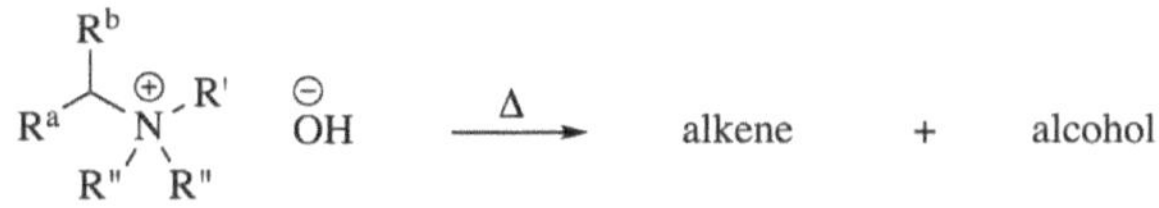

R^a	R^b	R'	R"	alkene	alcohol
Me	H	Me	Me	$H_2C{=}CH_2$	none
Et	H	Me	Me	$MeCH{=}CH_2$	MeOH (trace)
Me	Me	Me	Me	$MeCH{=}CH_2$	MeOH (trace)
i-Pr	H	Me	Me	trace	MeOH (trace)
i-Bu	H	Me	Me	*i*-$PrCH{=}CH_2$	MeOH ("small")
$H_2C{=}CH_2$	H	Me	Me	$MeCH{=}CH_2$	MeOH (trace)
Et	H	Me	Me	$Me_2CHCH{=}CH_2$	
HH	H	Ph	Me		MeOH only
Ph	H	Me	Me		$PhCH_2OH$
Me	H	Me	Et	$H_2C{=}CH_2$	
Me	H	Me	Me	$H_2C{=}CH_2$	
Ph	H	Me	Me		$PhCH_2OH$

give methanol. The data in Table 7.1 show that as the length of the alkyl chain increases, the relative amount of the product from this "anomalous" elimination rises from 0% to 75%.

Zaitsev elimination

One of the earliest reactions that proceeded with Zaitsev orientation to be subjected to extensive study was the dehydration of alcohols. Of particular interest to organic chemists of the era was the dehydration of isomeric amyl alcohols [1-pentanol (**202**), 3-methyl-1-butanol (**203**) and *S*-(−)-2-methyl-1-butanol (**204**)][11–13] from the fractional distillation of fusel oil from the fermentation of potatoes to give mixed isomers of pentenes (**205**) (Scheme 7.3).

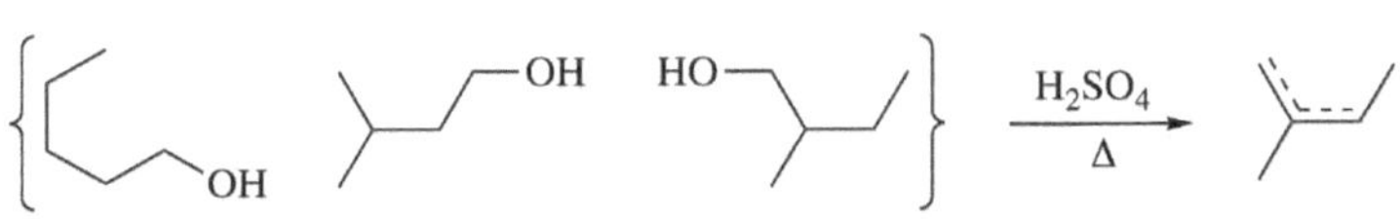

202 **203** **204** **205**

Scheme 7.3 Acid-catalyzed dehydration of isomeric amyl alcohols.

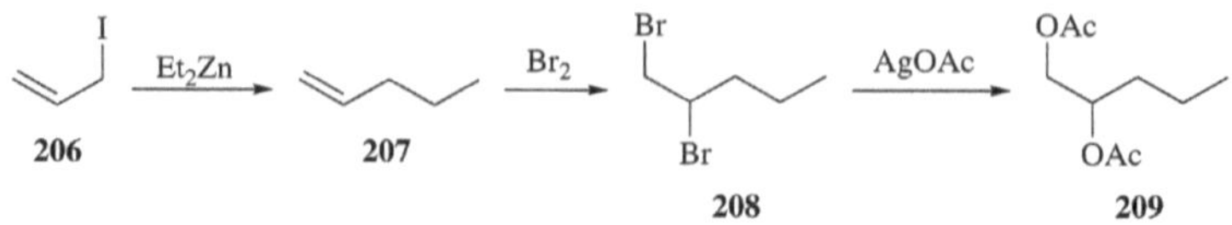

Scheme 7.4 The synthesis and derivatization of 1-pentene (207).

Following the literature in this area is often confusing because compounds based on two different carbon skeletons were named, "amylene." The first was obtained from the reaction of diethylzinc with allyl iodide (**206**), which makes it 1-pentene (**207**). The direct or indirect dehydration of the amyl alcohols referred to above (which the Russian chemists always referred to as "amyl alcohol from fermentation") gives alkenes based on the methylbutene carbon skeleton. The structures of the alkenes were often deduced from the structure of the addition product with hydrogen iodide or from the glycol formed by addition of bromine and subsequent displacement of the bromide with silver acetate followed by chromic acid cleavage of the diol.[14,15] For example, 1-pentene (**207**) reacts with bromine to give the dibromide **208**, which reacts with silver acetate to give the diacetate **209** (Scheme 7.4).

Addition reactions

The acid-catalyzed hydration of amylenes was one method used to try to prepare new isomers of $C_5H_{12}O$. Its results were complicated by the fact that at least one of the intermediate carbocations was prone to rearrangement, which led to difficulties in interpreting the results of the reaction. For example, the reaction between isoamyl alcohol (**203**) and concentrated hydriodic acid had been firmly established to give the primary isoamyl iodide (**210**).[16,17] The base elimination of this iodide can only give 3-methyl-1-butene (**211**), but the hydration of this alkene gave the tertiary alcohol **212** as the major (only) product of the reaction, instead of the expected Markovnikov adduct, the secondary alcohol **213**. The acid-catalyzed dehydration and rehydration of the alcohol **214** gave the same major final product as the hydration of alkene **176**, but in the absence of any knowledge about carbocation rearrangements (still two decades in the future[18–23]), the identity of the alkene produced by dehydration of the alcohol was deduced as most probably being **215**, which would, in fact, give **212** as the product of Markovnikov addition (Scheme 7.5).[24–26]

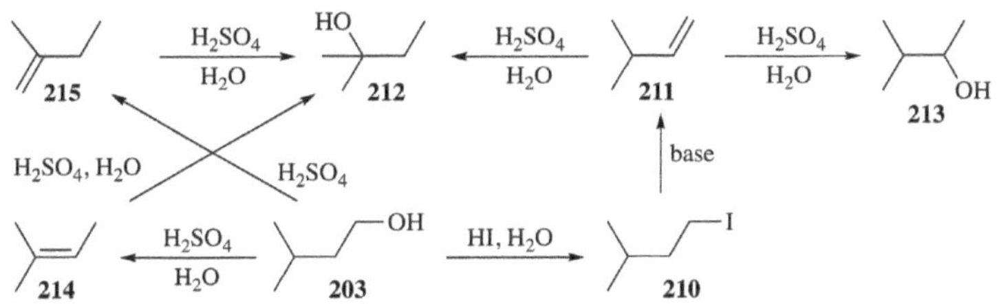

Scheme 7.5 The chemistry of isoamyl compounds developed by Wurtz, Bertholet, Kolbe, and others.

The beginnings of modern mechanistic studies

The rules relating to reaction regiochemistry (the Markovnikov, Hofmann, and Zaitsev Rules), as well as the observations in the Walden cycle, were all empirical rules governing those reactions. The earliest systematic mechanistic studies of nucleophilic substitution were carried out by Russian chemist, Nikolai Aleksandrovich Menshutkin (Fig. 4.8)[27–30] at St. Petersburg Imperial (now St. Petersburg State) University, who published an extensive series of papers in which he used kinetics to deduce the relative reactivities of alcohols and acids in esterification and alkyl halides in the quaternization of tertiary amines. His pioneering work in using kinetics paved the way for later, more sophisticated applications and was a key step in the development of physical organic chemistry, but even at the end of the 1920s, there was still no clear picture of the mechanisms of the four reactions that are the subject of this book.

The key contributions to understanding these mechanisms were made beginning in the 1930s by British chemists E. D. (Edward David) Hughes (1906–63)[31] and Sir Christopher Kelk Ingold (1893–970)[32–38] at University College, London. Their work led to a unified mechanistic rationalization of all these reactions. The Hughes and Ingold collaboration, which lasted a third of a century, helped to establish the modern subdiscipline of physical organic chemistry.

The key early players

Nikolai Aleksandrovich Menshutkin (1842–907)

Menshutkin is the pioneering physical organic chemist whose critical role in developing the field is most often overlooked. He was the sixth of eight children born to Aleksandr Nikolaevich Menshutkin (1803–76), a successful merchant and property owner in St. Petersburg. In 1848, Nikolai was sent to the Schlessner Boarding School until he was 10 years old, at which time he entered the St. Petersburg German School (the Sankt Peterschule, SPS, also known as the Petrischule).

Despite having done poorly in his first year, he graduated with honors from the SPS in 1857, but at 15 years of age, he was 11 and a half months younger than the statutory minimum age for university entrance. In fact, when he was finally admitted in the Fall of 1858, he was still under the official minimum age.

In St. Petersburg, he took the courses in analytical and inorganic chemistry from Abram Abramovich Voskresenskii (1808–80, Fig. 7.1) and the course in organic chemistry from Nikolai Nikolaevich Sokolov (1826–77, Fig. 7.1). The organic chemistry course had been started by Mendeleev, but he left St. Petersburg for a 2-year period abroad in 1860.

At the time, laboratory facilities in most Russian universities were almost nonexistent (the exception may have been Kazan), so as soon as he graduated, he left for Western Europe to get the practical laboratory training he lacked. He spent his first year abroad in the Tübingen laboratory of Adolph Strecker (1822–71, Fig. 7.2). In April 1864, he moved to the laboratory of Adolphe Wurtz (Fig. 4.4) in Paris. Here he began his studies of phosphorous acid, which he continued in the laboratory of Hermann Kolbe (Fig. 2.11).[39–43]

Figure 7.1 A. A. Voskresenskii (left), and N. N. Sokolov (right). *Public domain image of Voskresenskii from https://commons.wikimedia.org/wiki/File:Voskresensky_A.A.jpg. Public domain image of Sokolov from https://commons.wikimedia.org/wiki/File:Sokolov_NN_1870.jpg.*

Figure 7.2 Adolph Strecker. *Public domain image from https://en.wikipedia.org/wiki/Adolph_Strecker#/media/File:Adolph_Strecker.jpg.*

On August 30, 1866, he returned to St. Petersburg and single-mindedly set about completing his Magistr Khimii degree, which would qualify him to occupy a tenurable position at the university. Before the end of the year, he had completed the examinations, and he immediately began writing his dissertation based on the work he had done abroad. He submitted it and successfully defended it in March 1867.[44] He immediately petitioned the university for permission to deliver lectures as a Privat-docent, a nontenurable position that would permit him to deliver lectures to students in return for payment of fees. Instead, he was asked to deliver two trial lectures and then appointed Docent, a salaried, tenurable position, instead.

On January 5, 1868 (January 17, 1868 Gregorian), Menshutkin, along with Borodin, Beilstein, Mendeleev, Markovnikov, and Voskresenskii, among others (Fig. 7.3), attended the first Congress of Russian Naturalists and Physicians, where the Russian Chemical Society was first organized. Of the chemists pictured, only Markovnikov did not later sign the Charter. In 1869, Menshutkin became the Secretary of the Society, a post he held

Figure 7.3 Standing (L–R): F. R. Vreden, P. A. Lachinov, G. A. Shmidt, A. R. Shulyachenko, A. P. Borodin, N. A. Menshutkin, N. A. Sokovnin, F. F. Beil'shtein, K. I. Lisenko, D. I. Mendeleev, F. N. Savchenkov. Seated (L–R): V. Yu. Rikhter, S. I. Kovalevskii, N. P. Nechaev, V. V. Markovnikov (Morkovnikov), A. A. Voskresenskii, P. A. Il'yenkov, P. P. Alekseev, A. N. Engel'gardt. Courtesy of the Mendeleev Museum at St. Petersburg State University. *Public domain image downloaded from https://commons.wikimedia.org/wiki/File:RFCS_1868.jpg.*

until 1891, and he assumed the role of Editor of its journal, the *Zhurnal Russkago Khimicheskago Obshchestva [Journal of the Russian Chemical Society].* He continued as Editor for another 30 years. He was elected President of the Society in 1906. At the meeting of the society on March 6, 1869, he presented a paper to the Society, reported in the first volume of the Zhurnal as follows Fig. 7.4:

> *"On behalf of D. Mendeleev, N. Menshutkin communicates an experimental system of the elements based on their atomic weights and chemical similarities. In the absence of D. Mendeleev, the discussion of this communication was postponed until the next meeting."*

1. Н. Меншуткинъ сообщаетъ отъ имени *Д. Менделѣева*, опытъ системы элементовъ, основанной на ихъ атомномъ вѣсѣ и химическомъ сходствѣ. За отсутствіемъ Д. Менделѣева обсужденіе этаго сообщенія отложено до слѣдующаго засѣданія.

Figure 7.4 *Minutes of the meetings of the Society. Meeting of the 6th of March, 1869.* Zh. Russ. Khim. O-va. ***1869***, 1, 33–38. *The image is on p. 35.*

In 1869, Menshutkin changed his research direction again to study the preparations and properties of ureides.[45–47] He submitted and defended his Dr. Chem. Sci. dissertation[48] the same year and thus held the minimum qualification to hold a Professorial Chair in Russia. The chemistry of ureides remained a part of his research program for the remainder of his career. Just 6 days after his successful defense, he was unanimously elected Extraordinary Professor in the Chair of Technical Chemistry by the Physics—Mathematics Faculty. His appointment was confirmed a month later by the University Council. In 1872, A. M. Zaitsev, the Professor at Kazan University, offered him a Full Professorship there. He declined the offer, one of many he received. He was promoted to Full Professor in February 1876. Menshutkin was twice elected to 3-year terms as Secretary of the Physics—Mathematics Faculty, and in 1879, he became Dean of the Faculty, a post he held until 1887. When the new St. Petersburg Technical Institute was formed, Menshutkin was appointed Chair of Organic Chemistry and Dean of Mining in 1901, while still retaining his appointment at the University.

Menshutkin died just days after his good friend Mendeleev and a few months after the third eminent Russian chemist of this era, Friedrich Konrad Beilstein (Fyodor Fyodorovich Beil'shtein). In his career, he produced over 120 publications, and he frequently translated his papers in Russian into German or French. He also published a highly successful textbook, *Analiticheskaya Khimiya*,[49] which went through 16 Russian editions and was translated into both German[50] and English.[51] His encyclopedic textbook of organic chemistry[52] has never been translated from the Russian.

Edward David Hughes

Hughes was the youngest of nine children born to a North Welsh farmer, and his early upbringing was typical of a farm boy: when he came home after school each day, he had farm work to do. He entered University College at Bangor (Wales) in 1923 at the urging of his high school science teacher. He accepted a grant from the Board of Education, which was contingent on him becoming a secondary-school teacher after graduation. He later repaid this grant, when it became clear that he would not be following up on the required commitment. By then, he had support from his College with a gradual rise in salary until he obtained a Fellowship. He graduated with the B.Sc. (Hons.) in 1927 and undertook his training in secondary education the following year, graduating with the Dip. Ed. in 1928.

During his Dip. Ed. training, he began carrying out research on prototropy with H. B. Watson,[53–55] under whom he earned his Ph.D. from the University of Wales in 1930. That year, he took a postdoctoral position with Ingold. Ingold was at Leeds when Hughes accepted the offer of a postdoctoral appointment, but Ingold moved to University College of London (UCL) before the two could meet, and their first meeting was at UCL.

Thus began one of the great collaborations of physical organic chemistry, one that was to last the next 33 years. Hughes took more degrees at UCL: the M.Sc. in 1932 and the D.Sc. in 1938. He earned the Meldola Medal of the Royal Institute of Chemistry in 1936 and became a Fellow of the Institute in 1938.

In 1939, UCL was hastily evacuated from London to Wales, setting up at the University Colleges of Aberystwyth and Bangor. This first evacuation was poorly organized: the transport of equipment was not effective, meaning that much of the important equipment was left behind in London. A few months later, a return to London was attempted, but this time it was quashed by the London blitz. However, the second evacuation in 1940 was much more orderly and allowed the scientists to bring their important equipment with them.

As a leader of the relocation, Hughes settled in Aberystwyth. In 1943, the Chair became vacant at Bangor, and Hughes was elected Professor at his alma mater. He remained in Bangor until 1948, when he was appointed Professor of Chemistry at UCL. He remained at UCL until his death in 1963. At the time, he was Head of the Chemistry Department, and he worked in hospital almost to his dying day. It may be fitting that his last papers were published posthumously.[56–58]

Hughes was a gregarious individual and had been so since childhood. As a schoolboy, he preferred to study in rooms full of family members with the attendant noise, despite his parents having prepared a quiet place for him to study at home. Students everywhere he worked could always find a friendly ear, as could his younger colleagues. In his spare time, Hughes bred and raced greyhounds, being known at the racetrack as "The Prof." Hughes was an indefatigable worker, whose orderly approach to problems is evident in his over 200 research papers, indicative of his tendency to workaholism. He was elected a Fellow of the Royal Society in 1949; he died in 1963 at the young age of 57.

Hughes published over 30 papers with his students on mechanistic topics from 1933 to the year after his death and 84 papers jointly with Ingold on a range of mechanistic topics over the same time period.

Christopher Kelk Ingold

The other half of this extraordinarily productive team was Christopher Kelk Ingold, who stands alongside Hughes as one of the foundational giants of the discipline of physical organic chemistry. Along with Robert Robinson, Ingold was one of the first to apply modern electronic theory to organic problems.

Ingold was born in London to silk mercer, William Kelk Ingold, but due to his father's tuberculosis, the family moved to the Ise of Wight, where the Royal Ventnor Hospital for Consumptives[59] was located. There the treatment protocols of Dr. Emmet Densmore[60,61] were in effect. William changed his occupation on the Isle of Wight and became a confectioner. Ingold lost his father to tuberculosis when he was just 5 years old, and his mother took over her husband's business. His family situation, which probably required him to work to help to support the family, led him to take is B.Sc, in 1913 at the nearby Hartley University College, Southampton (now the University of Southampton) as an external student of the University of London under a local scholarship. He was a better physics student than a chemistry student, but physics at Southampton was taught as a dead subject with nothing new and exciting to offer. Chemistry was the antithesis of this, so Ingold moved his allegiance to chemistry. After he had completed his B.Sc. (Hons.) at Southampton, Ingold he moved to Imperial College in London.

At Imperial College, he studied with Sir Jocelyn Field Thorpe (1872–940). He immediately began producing important results, including the Thorpe–Ingold effect in 1915.[62–64] In 1917, he took a position as a research chemist in the Cassel Cyanide Company in Glasgow. This company (originally the Cassel Gold Extraction Company, which was eventually absorbed by Imperial Chemical Industries) was involved in the extraction of gold from ores by the cyanide leaching process. Ingold's work there was under the auspices of the War Office and involved the synthesis of new lachrymatory agents for use as chemical weapons.

He earned his M.Sc. from the University of London in 1919. In 1920, he returned to Imperial College and once again began working with Thorpe, first as Demonstrator, then as Lecturer in organic chemistry. Within 2 years, he had begun a research program that was characterized by its enormous productivity: he published 438 papers, reviews, monographs, and books between 1921 and 1969: an average of more than nine publications every year for nearly five decades. In 1921, he earned the D.Sc. from the University of London. In 1923, he moved to Leeds as professor of organic chemistry, and in 1924, he was elected a Fellow of the Royal Society, becoming one of the youngest chemists to attain that accolade. In 1930, Ingold returned to London as Professor at University College of the University of London, where he remained the rest of his career as Head and Director of Laboratories.[31] He maintained this position when the laboratories were evacuated to Aberystwyth during the Nazi blitz on London.

The year that he returned to UCL, Ingold was joined by E.D. Hughes. Their joint work established the currently accepted mechanisms of substitution and elimination reactions and continued until Hughes' death in 1963. Although Ingold received numerous medals and honors throughout his lifetime, including being knighted in 1958, he never received the Nobel Prize. The question why has been addressed by several authors: His first paper on the mechanism of electrophilic aromatic substitution began a long and bitter feud with another giant of British chemistry, Sir Robert Robinson, over claims of priority in conceiving the idea. This bitter feud was almost certainly a contributing factor in Ingold not receiving the Nobel, but the sheer breadth of his contributions may also have played a role—it would have been difficult to choose just one contribution to honor.[34] One can also infer this from the biography by Shoppee,[38] who states,

> *An adequate and detailed account of Ingold's scientific work is rendered very difficult by three circumstances: (i) he was a chemical genius, an excellent physicist, and a talented mathematician, able to employ and to combine all aspects of organic, inorganic, and physical chemistry; (ii) his enormous output of work; (iii) his long association with the late Professor E. D. Hughes, F.R.S., who was successively his post-doctoral research student, his junior colleague, his professorial colleague, and his partner.*

Ingold was a keen mountaineer and bird watcher; he was also something of a dare-devil—at the age of 71 he took up water-skiing!

The origin of physical organic chemistry: Menshutkin's kinetic studies

Effects of reactant structure on the rate of esterification

Menshutkin had begun his research career with studies of phosphorous acid and its reactions,[39–44] and his doctoral work focused on the chemistry of ureides.[45–48] These topics were displaced by kinetics: he reported his first kinetic study in 1877, on the effects of alcohol structure on the rates of esterification with acetic acid.[65,66] Unlike modern kinetic measurements,[65] where the reaction is carried out in a solvent so that concentrations can be accurately measured, Menshutkin mixed equimolar quantities of the acid and the alcohol and then heated them in a sealed tube. These early results all give excellent fits to a plot of 1/[A] versus t, consistent with first order in both reactants. He later revised his procedure, using benzene as solvent. During this phase of his research, he carried out an exhaustive, systematic study of esterification of mixtures of alcohols and carboxylic acids that extended from monohydric and polyhydric alcohols with mono-, poly-, and hydroxy-carboxylic acids. Much of the work on the effects of alcohol structure on the rate of acetylation was summarized in a 28-page article in 1891.[65] Some representative results that demonstrate the broad scope of these studies are shown in Table 7.2.

Menshutkin had difficulties in studying the esterification of tertiary alcohols, so he conducted a study of the thermolysis of *tert*-amyl acetate at 155°C, the same temperature at which he had studied the esterification reaction.[67–69] The plot of the mole fraction of acetic acid versus time is sigmoidal: evidence that the reaction is autocatalytic.

Table 7.2 Representative values of the relative rate constants of the esterification of alcohols with acetic acid.

Primary alcohols

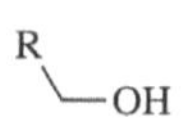

R	k_{rel}
Me	48.4
Et	42.9
n-Pr	41.6
i-Pr	35.9
n-C_6H_{13}	35.1
n-C_7H_{15}	33.7
n-$C_{13}H_{27}$	26.0
n-$C_{15}H_{31}$	24.0
n-$C_{17}H_{35}$	21.9
n-$C_{29}H_{59}$	16.5
H_2C=CH	25.9
H_2C=C(Me)	23.9
Ph	25.0
HC≡C	17.9
MeCH=$CHCH_2$	20.3
OAc	11.9
Cl	10.1
Cl	2.3
CN	8.0
CO_2Et	23.5
$ClCH_2CH$(Cl)	4.5

Secondary alcohols

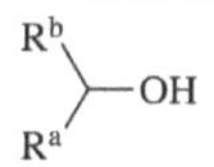

R^a	R^b	k_{rel}
Me	Me	13.2
Me	Et	11.0
Me	*n*-C_6H_{13}	8.1
Me	H_2C=$CHCH_2$	5.7
Et	Et	2.3
Ph	Ph	0.96
$PhCH_2$	$PhCH_2$	2.7
Me	CH_2CN	0.3
CH_2Cl	CH_2Cl	5.1
Me	MeCH=CH	2.9
Et	*i*-Pr	5.1
Me	CO_2Et	5.9

Tertiary alcohols

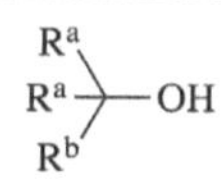

R^a	R^b	k_{rel}
Me	Me	0.8
Et	*i*-Pr	0.5

Data from Menshutkin, N. Vliyanie Sostava Alkogolei Na Skorost Eterifikatsii Ikh [The Influence of the Composition of Alcohols on the Rate of Their Esterification]. *Zh. Russ. Fiz.-Khim. O-va.* 1891, 23, 263–283 and Menschutkin, N. Sur La Decomposition de l'acétate d'amyle Tertiaire Par La Chaleur. *Compt. Rend. Hebd. Séances Acad. Sci.* **1882**, *95*, 648–651.

The Menshutkin reaction

In 1882, Menshutkin moved his focus to the formation of amides (especially anilides), and he began to study the thermal decomposition of esters. In 1890, he made his initial report of the reaction that bears his name, the S_N2 displacement of the halide from alkyl halides by triethylamine. In the last few years of the century, he reported the study of the effects of solvent on reaction rate in a study of the quaternization of triethylamine now known by his name.[70–73] The results of these studies, gathered in Table 7.3, revealed that there is an increase of more than two orders of magnitude in

Table 7.3 Solvent effects on the rate of the Menshutkin reaction between ethyl iodide and triethylamine.

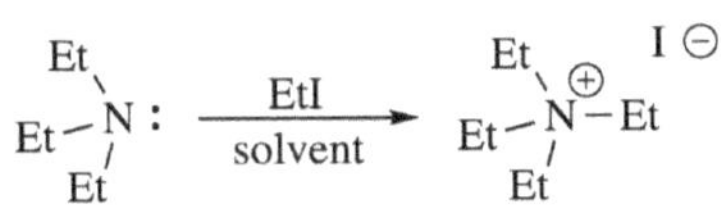

Solvent	k_{rel}	ε	Solvent	k_{rel}	ε
hydrocarbons			*alcohols*		
n-C_6H_{14}	0.93	1.89	i-BuOH	19.4	18.7
n-C_7H_{16}	0.17	1.92	EtOH	27.5	24.6
$C_6H_4Me_2$	2.2	2.4	$H_2C{=}CHCH_2OH$	32.5	22
C_6H_6	4.4	2.28	MeOH	38.0	32.6
organic halides			$PhCH_2OH$	100.0	13
n-PrCl	4.0	8.53	*ketones*		
PhCl	17.4	5.62	Me_2CO	45.7	21.0
PhBr	20.3	5.4	Me_2CO-H_2O 95:5	66.9	
α-$C_{10}H_7Br$	84.9	5.1	PhCOMe	97.3	17.3
ethers					
$C_5H_{11}OEt$	0.47	3.6			
Et_2O	0.57	4.27			
PhOEt	16.0	4.2			
PhOMe	30.3	4.3			

Data from Hughes, E. D.; Watson, H. B. The Reaction of Bromine with Aliphatic Acids. Part IV. Succinic Acid. *J. Chem. Soc.* **1930**, 1733–1740. https://doi.org/10.1039/JR9300001733. Hughes, E. D.; Watson, H. B.; Yates, E. D. Constitutional Factors Controlling Prototropic Changes in Carbonyl Compounds. Part I. The Relative Speeds of Enolisation of Acetone and Bromoacetone and the Effect of Acid Catalyst. *J. Chem. Soc.* **1931**, 3318–3324. https://doi.org/10.1039/JR9310003318. Dielectric constants from https://www.kabusa.com/Dilectric-Constants.pdf.

the relative rate constants on replacing hexane as the solvent with methanol or benzyl alcohol. Although there is a general increase in rate with increasing polarity of the solvent, it is not straightforward.

Classifying substitution and elimination mechanisms: Hughes and Ingold

Beginning in the 1930s, Hughes and Ingold began an extensive program of research directed toward elucidating the mechanism of substitution and elimination reactions. By 1940, their careful study of the kinetics and stereochemistry of nucleophilic substitution and base elimination reactions of a wide range of alkyl compounds allowed them to identify four classes of these reactions,[74,75] based on mechanism:

- S_N1, stepwise unimolecular substitution reactions that occur through a carbocation intermediate and typically exhibit first-order kinetics;
- S_N2, concerted bimolecular substitution reactions that typically exhibit second-order kinetics (first order in each reactant);
- E1, stepwise unimolecular elimination reactions that occur through a carbocation intermediate and typically exhibit first-order kinetics; and
- E2, concerted bimolecular elimination reactions that typically exhibit second-order kinetics (first order in each reactant).

S_N2 substitution and the Walden inversion: a stereochemical enigma

Almost two decades after it was first reported, the Walden inversion remained a mechanistic riddle. The Walden cycle required two reactions proceeding with inversion of configuration, and two with retention, but the exact details were still very much the province of speculation.

Prior to 1916, when Lewis published his model for covalent bonding,[76] the consensus appears to have been that the displacement with inversion proceeded in two steps: the addition of the nucleophile opposite the leaving group, and then the loss of the leaving group: a form of "associative" mechanism. This was the model adopted by Emil Fischer,[77,78] Paul Pfeiffer,[79] and Johannes Gadamer[80]; a similar mechanism was proposed for substitution in cobalt complexes by Werner.[81,82] The Presidential Address presented to the Chemical Society (London) by Percy Faraday Frankland in 1913[83] gave an excellent overview of the state of knowledge at the time (including some empirical observations), but it did not contain any mechanistic rationalizations, and information was still being gleaned from new experiments.[84]

Lewis' 1916 paper removed all stepwise associative mechanisms through a stable pentavalent adduct from consideration because such a species

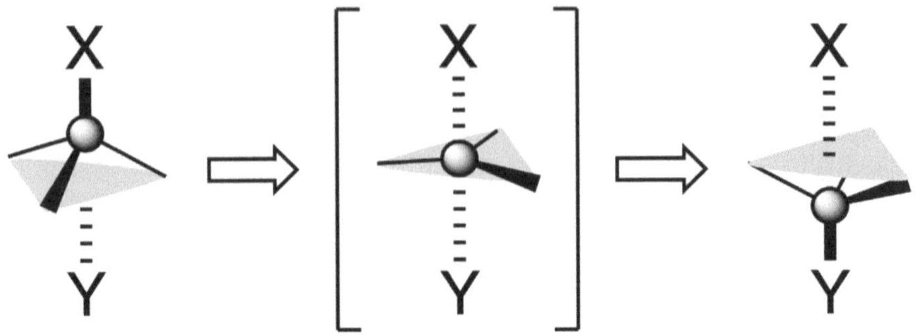

Figure 7.5 G. N. Lewis' model for the Walden inversion.

violated the octet rule. However, in his 1923 book,[85] he did postulate that transient pentavalent species with the incoming group attacking the face of the tetrahedron opposite the departing group (Fig. 7.5) were not precluded from existing.

In Lewis' view, the reaction is envisaged as a transfer of the carbon from the leaving group to the nucleophile by passing the carbon atom through the center of the plane defined by the three other substituents on the basal plane of the tetrahedron. This results in a transient pentacoordinate carbon atom at the midpoint. However, Lewis provided little evidence for this position beyond one paragraph on p. 113 of his book. So, it was still necessary to develop new ideas about the possible mechanism of the reaction, especially about the reasons why the incoming group should approach the substrate from the back side, which would take another decade and more.

Hughes and Ingold undertook an extensive program of kinetic and stereochemical studies to address this question. In a series of papers beginning in 1937 and culminating with papers published after Hughes' death,[86–93] the research group at University College, London, studied the kinetics of the Walden inversion. They subdivided the reactions into four subclasses: (a) reactions where the leaving group is uncharged and the nucleophile is anionic; (b) reactions where the leaving group is positively charged and the nucleophile is anionic; (c) reactions where the leaving group and the nucleophile are both uncharged; and (d) reactions where the leaving group is positively charged and the nucleophile is uncharged (Fig. 7.6).

Their results were clear: Walden inversion was universal for all bimolecular nucleophilic substitutions at carbon. With the exception of the reaction between secondary and tertiary alcohols and anhydrous hydrogen iodide, these nucleophilic substitution reactions proceeded with a second-order rate law—Rate = k $[RX][Nu^-]$; the reaction of alcohols with hydrogen

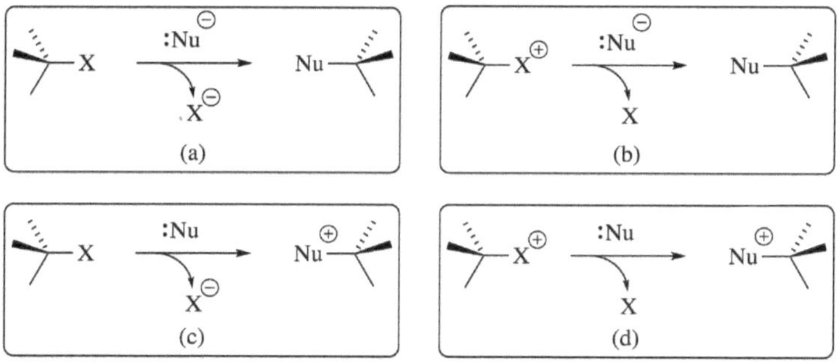

Figure 7.6 The four classes of bimolecular nucleophilic substitution studied by Hughes and Ingold.

iodide exhibited a third-order rate law (Rate = k [ROH][H^+][I^-]) consistent with what we now know is rapid preequilibrium protonation of the alcohol as the first step of the mechanism. They interpreted the kinetic results of all their studies of the Walden inversion in terms of a bimolecular displacement reaction through a transition state with the geometry of Lewis' transient pentavalent species. This allowed the complications from the change in kinetic order due to preequilibrium reactions to be sidestepped where the molecularity of the reaction remained unchanged.

Hughes and Ingold designated the mechanism of the Walden inversion by the mnemonic, S_N2 (Substitution, Nucleophilic, Bimolecular [Molecularity 2]). The absolute stereochemical control provided by this mechanism made the S_N2 reaction a cornerstone synthetic method as organic synthesis advanced from control of simple chemo- and regioselectivity to control of diastereoselectivity.

By 1946,[94–100] the UCL chemists had elucidated the effects of steric hindrance on the relative rates of S_N2 reactions of primary bromides (Fig. 7.7) and on the effects of α-substitution on the relative rates of the S_N1, S_N2, E1, and E2 reactions (Fig. 7.8).

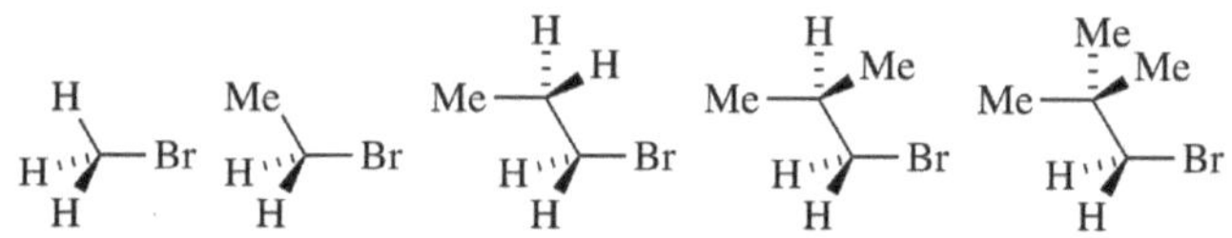

55 °C, EtOH	17.6	1	0.28	0.030	4.2×10^{-6}
95 °C, EtOH	15	1		0.041	1.0×10^{-5}
95 °C, Me_2CO		1			6.2×10^{-5}

Figure 7.7 The effects of steric hindrance on the rate of the S_N2 substitution of primary bromides relative to the rate of for ethyl bromide defined as 1.

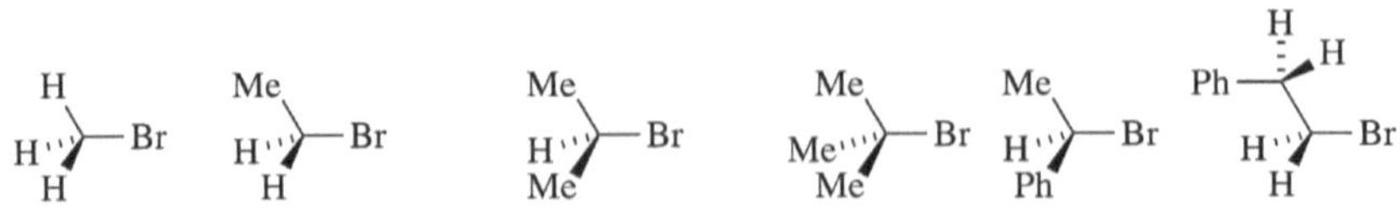

S_N1	(0.002)*	n/a ($8.1x10^{-4}$)*	$8.5x10^{-4}$ ($14x10^{-4}$)*	0.7 (5.9)*	0.95	0.01
S_N2	(12.5)*	1 (1)*	0.025 (0.028)*	n/a (n/a)*	n/a	0.15
E1		n/a	$<5x10^{-5}$	0.275	<0.025	n/a
E2		n/a	0.075	0.7	n/a	3.0

Figure 7.8 The effects α-substitution on the rate of nucleophilic substitution and elimination of alkyl bromides in anhydrous ethanol at 65°C and 80% aqueous ethanol at 55°C*, relative to the rate for ethyl bromide defined as 1.

Elimination reactions with base

The elimination of hydrogen halides from alkyl halides by bases was studied by both Markovnikov and Zaitsev. Markovnikov observed that the base elimination of hydrogen iodide from isobutyl iodide gave isobutylene, as did the elimination from *tert*-butyl iodide. He interpreted these observations in conformity with the conventional wisdom of the time, which held that the hydrogen removed in an elimination reaction would come from the less hydrogenated carbon. Of course, the eliminations that Markovnikov studied cannot have any other outcome. The experiments where more than one outcome could be expected were done, instead, by Zaitsev and his students.[101–106] Hofmann, on the other hand, had studied the pyrolysis of quaternary ammonium hydroxides, reactions with regiochemistry opposite to the halide elimination.[1,2,107]

Hughes, Ingold and their students conducted some 60 studies on the mechanism of eliminations that were published in the *Journal of the Chemical Society* between 1933 and 1964. In the course of these studies, they identified a number of the critical factors affecting the reaction, including the leaving group, the structure of the alkyl group, and the effect of solvent. The eliminations studied by Hughes and Ingold are readily separated into two classes, based on the reaction kinetics and molecularity.

E2 elimination

The first proposal of the bimolecular mechanism was for the Hofmann elimination by pyrolysis of quaternary ammonium hydroxides.[108–114] The study of this reaction and the related eliminations of trialkylsulfonium ions,[115–123] alkyl bromides,[115–119] along with the "anomalous" hydrolysis reaction that frequently accompanied both provided much of the basis for the early mechanistic work of Hughes and Ingold. During these studies, the stereochemistry and mechanism of the E2 reaction were probed.[124,125] One of these studies,[125] the E2 eliminations from menthyl chloride (**216**) and neomenthyl chloride (**219**), has found its way into organic chemistry textbooks at both the graduate and undergraduate levels. The E2 elimination of menthyl chloride, which can only eliminate with *anti* stereochemistry by attaining the high-energy conformation **217**, where the leaving group and a single β-hydrogen are both axial, gave 2-menthene (**218**) as the only product of a slow reaction; the E2 elimination of neomenthyl chloride, where there are two β-hydrogen atoms anti to the leaving group in the low-energy conformation, gave mainly the Zaitsev regioisomer, 3-menthene (**220**), and proceeds some 200 times faster (Scheme 7.6). These experiments provided clear evidence that E2 elimination must occur with *anti* stereochemistry.

The reason for the differing regiochemistry of Zaitsev and Hofmann eliminations is traced to the different leaving groups. In the Zaitsev elimination, the leaving group departs as the conjugate base of a strong Lowry–Brønsted acid: Cl^- (HCl), Br^- (HBr), I^- (HI), OTs^- (TsOH), OMs^- (MsOH), H_2O (H_3O^+), etc. These are usually described as good leaving groups. In contrast to this, the leaving group in Hofmann eliminations departs as a weak base itself or the conjugate base of a weak acid: R_3N (R_3NH^+), R_2S (R_2SH^+), F^- (HF), RCO_2^- (RCO_2H), etc. These are usually not described as good leaving groups, but rather as poor or modest leaving groups.

The elimination of a poor leaving group is facilitated by generation of negative charge density at the β-carbon. This leads to another mechanism

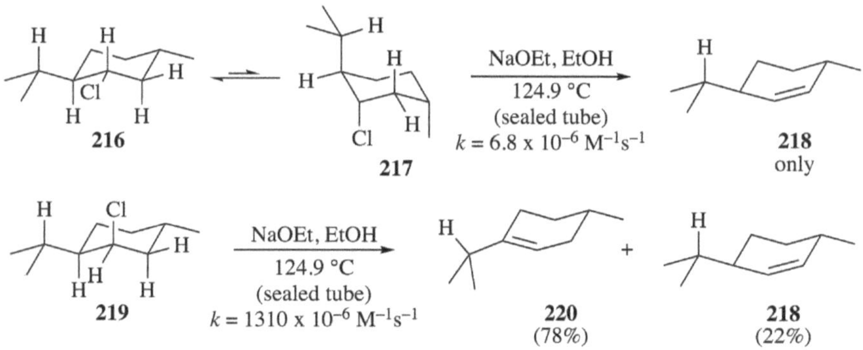

Scheme 7.6 The E2 eliminations from menthyl chloride and neomenthyl chloride provided the evidence for the mandated antistereochemistry of the reaction.

for elimination: E1cB (Elimination, Unimolecular, from conjugate base).[126] Because the conjugate base of the starting compound is involved, the reaction proceeds most rapidly when the most acidic β-hydrogen is abstracted. The acidity of hydrogens in an alkyl halide decreases as the substitution of the hydrogens at the β-carbon by alkyl increases: the hydrogens on the less substituted carbon are the most acidic, leading to the least substituted alkene as the major product.

Electrophilic addition to alkenes

Lowry-Brønsted acids

Beginning with Markovnikov's observations, the addition of protic acids to alkenes has been a fruitful area of research for over 150 years. Markovnikov himself studied the acid-catalyzed hydration of alkenes, as well as the addition of hydrogen halides in aqueous solution. Chemists following in his footsteps confirmed his results for hydration and the addition of hydrogen chloride and hydrogen iodide, but the addition of hydrogen bromide proved to be infuriatingly inconsistent—part of the time, the Markovnikov regioisomer was the major product, but at other times, the anti-Markovnikov isomer was the sole product. The solution to this enigma was found by Morris S. Kharasch at the University of Chicago.

The addition of Lowry–Brønsted acids to alkenes is often subject to Wagner–Meerwein rearrangements.[22,23,127–129] Representative reactions of this type,[130–132] which proceed through cations, are shown in Scheme 7.7. Thus, 3-methyl-1-butene (**221**) reacts with hydrogen chloride to give an approximately equal mixture of the unrearranged, secondary

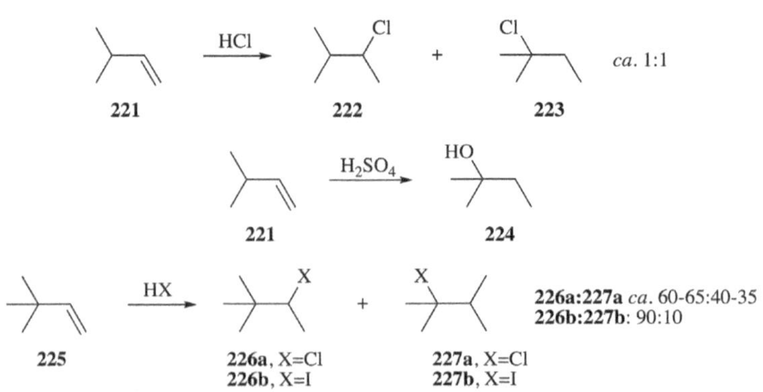

Scheme 7.7 Wagner–Meerwein rearrangements in the addition of Lowry–Brønsted acids to alkenes.

chloride **222**, and the rearranged, tertiary chloride **223**. The analogous acid-catalyzed hydration gives the rearranged alcohol **224** as the only product of the reaction. The analogous addition of hydrogen halides to 3,3-dimethyl-1-butene (**225**) gives a mixture of rearranged and unrearranged products. With hydrogen chloride, the reaction gives an approximately 2:1 mixture of the unrearranged chloride, **226a**, and the rearranged chloride, **227a**. When hydrogen iodide is used, the unrearranged iodide, **226b**, dominates the product mixture by a ratio of 9:1.

Markovnikov hydration without rearrangement

Acid-catalyzed hydration was the other reaction studied by Markovnikov in the work leading up to his Rule. In this case, the reaction was occasionally complicated by Wagner–Meerwein rearrangement, but in those cases where rearrangement did not occur, it was clear that there was a strong preference for the oxygen to end up on the more substituted carbon of the alkene—for the Markovnikov product to form. Thus, the acid-catalyzed hydration of 3-methyl-1-butene (**221**) gives the rearranged alcohol **224** as the major product (Scheme 7.7).

Suppressing the Wagner–Meerwein rearrangement was one major task for improving the usefulness of the alkene hydration reaction. This was accomplished by oxymercuration–demercuration.[133–139] Its effectiveness is nicely illustrated by the hydration of 3,3-dimethyl-1-butene (**225**), to the alcohols **228** and **229**, which gives a high ratio of the Markovnikov (**228**) to anti-Markovnikov (**229**) isomers, but no amounts of the rearranged alcohol (**230**) above the limits of detection (Scheme 7.8).[136]

Kharasch and the "peroxide effect"

The work of Kharasch in rationalizing the effects of oxygen and peroxides on the orientation of the addition of hydrogen bromide to alkenes was discussed at length in Chapter 3.[140–142] In short, Kharasch was able to show that in the presence of peroxides, the reaction proceeded by a free radical,

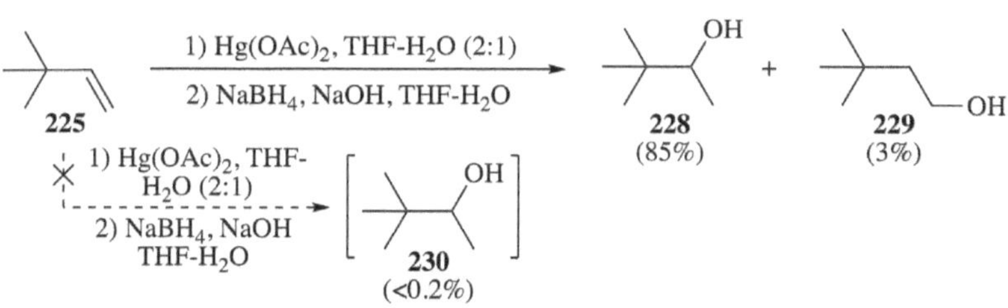

Scheme 7.8 Oxymercuration–demercuration of 3,3-dimethyl-1-butene.

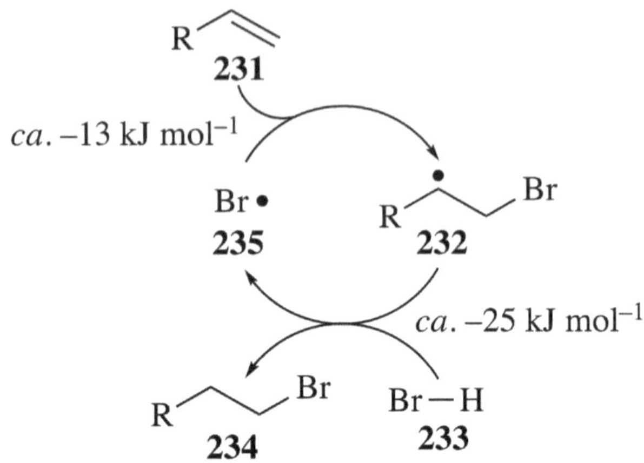

Scheme 7.9 Both propagation steps of the radical addition of HBr to an alkene are exothermic.

rather than an ionic mechanism. This reversed the order of the addition of the elements of H—Br, so the reaction proceeding through the lower-energy intermediate now gave the anti-Markovnikov product. Because both propagation steps in the radical addition of HBr to the alkene are exothermic (Scheme 7.9), in contrast to HCl or HI, only HBr addition can occur by this mechanism—explaining another of the observations of the "strange" behavior of HBr in the addition reaction.

Halogens and similar nonacidic electrophiles

The addition of bromine in a halogenated solvent to an alkene results in the formation of a vicinal dibromide by a stepwise mechanism involving an open-chain cationic intermediate in concord with Markovnikov's Rule.[143,144] This cation could be intercepted by competing nucleophiles, so the addition of bromine in aqueous solutions of sodium chloride, sodium iodide, and sodium nitrate gave 2-bromo-1-chloroethane, 2-iodo-1-bromoethane and 2-bromoethanol nitrate; the addition of bromine in concentrated aqueous sodium bromide suppressed the formation of the bromohydrin[145] Reactions in polar, nucleophilic solvents such as methanol give crossed addition products in which the reactive intermediate is intercepted by the solvent.[146] These results were interpreted in terms of a cationic intermediate (**237**) (Scheme 7.10), but the origin of the stereochemistry was not rationalized until 1937.[147]

The stereochemistry of the addition in polar solvents is anti, and this was interpreted by Roberts and Kimball in terms of a three-membered cyclic

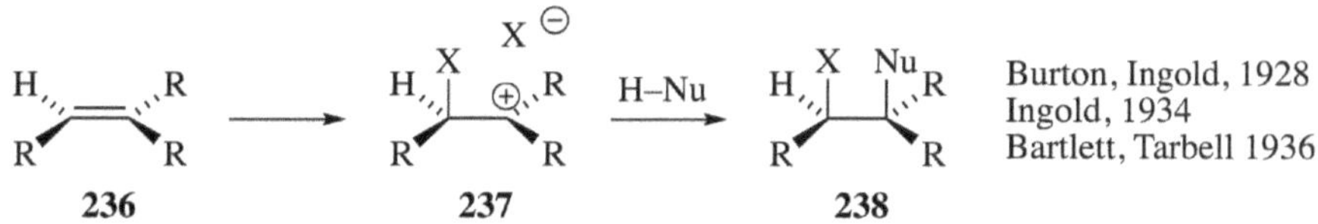

Scheme 7.10 The early mechanistic interpretations of the addition of halogen-like electrophiles to alkenes.

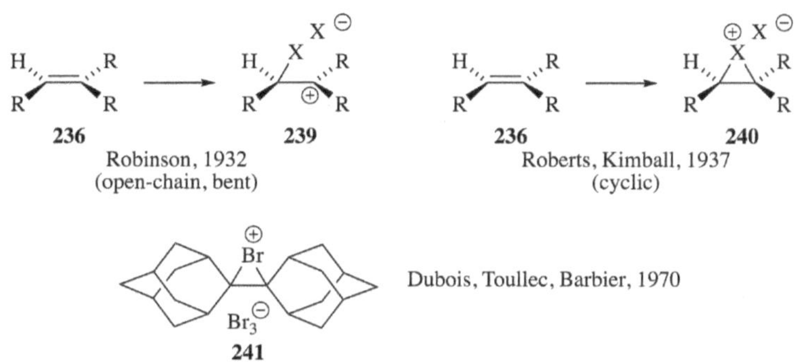

Scheme 7.11 The evolution of the three-membered cyclic heteronium ion intermediate and the first stable bromonium ion.

halonium ion.[147] In 1932, Robinson had proposed an open-chain cation (**239**) in which one face of the cation is shielded. In 1937, Roberts and Kimball formalized the three-membered heteronium ion **240** (Scheme 7.11). All atoms in the three-membered heteronium ion possess complete octets, so there is no thermodynamic advantage to a rearrangement, which is suppressed. In addition, all three-membered heteronium ions open by an S_N1-like S_N2 mechanism, leading to stereospecific antiaddition and Markovnikov regiochemistry. Unequivocal evidence for the existence of stable three-membered heteronium ions was provided in 1970 with the isolation of the three-membered cyclic bromonium salt **241**.[148]

Isotope effect evidence for the mechanisms of elimination

The advent of kinetic isotope effects gave a tool for probing the intimate details of the activated complex structure and the reaction coordinate motion in organic reactions.[149–155] This is especially so when compounds isotopically labeled at several different locations ("successive labeling"[156]) are studied under the same conditions. Successive labeling and a combined experimental—theoretical approach have been advocated as being particularly effective.[157] Stated most simply, a kinetic isotope effect will occur only if there is a net breaking or formations of bonds to the labeled atom; net bond rupture leads to $^{L}k/^{H}k > 1$, where ^{L}k refers to the rate constant of the lighter isotope, and ^{H}k refers to the heavier isotope. When $^{L}k/^{H}k > 1$, this is referred to as a "normal" isotope effect. When $^{L}k/^{H}k < 1$, this is referred to as an "inverse" isotope effect. The more ready availability of enriched heavy isotopes has permitted kinetic isotopes to be measured for elements other than hydrogen (e.g., $^{12}k/^{14}k$ for carbon, $^{14}k/^{15}k$ for nitrogen, $^{16}k/^{18}k$ for oxygen, $^{32}k/^{34}k$ for sulfur, and $^{35}k/^{37}k$ for chlorine, etc.), and these heavy-atom isotope effects allow us to monitor bonding changes at positions other than hydrogen. In general, a normal isotope effect is associated with a general "loosening" of the vibrations at the labeled atom, which usually correlates with bond rupture and a change in hybridization from sp^3 to sp^2, as in the solvolysis of *tert*-butyl chloride.[158] Conversely, an inverse isotope effect is associated with a general "tightening" of the vibrations at the labeled atom, which usually correlates with bond formation at the labeled atom. When bond formation compensates for bond rupture (as in an S_N2 reaction), the isotope is often very small—a "null" isotope effect. Representative theoretical maximum values of $^{L}k/^{H}k$ are collected in Table 7.4.

An excellent example of the use of successive labeling in the elucidation of reaction mechanism was provided by the 1973 study of the Hofmann

Table 7.4 Theoretical maximum values for kinetic isotope effects.

Isotope Exchange	Theoretical Maximum kie	Isotope Exchance	Theoretical Maximim kie
H→D	8.3	$^{12}C \rightarrow ^{14}C$	1.3
$^{14}N \rightarrow ^{15}N$	1.12	$^{16}O \rightarrow ^{18}O$	1.19
$^{32}S \rightarrow ^{34}S$	1.08	$^{35}Cl \rightarrow ^{37}Cl$	1.05

Data from Lewis, D. E., *Advanced Organic Chemistry*; Oxford: New York, 2016; p. 295. ISBN 9780199758975.

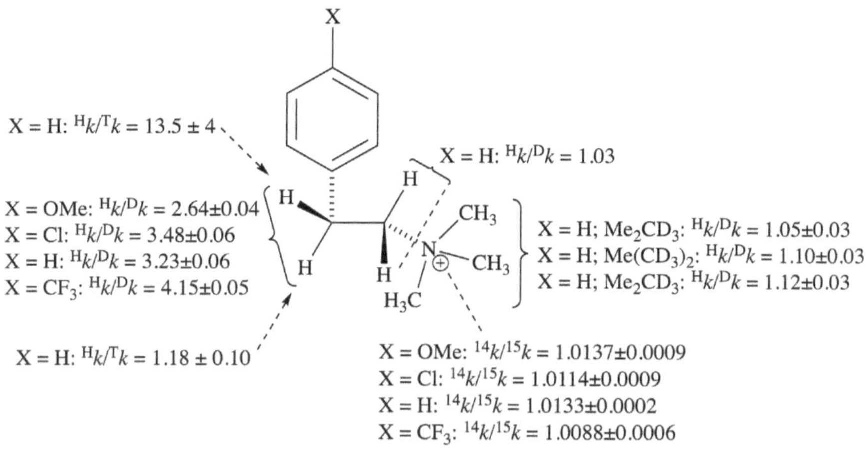

Figure 7.9 Kinetic isotope effects at different locations in several substituted 2-phenylethyltrimethylammonium ions.

elimination of 2-arylethyltrimethylammonium ions by Smith and Bourns.[159] The isotope effects for labeling at several different sites in the cation are shown in Fig. 7.9.

The primary deuterium isotope effects at the benzyl methylene groups (CH_2 vs. CD_2) indicate that there is rupture of the C—H bond at the benzyl carbon in all four cations, and that there is partial C—N bond rupture also. The isotope effects in the parent ion were used to show that the reaction passes through an E1cb-like E2 transition state, with a charge near −0.3 at the benzyl carbon, and that the more electron withdrawing the substituent on the aromatic ring, the more reactant-like the activated complex.[160]

References

1. Hofmann, A. W. Beiträge zur Kenntniss der flüchtigen organischen Basen. *Justus Liebigs Ann. Chem.* **1851,** *79* (1), 11—39. https://doi.org/10.1002/jlac.18510790103.
2. Hofmann, A. W. Beiträge zur Kenntniss der flüchtigen organischen Basen. *Justus Liebigs Ann. Chem.* **1851,** *78* (3), 253—286. https://doi.org/10.1002/jlac.18510780302.
3. Lossen, W. Ueber Ammoniumverbindungen. *Justus Liebigs Ann. Chem.* **1876,** *181* (3), 364—383. https://doi.org/10.1002/jlac.18761810303.
4. Collie, N.; Schryver, S. B. LIII.—The Action of Heat on the Chlorides and Hydroxides of Mixed Quaternary Ammonium Compounds. *J. Chem. Soc. Trans.* **1890,** *57,* 767—782. https://doi.org/10.1039/CT8905700767.
5. von Braun, J. Über den Zerfall quartärer Ammoniumhydroxyde. *Justus Liebigs Ann. Chem.* **1911,** *382* (1), 1—49. https://doi.org/10.1002/jlac.19113820102.

6. von Braun, J.; Cahn, R. S. Über den Zerfall quartärer Ammoniumhydroxyde. *Justus Liebigs Ann. Chem.* **1924,** *436* (1), 262–273. https://doi.org/10.1002/jlac.19244360115.
7. von Braun, J.; Buchman, E. R. Über Den Zerfall Quartärer Ammoniumhydroxyde (V. Mitteil.). *Ber. Dtsch. Chem. Ges.* **1931,** 2610–2617. https://doi.org/10.1002/cber.19310641007.
8. von Braun, J. Über den Zerfall quartärer Ammoniumhydroxyde. *Justus Liebigs Ann. Chem.* **1912,** *386* (3), 273–303. https://doi.org/10.1002/jlac.19123860302.
9. von Braun, J.; Anton, E. Über Den Zerfall Quartärer Ammoniumhydroxyde (VI. Mitteil.). *Ber. Dtsch. Chem. Ges.* **1931,** 2865–2869. https://doi.org/10.1002/cber.19310641122.
10. von Braun, J.; Teuffert, W.; Weißbach, K. Über den Zerfall quartärer Ammonium- und Sulfoniumhydroxyde. IV. *Justus Liebigs Ann. Chem.* **1929,** *472* (1), 121–142. https://doi.org/10.1002/jlac.19294720104.
11. Flavitskii, F., Ob izomerii amilenov iz amil'nago brozheniya [On the isomers of the amylenes from the alcohol from fermentation]. Zh. Russ. Fiz.-Khim. O-va. 1875, 7, 103–130.
12. Vyshnegradskii, A. Vyshnegradskii Ob izomerii amilenov,. Stat'ya vtoraya [On the isomers of the amylenes. Second paper]. *Zh. Russ. Khim. O-va. i Fiz. O-va* **1877,** *9,* 192–208.
13. Wischnegradsky, A. Ueber verschiedene Amylene und Amylalkohole. *Justus Liebigs Ann. Chem.* **1878,** *190* (3), 328–366. https://doi.org/10.1002/jlac.18781900307.
14. Berthelot, M. Nouvelle Méthode Pour la Synthèse des Acides Organiques. *Ann. Chim. Phys.* **1871,** *23,* 212–222.
15. Wurtz, A. Mémoire sur les Glycols ou Alcools Diatomiques. *Ann. Chim. Phys.* **1859,** *55,* 400–478.
16. Bauer, A. Ueber das Amylen end einige damit Isomere Substanzes. *Z. Chem. Pharm.* **1861,** *4,* 645–658.
17. Wurtz, A. Sur un isomère de l'alcool amylique. *Compt. Rend. Hebd. Séances Acad. Sci.* **1862,** *55,* 370–375.
18. Meerwein, H. Über den Reaktionsmechanismus der Umwandlung von Borneol in Camphen; (Dritte Mitteilung über Pinakolinumlagerungen). *Justus Liebigs Ann. Chem.* **1914,** *405* (2), 129–175. https://doi.org/10.1002/jlac.19144050202.
19. Meerwein, H.; van Emster, K. Untersuchungen in der Camphen-Reihe, I.: Über den Reaktionsmechanismus der Isoborneol $\rightleftharpoons$ Camphen-Umlagerung. *Ber. Dtsch. Chem. Ges.* **1920,** 1815–1829. https://doi.org/10.1002/cber.19200530928.
20. Meerwein, H.; van Emster, K. Über die Gleichgewichts-Isomerie zwischen Bornylchlorid, Isobornylchlorid und Camphen-Chlorhydrat. *Ber. Dtsch. Chem. Ges.* **1922,** 2500–2528. https://doi.org/10.1002/cber.19220550829.
21. Y., Vagner, Ye. Ot imeni Ye. Ye. Vagnera «o stroenii kamfena» [On behalf of Ye. Ye. Vagner, "On the structure of camphene". Zh. Russ. Fiz.-Khim. O-va. 1899, 31, 680–684.
22. Wagner, G.; Brickner, W. Ueber die Beziehung der Pinenhaloïdhydrate zu den Haloïdanhydriden des Borneols. *Ber. Dtsch. Chem. Ges.* **1899,** *32* (2), 2302–2325. https://doi.org/10.1002/cber.189903202168.
23. Wagner, G.; Brykner, W. Bornylen, ein neues Terpen. *Ber. Dtsch. Chem. Ges.* **1900,** *33* (2), 2121–2125. https://doi.org/10.1002/cber.190003302117.
24. M., Osipov Poluchenie amilovykh spirtov deistviem sernoi kisloty na amilen [The preparation of amyl alcohols by the action of sulfuric acid on amylene]. Zh. Russ. Khim. O-va. i Fiz. O-va. 1875, 214–221.
25. A., Vyshnegradskii Ob uplotenii izoamilena [On the condensation of isoamylene]. Zh. Russ. Khim. O-va. i Fiz. O-va. 1875, 7, 165–172.

26. Vyshnegradskii, A., Ob izomerii amilenov iz amil'nago alkogolya brozheniya [On the isomers of the amylenes from the amyl alcohol from fermentation]. Zh. Russ. Khim. O-va. i Fiz. O-va. 1877, 9, 150–163.
27. Menschutkin, B. Nikolai Alexandrowitsch Menschutkin. *Ber. Dtsch. Chem. Ges.* **1907,** *40* (4), 5087–5098. https://doi.org/10.1002/cber.190704004182.
28. Menshutkin, B. N. 25 Pages of Testimonials and Condolence Messages from All over Europe. *Zh. Russ. Fiz.-Khim. O-va* **1907,** *39,* 278–304.
29. Menshutkin, B.N., Zhizn i Deyatel'nost Nikolaya Alekdandra Menshutkina [The life and works of Nikolai Aleksandrovich Menshutkin]. M. Frolova Press: St. Petersburg, 1908.
30. Witt, O. N. Obituary Notices: Friedrich Konrad Beilstein, 1838–1906; Emil Erlenmeyer, 1825–1909; Rudolph Fittig, 1835–1910; Hans Heinrich Landolt, 1831–1910; Nikolai Alexandrovitsch Menschutkin, 1842–1907; Sir Walter Palmer, Bart., 1858–1910. *J. Chem. Soc. Trans.* **1911,** 1646–1668. https://doi.org/10.1039/CT9119901646.
31. Ingold, C. K. Edward David Hughes. 1906-1963. *Biogr. Mem. Fellows R. Soc.* **1964,** *10,* 147–182.
32. Barton, D. H. R. Ingold, Robinson, Winstein, Woodward, and I. *Bull. Hist. Chem.* **1996,** *19,* 43–47.
33. Multiple authors. C. K. Ingold: Master and Mandarin of Physical Organic Chemistry (Special Issue). *Bull. Hist. Chem.* **1996,** *19,* 1–93.
34. Ridd, J. H. Christopher Ingold: The Missing Nobel Prize. In *The Posthumous Nobel Prize in Chemistry;* Strom, E. T., Mainz, V. V., Eds.; *Correcting the Errors and Oversights of the Nobel Prize Committee. ACS Symp. Ser.*; American Chemical Society: Washington, D.C, 2017, Vol. 1262; pp 207–218.
35. Roberts, G. K. Ingold at University College London: Educator and Department Head. *Bull. Hist. Chem* **1996,** *19,* 2–12.
36. Roberts, G. K. Christopher Kelk Ingold. In New Dictionary of Scientific Biography; Koertge, N., Ed.; Charles Scribner's Sons/Thomas Gale: Farmington Hills, MI, USA, 2007.
37. Saltzman, M. D. Ingold's Development of the Concept of Mesomerism. *Bull. Hist. Chem.* **1996,** *19,* 25–32.
38. Shoppee, C. W. Christopher Kelk Ingold, 1893-1970. *Biogr. Mem. Fellows R. Soc.* **1972,** *18,* 349–411.
39. Menschutkin, N. Sur l'action Du Chlorure d'acétyle Sur l'acide Phosphoreux. *Compt. Rend. hebd. Séances Acad. Sci.* **1864,** *59,* 295–298.
40. Menschutkin, N. Sur Les Acétopyrophosphates. *Compt. Rend. hebd. Séances Acad. Sci.* **1865,** *60,* 532–534.
41. Menschutkin, N. Untersuchungen aus dem chemischen Laboratorium des Professor Kolbe. XLII. Ueber die Producte der Einwirkung wasserfreier Alkohole auf Dreifach-Chlorphosphor. *Ann. Chem. Pharm.* **1866,** *139,* 343–354. https://doi.org/10.1002/jlac.18661390309.
42. Menschutkine, N. Sur l'action du chlorure d'acétyle sur l'acide phosphoreux. – Acide acétopyrophosphoreux. *Bull. Soc. Chim. Paris* **1864,** *2,* 241–244.
43. Menschutkine, N. Sur l'action du chlorure d'acétyle sur l'acide phosphoreux. *Bull. Soc. Chim. Paris* **1864,** *2,* 122–123.
44. Menshutkin, N., O vodorode fosforistoi kisloty ne sposobnom k metallicheskomu zameshcheniyu pri obyknovennykh usloviyakh dlya kislot [About the hydrogen of phosphorous acid, which is not capable of substitution by metals under ordinary conditions for acids]. Magistr Khimii diss. St. Petersburg University, 1866.
45. Menschutkin, N. Zur Kenntniss der Harnstoffverbindungen. Erste Abhandlung. Die Ureïde. *Ann. Chem. Pharm.* **1870,** *153,* 83–110. https://doi.org/10.1002/jlac.18701530104.

46. Menshutkin, N., O deistvii tsianovokislago kaliya na amidokisloty i ikh proizvodnyya (1 Chast) [On the action of potassium cyanate on amino acids and their derivatives (Part 1)]. Zh. Russ. Khim. O-va. 1869, 1, 38–53.
47. Menshutkin, N., Razlichie amidnykh grupp v gidantoinovoi kisloty. Klassifikatsiya ureidov [The difference in the amide groups in hydantoic acid. Classification of ureides]. Zh. Russ. Khim. O-va. 1869, 1, 134–142.
48. Menshutkin, N., Sintez i svoystva ureidov [Synthesis and properties of ureides]. Doktor Khimii diss. St. Petersburg University, 1869.
49. Menshutkin, N. *Analiticheskaya Khimiya* (*Analytical Chemistry*), 1st ed.; V. Demanova: St. Petersburg, 1871.
50. Menschutkin, N., *Analytische Chemie, Deutsche Ausgabe Unter Mitwirgung Des Verfassers Übersetzt von Dr. O. Bach,* 1st;; Quandt und Handel: Leipzig, 1871.
51. Menshutkin, N.; Locke, J. In *Analytical Chemistry, by N. Menschutkin. Translated from the Third German Edition under the Supervision of the Author by James Locke;* MacMillan & Co: London and New York, 1895.
52. Menshutkin, N., Lektsii organicheskoi khimii [Lectures in Organic Chemistry]. Tipografiya V. Demakova: St. Petersburg, 1884.
53. Hughes, E. D.; Watson, H. B. The Reaction of Bromine with Aliphatic Acids. Part III.α- and γ-Ketonic Acids. *J. Chem. Soc.* **1929,** 1945–1954. https://doi.org/10.1039/JR9290001945.
54. Hughes, E. D.; Watson, H. B. The Reaction of Bromine with Aliphatic Acids. Part IV. Succinic Acid. *J. Chem. Soc.* **1930,** 1733–1740. https://doi.org/10.1039/JR9300001733.
55. Hughes, E. D.; Watson, H. B.; Yates, E. D. Constitutional Factors Controlling Prototropic Changes in Carbonyl Compounds. Part I. The Relative Speeds of Enolisation of Acetone and Bromoacetone and the Effect of Acid Catalyst. *J. Chem. Soc.* **1931,** 3318–3324. https://doi.org/10.1039/JR9310003318.
56. Hofmann, H. M. R.; Hughes, E. D. Reaction Kinetics and the Walden Inversion. Part XII. Steric Course of Bimolecular Nucleophilic Substitution of an Initially Positively Charged Leaving Group by a Neutral Nucleophile. *J. Chem. Soc.* **1964,** 1259–1264. https://doi.org/10.1039/jr9640001259.
57. Hofmann, H. M. R.; Hughes, E. D. Reaction Kinetics and the Walden Inversion. Part IX. The Rotation of Optically Pure 1-Phenylethyl Chloride and 1-Phenylethyl Bromide. *J. Chem. Soc.* **1964,** 1244–1249. https://doi.org/10.1039/JR9640001244.
58. Hofmann, H. M. R.; Hughes, E. D. Reaction Kinetics and the Walden Inversion. Part XI. Steric Course of Bimolecular Nucleophilic Substitution by a Neutral Nucleophile in a Neutral Substrate. *J. Chem. Soc.* **1964,** 1252–1258. https://doi.org/10.1039/JR9640001252.
59. Mitchell, S.; Trollope, A. (1815-1882). In *Victorian. Britain. An Encyclopedia;* Mitchell, S., Ed.; Routledge: New York, 1988; pp 822–823. https://doi.org/10.4324/9780203815595.
60. Densmore, E. *Consumption and Chronic Diseases: A Hygienic Cure, at Patient's Home of Incipient and Advanced Cases;* Swan Sonnenschein & Co.: London, 1899.
61. Densmore, H. The Cure of Consumption. *Med. Leg. J.* **1900,** *18,* 282–291.
62. Beesley, R. M.; Thorpe, J. F. The Formation and Stability of Associated Alicyclic Systems. Part I. A System of Nomenclature, and Some Derivatives of Methane-II-cyclopropane and of Methane-III-cyclopropane. (This paper contains an insert, pp. 603–610: "[The following note has been written by Mr. C. K. Ingold.]"). *J. Chem. Soc. Trans.* **1920,** *117,* 591–620. https://doi.org/10.1039/CT920170591B.
63. Beesley, R. M.; Ingold, C. K.; Thorpe, J. F. The Formation and Stability of *spiro*-Compounds. Part I. *spiro*-Compounds from *cyclo*-Hexane. *J. Chem. Soc. Trans.* **1915,** *107.* https://doi.org/10.1039/CT9150701080.

64. Ingold, C. K.; Thorpe, J. F. The Formation and Stability of *spiro*-Compounds. Part II. Bridged *spiro*-Compounds Derived from *cyclo*-Hexane. *J. Chem. Soc., Trans.* **1919,** *115,* 320–383. https://doi.org/10.1039/CT9191500320.
65. Menschutkin, N. Sur la vitesse de formation des éthers composés. *Bull. Acad. Roy. Sci. Lett. Beaux-arts Belgique* **1891,** *21,* 559–587.
66. N., Menshutkin Vliyanie sostava alkogolei na skorost eterifikatsnn ikh [The influence of the composition of alcohols on the rate of their esterification]. Zh. Russ. Fiz.-Khim. O-va. 1891, 23, 263–283.
67. Menschutkin, N. Sur la décomposition de l'acétate d'amyle tertiaire par la chaleur. *Compt. Rend. Hebd. Séances Acad. Sci.* **1882,** *95,* 648–651.
68. Menschutkin, N. Ueber die Zersetzung des tertiären Amylacetats durch Wärme. *Ber. Dtsch. Chem. Ges* **1882,** *15,* 2512–2518. https://doi.org/10.1002/cber.188201502211.
69. Menshutkin, N., Issedovanie raspadeniya uksusnago tretichnago amila pri nagrevanii [An investigation of the decomposition of tertiary amyl acetate on heating]. Zh. Russ. Fiz.-Khim. O-va. 1882, 14, 292–300.
70. Menschutkin, N. Beiträge zur Kenntnis der Affinitätskoeffizienten der Alkylhaloide und der organischen Amine. *Z. Phys. Chem.* **1890,** *5,* 589–600. https://doi.org/10.1515/zpch-1890-0546.
71. Menschutkin, N. Über die Affinitätskoeffizienten der Alkylhaloide und der Amine. Zweiter Teil. Über den Einfluss des chemisch indifferenten flüssigen Mediums Auf die Geschwindigkeit der Verbindung des Triäthylamins mit den Alkyljodiden. *Z. Phys. Chem.* **1890,** *6,* 41–57. https://doi.org/10.1515/zpch-1890-0607.
72. Menshutkin, N., O koeffitsientakh srodstva galoidgidrinov i aminov. Chast vtoraya. [On the affinity coefficients of alkyl halides and amines. Second paper]. Zh. Russ. Fiz.-Khim. O-va. 1890, 22, 393–409.
73. Menshutkin, N., Vasil'ev, M., O koeffitsientakh srodstva galoidgidrinov i aminov. Chast pervaya. [On the affinity coefficients of alkyl halides and amines. First paper]. Zh. Russ. Fiz.-Khim. O-va. 1890, 22, 346–358.
74. Hughes, E. D.; Ingold, C. K.; Masterman, S.; MacNulty, B. J. Mechanism of Elimination Reactions. Part V. Kinetics of Olefin Elimination from Ethyl, Isopropyl, Tert.-Butyl and α- and β-Phenylethyl Bromides in Acidic and in Alkaline Alcoholic Solution. Effects Due to, and Factors Influencing, the Two Mechanisms of Elimination. *J. Chem. Soc.* **1940,** 899–912. https://doi.org/10.1039/JR9400000899.
75. Ingold, C. K. *Structure and Mechanism in Organic Chemistry;* Cornell University Press: Ithaca, NY, 1953; pp 306–472.
76. Lewis, G. N. The Atom and the Molecule. *J. Am. Chem. Soc.* **1916,** *38* (4), 762–785. https://doi.org/10.1021/ja02261a002.
77. Fischer, E. Waldensche Umkehrung und Substitutionsvorgang. *Justus Liebigs Ann. Chem.* **1911,** *381* (1), 123–141. https://doi.org/10.1002/jlac.19113810108.
78. Fischer, E. Nachtrag zu der Abhandlung: Waldensche Umkehrung und Substitutionsvorgang. *Justus Liebigs Ann. Chem.* **1912,** *386* (3), 374–386. https://doi.org/10.1002/jlac.19123860307.
79. Pfeiffer, P. Beitrag zur Theorie der Halochromieerscheinungen II. Unter experimenteller Mitarbeit von B. Friedmann, Z. Goldberg, E. Pros und V. Schwarzkopf. *Justus Liebigs Ann. Chem.* **1911,** *383* (1), 92–155. https://doi.org/10.1002/jlac.19113830106 (esp. pp. 123–130).
80. Gadamer, J. Über die Racemisation der Tropasäure und ihrer Ester; zugleich eine Theorie der Racemisation, Substitution und Waldenschen Umkehrung. *J. Prakt. Chem.* **1913,** 312–392. https://doi.org/10.1002/prac.19130870122.

81. Werner, A. Über den räumlichen Stellungswechsel bei Umsetzungen von raumisomeren Verbindungen. *Ber. Dtsch. Chem. Ges* **1911,** *44* (1), 873–882. https://doi.org/10.1002/cber.191104401130.
82. Werner, A. Über die raumisomeren Kobaltverbindungen. *Justus Liebigs Ann. Chem.* **1912,** *386* (1–2), 1–272. https://doi.org/10.1002/jlac.19123860102 (esp. 70 ff.).
83. Frankland, P. F. Presidential Address. The Walden Inversion. *J. Chem. Soc. Trans.* **1913,** *103,* 713–742. https://doi.org/10.1039/CT9130300713.
84. McKenzie, A.; Clough, G. W. Experiments on the Walden Inversion. Part IX. The Interconversion of the Optically Active Phenylmethylcarbinols. *J. Chem. Soc. Trans.* **1913,** *103,* 687–699. https://doi.org/10.1039/CT9130300687.
85. Lewis, G. N. *Valence and the Structure of Atoms and Molecules;* The Chemical Catalog Company: New York, 1923.
86. Cowdrey, W. A.; Hughes, E. D.; Ingold, C. K.; Masterman, S.; Scott, A. D. Reaction Kinetics and the Walden Inversion. Part VI. Relation of Steric Orientation to Mechanism in Substitutions Involving Halogen Atoms and Simple or Substituted Hydroxyl Groups. *J. Chem. Soc.* **1937,** 1252–1271. https://doi.org/10.1039/jr9370001252.
87. Cowdrey, W. A.; Hughes, E. D.; Nevell, T. P.; Wilson, C. L. Aliphatic Substitution and the Walden Inversion. Part III. Comparison, Using Radioactive Bromine, of the Rates of Inversion and Substitution in the Reaction of Bromide Ions with α-Bromopropionic Acid. *J. Chem. Soc.* **1938,** 209–211. https://doi.org/10.1039/JR9380000209.
88. Harvey, S. H.; Hoye, P. A. T.; Hughes, E. D.; Ingold, C. K. Reaction Kinetics and the Walden Inversion. Part VIII. Steric Course of Bimolecular Nucleophilic Substitution by an Anionic Substituting Agent in a Cationic Substrate. *J. Chem. Soc.* **1960,** 800–806. https://doi.org/10.1039/jr9600000800.
89. Hiron, F.; Hughes, E. D. Reaction Kinetics and the Walden Inversion. Part VII. The Configurations of Some Amino-Hydrocarbons. *J. Chem. Soc.* **1960,** 795–799. https://doi.org/10.1039/jr9600000795.
90. Hughes, E. D. Aliphatic Substitution and the Walden Inversion. *Trans. Faraday Soc.* **1938,** *34,* 202–221. https://doi.org/10.1039/TF9383400202.
91. Hughes, E. D.; Whittingham, D. J. Mechanism of Substitution at a Saturated Carbon Atom. Part LX. Nucleophilic Substitutions of the "Missing" Charge Type. *J. Chem. Soc.* **1960,** 806–809. https://doi.org/10.1039/JR9600000806.
92. Hughes, E. D.; Juliusburger, F.; Masterman, S.; Topley, B.; Weiss, J. Aliphatic Substitution and the Walden Inversion. Part I. *J. Chem. Soc.* **1935,** 1525–1529. https://doi.org/10.1039/jr9350001525.
93. Hughes, E. D.; Juliusburger, F.; Scott, A. D.; Topley, B.; Weiss, J. Aliphatic Substitution and the Walden Inversion. Part II. *J. Chem. Soc.* **1936,** 1173–1175. https://doi.org/10.1039/JR9360001173.
94. Dostrovsky, I.; Hughes, E. D. Mechanism of Substitution at a Saturated Carbon Atom. Part XXIX. The Rôle of Steric Hindrance. (Section D) the Mechanism of the Reaction of Neopentyl Bromide with Aqueous Ethyl Alcohol. *J. Chem. Soc.* **1946,** 166–169. https://doi.org/10.1039/JR9460000166.
95. Dostrovsky, I.; Hughes, E. D. Mechanism of Substitution at a Saturated Carbon Atom. Part XXVI. The Rôle of Steric Hindrance. (Section A) Introductory Remarks, and a Kinetic Study of the Reactions of Methyl, Ethyl, n-Propyl, Isobutyl, and Neopentyl Bromides with Sodium Ethoxide in Dry Ethyl Alcohol. *J. Chem. Soc.* **1946,** 157–161. https://doi.org/10.1039/JR9460000157.
96. Dostrovsky, I.; Hughes, E. D. Mechanism of Substitution at a Saturated Carbon Atom. Part XXVII. The Rôle of Steric Hindrance. (Section B) a Comparison of the Rates of Reaction of Ethyl and *Neo*pentyl Bromides with Inorganic Iodides in Acetone Solution. *J. Chem. Soc.* **1946,** 161–164. https://doi.org/10.1039/JR9460000161.

97. Dostrovsky, I.; Hughes, E. D. Mechanism of Substitution at a Saturated Carbon Atom. Part XXVIII. The Rôle of Steric Hindrance. (Section C) a Comparison of the Rates of Reaction of Methyl, Ethyl, *n*-Propyl, *Iso*butyl, and *Neo*pentyl Bromides with Aqueous Ethyl Alcohol. *J. Chem. Soc.* **1946,** 164–165. https://doi.org/10.1039/JR9460000164.
98. Dostrovsky, I.; Hughes, E. D. Mechanism of Substitution at a Saturated Carbon Atom. Part XXX. The Rôle of Steric Hindrance. (Section E) a Comparison of the Rates of Reaction of Methyl, Ethyl, *n*-Propyl, *Iso*butyl, and *Neo*pentyl Halides with Silver Nitrate in Aqueous Ethyl Alcohol. *J. Chem. Soc.* **1946,** 169–171. https://doi.org/10.1039/JR9460000169.
99. Dostrovsky, I.; Hughes, E. D. Mechanism of Substitution at a Saturated Carbon Atom. Part XXXI. The Rôle of Steric Hindrance. (Section F) a Comparison of the Rates of Reaction of Methyl, Ethyl, *n*-Propyl, and *Neo*pentyl Bromides with Wet Formic Acid. *J. Chem. Soc.* **1946,** 171–173. https://doi.org/10.1039/JR9460000171.
100. Dostrovsky, I.; Hughes, E. D.; Ingold, C. K. Mechanism of Substitution at a Saturated Carbon Atom. Part XXXII. The Rôle of Steric Hindrance. (Section G) Magnitude of Steric Effects, Range of Occurrence of Steric and Polar Effects, and Place of the Wagner Rearrangement in Nucleophilic Substitution and Elimination. *J. Chem. Soc.* **1946,** 173–194. https://doi.org/10.1039/JR9460000173.
101. Grabovskii, N.; Zaitsev, A. Bromistom Butilene i Butilovom Glikole Iz Normalnago Butil'nago Alkogolya (On Butylene Bromide and Butyl Glycol from Normal Butyl Alcohol). *Zh. Russ. Khim. O-va. i Fiz. O-va* **1875,** *7,* 317–328.
102. Grabowsky, N.; Saytzeff, A. Ueber Butylenbromür Und Butylglycol Aus Dem Normalen Butylalkohol. *Justus Liebigs Ann. Chem.* **1875,** *179* (3), 325–338. https://doi.org/10.1002/jlac.18751790307.
103. Vagner, Y., Zaitsev A., Prevrashenie dietilkarbinola v metilpropilkarbinol [The conversion of diethyl carbinol to methyl propyl carbinol]. Zh. Russ. Khim. O-va. i Fiz. O-va. 1875, 7, 302–312.
104. Wagner, G.; Saytzeff, A. Ueber Amylenbromür und Amylglycol aus Diäthylcarbinol. *Justus Liebigs Ann. Chem.* **1875,** *179* (3), 302–313. https://doi.org/10.1002/jlac.18751790305.
105. Saytzeff, A. Zur Kenntniss der Reihenfolge Der Analgerung und Ausscheidung der Jodwasserstoffelemente in organischen Verbindungen. *Justus Liebigs Ann. Chem.* **1875,** *179* (3), 296–301. https://doi.org/10.1002/jlac.18751790304.
106. Zaitsev, A., Voprosu poryadke prisoedineniya i vydeleniya elementov iodistago vodoroda v organicheskikh soedineniyakh [On the question of the order of the addition and elimination of the elements of hydrogen iodide in organic compounds]. Zh. Russ. Khim. O-va. i Fiz. O-va. 1875, 7, 289–293.
107. Hofmann, A. W. Researches into the Molecular Constitution of the Organic Bases. *Phil. Trans. Roy. Soc. Lond.* **1851,** *141,* 357–398. https://doi.org/10.1098/rstl.1851.0017.
108. Hughes, E. D.; Ingiold, C. K. Influence of Poles and Polar Linkings on the Course Pursued by Elimination Reactions. Part XV. Dynamics of the Elimination of Olefins from Quaternary Ammonium Compounds. *J. Chem. Soc.* **1933,** 523–526. https://doi.org/10.1039/JR9330000523.
109. Hughes, E. D.; Ingold, C. K.; Patel, C. S. Influence of Poles and Polar Linkings on the Course Pursued by Elimination Reactions. Part XVI. Mechanism of the Thermal Decomposition of Quaternary Ammonium Compounds. *J. Chem. Soc.* **1933,** 526–530. https://doi.org/10.1039/JR9330000526.

110. Ingold, C. K.; Jessop, J. A. Influence of Poles and Polar Linkings on the Course Pursued by Elimination Reactions. Part VI. 1:1-Elimination in the Degradation of Quaternary Ammonium Hydroxides. *J. Chem. Soc.* **1929,** 2357–2361. https://doi.org/10.1039/jr9290002357.
111. Ingold, C. K.; Kuriyan, K. I. Influence of Poles and Polar Linkings on the Course Pursued by Elimination Reactions. Part XX. Elimination of Saturated Primary Alcohols by Thermal Decomposition of Sulphonium Hydroxides. *J. Chem. Soc.* **1933,** 991–993. https://doi.org/10.1039/JR9330000991.
112. Ingold, C. K.; Patel, C. S. Influence of Poles and Polar Linkings on the Course Pursued by Elimination Reactions. Part XI. Decomposition of Quaternary Ammonium Hydroxides Containing the *tert.*-Butylcarbinyl Group. *J. Chem. Soc.* **1933,** 67–68. https://doi.org/10.1039/JR9330000067.
113. Ingold, C. K.; Patel, C. S. Influence of Poles and Polar Linkings on the Course Pursued by Elimination Reactions. Part XII. Decomposition of Quaternary Ammonium Alkoxides. *J. Chem. Soc.* **1933,** 68–69. https://doi.org/10.1039/JR9330000068.
114. Ingold, C. K.; Vass, C. C. N. Influence of Poles and Polar Linkings on the Course Pursued by Elimination Reactions. Part II. Mechanism of Exhaustive Methylation (Continued). *J. Chem. Soc.* **1928,** 3125–3127. https://doi.org/10.1039/JR9280003125.
115. Dhar, M. L.; Highes, E. D.; Ingold, C. K.; et al. Mechanism of Elimination Reactions. Part IX. Kinetics of Olefin Elimination from *n*-Propyl, *n*-Butyl, *n*-Amyl, and *iso*Butyl Bromides in Alkaline Alcoholic Media. *J. Chem. Soc.* **1948,** 2055–2057. https://doi.org/10.1039/JR9480002055.
116. Dhar, M. L.; Hughes, E. D.; Ingold, C. K. Mechanism of Elimination Reactions. Part X. Kinetics of Olefin Elimination from *iso*Propyl, *sec.*-Butyl, 2-*n*-Amyl, and 3-*n*-Amyl Bromides in Acidic and Alkaline Alcoholic Media. *J. Chem. Soc.* **1948,** 2058–2065. https://doi.org/10.1039/JR9480002058.
117. Dhar, M. L.; Hughes, E. D.; Ingold, C. K. Mechanism of Elimination Reactions. Part XI. Kinetics of Olefin Elimination from *tert.*-Butyl and *tert.*-Amyl Bromides in Acidic and Alkalinc Alcoholic Media. *J. Chem. Soc.* **1948,** 2065–2072. https://doi.org/10.1039/JR9480002065.
118. Dhar, M. S.; Hughes, E. D.; Ingold, C. K.; Mandour, A. M. M.; Maw, G. A.; Woolf, L. I. Mechanism of Elimination Reactions. Part XVI. Constitutional Influences in Elimination. A General Discussion. *J. Chem. Soc.* **1948,** 2093–2119. https://doi.org/10.1039/JR9480002093.
119. Hughes, E. D.; Ingold, C. K. The Mechanism and Kinetics of Elimination Reactions. *Trans. Faraday Soc.* **1941,** *37,* 657–685. https://doi.org/10.1039/TF9413700657.
120. Hughes, E. D.; Ingold, C. K.; Woolf, L. I. Mechanism of Elimination Reactions. Part XIV. Kinetics of Olefin Elimination from *tert.*-Butyl- and *tert.*-Amyl-Dimethylsulphonium Salts in Alkaline and Acidic Alcoholic Media. *J. Chem. Soc.* **1948,** 2084–2090. https://doi.org/10.1039/JR9480002084.
121. Hughes, E. D.; Ingold, C. K.; Maw, G. A.; et al. Mechanism of Elimination Reactions. Part XIII. Kinetics of Olefin Elimination from *iso*Propyl-, *sec.*-Butyl-, and 1-Phenylethyl-Dimethylsulphonium Salts in Alkaline Alcoholic Media. *J. Chem. Soc.* **1948,** 2077–2083. https://doi.org/10.1039/JR9480002077.
122. Cooper, K. A.; Dhar, M. L.; Hughes, E. D.; Ingold, C. K.; MacNulty, B. J.; Woolf, L. I. Mechanism of Elimination Reactions. Part VII. Solvent Effects on Rates and Product-Proportions in Uni- and Bi-molecular Substitution and Elimination Reactions of Alkyl Halides and Sulphonium Salts in Hydroxylic Solvents. *J. Chem. Soc.* **1948,** 2043–2049. https://doi.org/10.1039/JR9480002043.

123. Cooper, K. A.; Hughes, E. D.; Ingold, C. K.; MacNulty, B. J. Mechanism of Elimination Reactions. Part VI. Introduction to a Group of Papers. Unimolecular Olefin Formation from Tert.-Butyl- and Tert.-Amyl-Sulphonium Salts. *J. Chem. Soc.* **1948,** 2038–2042. https://doi.org/10.1039/JR9480002038.
124. Hughes, E. D.; Ingold, C. K.; Pasternak, R. Mechanism of Elimination Reactions. Part XVIII. Kinetics and Steric Course of Elimination from Isomeric Benzene Hexachlorides. *J. Chem. Soc.* **1953,** 3832–3839. https://doi.org/10.1039/jr9530003832.
125. Hughes, E. D.; Ingold, C. K.; Rose, J. B. Mechanism of Elimination Reactions. Part XIX. Kinetics and Steric Course of Elimination from Isomeric Menthyl Chlorides. *J. Chem. Soc.* **1953,** 3839–3845. https://doi.org/10.1039/jr9530003839.
126. McLennan, D. J. The Carbanion Mechanism of Olefin-Forming Elimination. *Q. Rev.* **1967,** *21,* 490–506. https://doi.org/10.1039/QR9672100490.
127. Meerwein, H.; Schäfer, J. Über die wechselseitige Umwandlung von Verbindungen mit sechs- und siebengliedrigem Kohlenstoffring. *J. Prakt. Chem.* **1922,** *104,* 289–310. https://doi.org/10.1002/prac.19221040121.
128. Vagner, Y.Y., Protokol zasedaniya Otdeleniya Khimii R. F. Khimicheskago Obshchestva. 9-go sentyabrya 1899 goda [Minutes of the meeting of the Chemistry Section of the Russian Physical-Chemical Society. 9th September 1899]. Zh. Russ. Fiz.-Khim. O-va. 1899, 31, 680–684.
129. Meerwein, H. Über Ringveränderungen bei der Wasserabspaltung aus alicyclic Alkoholen. *Justus Liebigs Ann. Chem.* **1918,** *417,* 255–277. https://doi.org/10.1002/jlac.19184170205.
130. Ecke, G. C.; Cook, N. C.; Whitmore, F. C. Some Reactions of *t*-Butylethylene. *J. Am. Chem. Soc.* **1950,** *72,* 1511–1513. https://doi.org/10.1021/ja01160a023.
131. Noyes, W. A.; Potter, R. S. Molecular Rearrangements in the Camphor Series. X. Campholytic and Related Compounds. Walden's Rearrangement. *J. Am. Chem. Soc.* **1912,** *34,* 1067–1080. https://doi.org/10.1021/ja02209a015.
132. Whitmore, F. C.; Johnston, F. Secondary Isoamyl Chloride, 3-Chloro-2-Methylbutane. *J. Am. Chem. Soc.* **1933,** *55,* 5020–5022. https://doi.org/10.1021/ja01339a053.
133. Chatt, J. The Addition Compounds of Olefins with Mercuric Salts. *Chem. Rev.* **1950,** *48,* 7–43. https://doi.org/10.1021/cr60149a002.
134. Henbest, H. B.; Nicholls, B. Aspects of Stereochemistry. Part XII. A Specific Directing Effect in the Mercuration of Some 4-Substituted *cyclo*Hexenes and *cis*-Hex-3-Enol. *J. Chem. Soc.* **1959,** 227–236. https://doi.org/10.1039/JR9590000227.
135. Sand, J.; Genssler, O. Mercuriverbindungen aus Ketonen. *Ber. Dtsch. Chem. Ges* **1903,** *36.* https://doi.org/10.1002/cber.190303603175, 3799–3706.
136. Brown, H. C.; Geoghegan, P., Jr. The Oxymercuration-Demercuration of Representative Olefins. A Convenient, Mild Procedure for the Markovnikov Hydration of the Carbon-Carbon Double Bond. *J. Am. Chem. Soc.* **1967,** *89,* 1522–1524. https://doi.org/10.1021/ja00982a043.
137. Brown, H. C.; Geoghegan, P., Jr. Solvomercuration-Demercuration. I. Oxymercuration-Demercuration of Representative Olefins in an Aqueous System. Mild Procedure for the Markovnikov Hydration of the Carbon-Carbon Double Bond. *J. Org. Chem.* **1970,** *35,* 1844–1850. https://doi.org/10.1021/jo00831a028.
138. Brown, H. C.; Hammar, W. J. Stereoselectivity of the Hydration of Representative Cyclic and Bicyclic Olefins by the New Oxymercuration-Demercuration Procedure. *J. Am. Chem. Soc.* **1967,** *1989,* 1524–1525. https://doi.org/10.1021/ja00982a044.

139. Brown, H. C.; Kawakami, J. H.; Ikegami, S. A Remarkable Stereoselectivity in the Oxymercuration-Demercuration of Norbornene, 7,7-Dimethylnorbornene, and Related Derivatives. *J. Am. Chem. Soc.* **1967,** *89,* 1525–1526. https://doi.org/10.1021/ja00982a045.
140. Kharasch, M. S.; Mayo, F. R. The Peroxide Effect in the Addition of Reagents to Unsaturated Compounds. I. The Addition of Hydrogen Bromide to Allyl Bromide. *J. Am. Chem. Soc.* **1933,** *55,* 2468–2496. https://doi.org/10.1021/ja01333a041.
141. Kharasch, M. S.; McNab, M. C.; Mayo, F. R. The Peroxide Effect in the Addition of Reagents to Unsaturated Compounds. II. The Addition of Hydrogen Bromide to Propylene. *J. Am. Chem. Soc.* **1933,** *55,* 2521–2530. https://doi.org/10.1021/ja01333a048.
142. Kharasch, M. S.; McNab, M. C.; Mayo, F. R. The Peroxide Effect in the Addition of Reagents to Unsaturated Compounds. III. The Addition of Hydrogen Bromide to Vinyl Bromide. *J. Am. Chem. Soc.* **1933,** *55,* 2531–2533. https://doi.org/10.1021/ja01333a049.
143. Burton, H.; Ingold, C. K. Mobile-anion Tautomerism. Part I. A Preliminary Study of the Conditions of Activation of the Three-Carbon System, and a Discussion of the Results in Relation to the Modes of Addition to Conjugated Systems. *J. Chem. Soc.* **1928,** 904–921. https://doi.org/10.1039/JR9280000904.
144. Ingold, C. K. Principles of an Electronic Theory of Organic Reactions. *Chem. Rev.* **1934,** *15,* 225–274. https://doi.org/10.1021/cr60051a003.
145. Francis, A. W. Studies on the Directive Influence of Substituents in the Benzene Ring. III. The Active Agent in Aqueous Bromination. *J. Am. Chem. Soc.* **1925,** *47,* 2340–2348. https://doi.org/10.1021/ja01686a011.
146. Bartlett, P. D.; Tarbell, D. S. The Mechanism of Addition Reactions. A Kinetic Study of the Addition of Methyl Hypobromite to Stilbene. *J. Am. Chem. Soc.* **1936,** *58,* 466–474. https://doi.org/10.1021/ja01294a023.
147. Roberts, I.; Kimball, G. E. The Halogenation of Ethylenes. *J. Am. Chem. Soc.* **1937,** *59,* 947–948. https://doi.org/10.1021/ja01284a507.
148. Dubois, J.-E.; Toullec, J.; Barbier, G. Attenuation of α- and β-Methyl Effects in the Bromination of Mono-Olefins with Conjugatively Electron-Donating Substituents. *Tetrahedron Lett.* **1970,** *11,* 4485–4488. https://doi.org/10.1016/S0040-4039(01)83956-3.
149. Collins, C. J., Bowman, N. S., Eds. *Isotope Effects in Chemical Reactions;* Van Nostrand Reinhold: New York, 1970.
150. Melander, L. *Isotope Effects on Reaction Rates;* Ronald Press: New York, 1960.
151. Melander, L.; Saunders, W. H., Jr. *Reaction Rates of Isotopic Molecules;* John Wiley: New York, 1980.
152. Westheimer, F. H.; Westheimer, F. H. The Magnitude of the Primary Isotope Effect for Compounds of Hydrogen and Deuterium. *Chem. Rev.* **1961,** *61,* 265–273. https://doi.org/10.1021/cr60211a004.
153. Wiberg, K. B. The Deuterium Isotope Effect. *Chem. Rev.* **1955,** *55,* 713–743. https://doi.org/10.1021/cr50004a004.
154. Willi, A. V. *Isotopeneffekte bei Chemischen Reaktionen;* Thieme Verlag: Stuttgart, 1983.
155. Kresge, A. J. Correlation of Kinetic Isotope Effects with Free Energies of Reaction. *J. Am. Chem. Soc.* **1980,** *102,* 7797–7798. https://doi.org/10.1021/ja00546a036.
156. Fry, A. Application of the Successive Labelling Technique to Some Carbon, Nitrogen and Chlorine Isotope Effect Studies of Organic Reaction Mechanisms. *Pure Appl. Chem.* **1964,** *8,* 409–419. https://doi.org/10.1351/pac196408030409.
157. Sims, L. B.; Lewis, D. E. Bond Order Methods for Calculating Isotope Effects in Organic Reactions. In *Isotopes in Organic Chemistry;* Buncel, E., Lee, C. C., Eds.; Elsevier: Amsterdam, 1984; pp 161–259.

158. Burton, G. W.; Sims, L. B.; Wilson, J. C.; et al. Calculation of Carbon-14, Chlorine-37, and Deuterium Kinetic Isotope Effects in the Solvolysis of *tert*-Butyl Chloride. *J. Am. Chem. Soc.* **1977,** *99* (10), 3371–3379. https://doi.org/10.1021/ja00452a031.
159. Smith, P. J.; Bourns, A. N. Isotope Effect Studies on Elimination Reactions. The Nature of the Transition State for the E2 Reaction of 2-Arylethylammonium Ions with Ethoxide in Ethanol. *Can. J. Chem.* **1974,** *52,* 749–760. https://doi.org/10.1139/v74-118.
160. Lewis, D. E.; Sims, L. B.; Yamataka, H.; et al. Calculations of Kinetic Isotope Effects in the Hofmann Eliminations of Substituted (2-Phenylethyl)trimethylammonium Ions. *J. Am. Chem. Soc.* **1980,** *102,* 7411–7419. https://doi.org/10.1021/ja00545a001.

CHAPTER EIGHT

Development of highly regioselective addition reactions of alkenes and alkynes

Its synthetic versatility has made electrophilic addition to alkenes an extremely useful synthetic method,[1] and considerable effort has been expended in developing highly regio- and stereoselective versions of the reaction. However, one of the major problems with the addition of Lowry—Brønsted acids to alkenes is that the intermediate cation is susceptible to rearrangement.[2] This was clearly demonstrated by Whitmore in the reaction of *tert*-butylethylene (**242**) with hydrogen chloride in the absence of a solvent.[3] His results are compared with those of Fahey,[4] who used acetic acid as solvent, in Table 8.1.

Table 8.1 The addition of hydrogen chloride to *tert*-butylethylene.

Whitmore (1950)

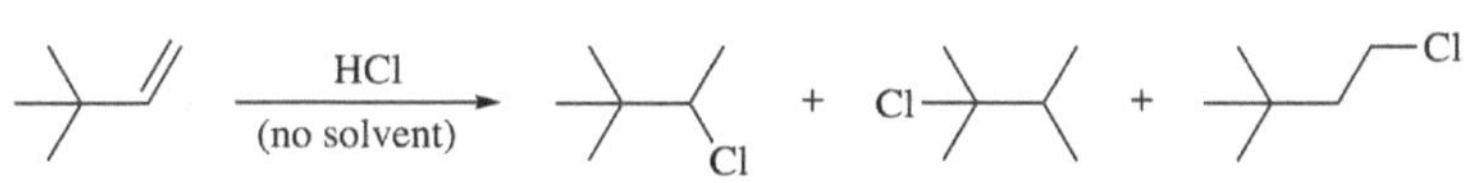

	243	244	245
[HCl]/[alkene] = 0.23	42%	53%	2%
[HCl]/[alkene] = 0.37	26%	67%	4%
[HCl]/[alkene] = 0.87	38%	61%	0.3%

Fahey (1969)

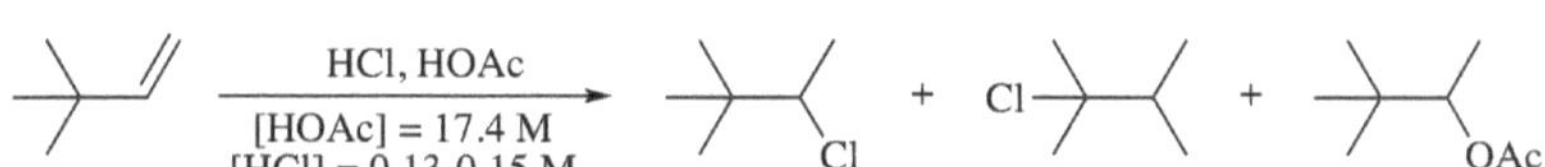

	243	244	246
25 °C	37%	44%	19%
50 °C	33%	48%	17%
125 °C	39%	47%	14%

The results of the Whitmore experiments, and the results of his study of the reaction of 3-methyl-1-butene,[5] show that in the absence of a competing nucleophile, the ratio of rearranged (**244**) to unrearranged (**243**) chloride varied from 1.25:1 to 2.6:1. The amount of the anti-Markovnikov product (**245**) in these studies was uniformly small and never

Addition, Elimination and Substitution: Markovnikov, Hofmann, Zaitsev and Walden
ISBN: 978-0-12-821027-7
https://doi.org/10.1016/B978-0-12-821027-7.00009-8

more than 4%. For the reaction in acetic acid, the solvent introduces a competing nucleophile. Three products were formed in this reaction: the rearranged chloride (**244**), the unrearranged product (**243**), and the product (**246**) formed by trapping of the intermediate cation by the solvent. It is noteworthy that the ratio **244**:**253** remained close to 1.4 at every temperature that the reaction was carried out. The outcome of the reaction in acetic acid is also shown in Table 8.1.

Factors affecting reaction regiochemistry: the alkene

Mechanisms of carbocation stabilization

Ultimately, the regiochemistry of the addition reaction depends on the substituents on the alkene and the nature of the electrophile. Reactions with Markovnikov regioselectivity are most strongly favored by alkenes where the substituent strongly stabilizes the carbocation, and by strong electrophiles, which add rapidly to most alkene π bonds. Historically, three modes of carbocation stabilization have been considered. In approximate descending order of general effectiveness, they are hyperconjugation, conjugation with carbon-carbon π bonds or aromatic rings, and conjugation with adjacent heteroatoms carrying lone pairs. These mechanisms of stabilization are summarized in Fig. 8.1.

Thus, the alkene **247**, which carries a substituent that can only stabilize the adjacent cationic center by hyperconjugation, gives the cation **248**, where the major canonical form has no π bond. The diene **249** holds the possibility of protonation at the terminus of the diene or in the middle; protonation at the less substituted terminal atom gives the allylic cation, **250**,

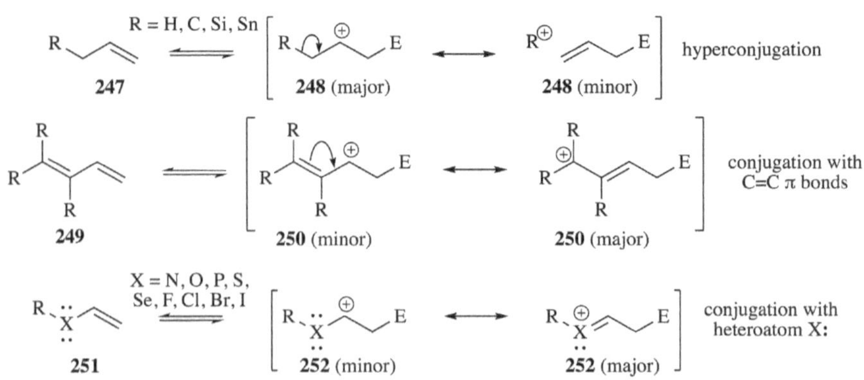

Figure 8.1 Mechanisms of stabilization of a carbocation.

that is most strongly stabilized by resonance. The enamine **251** reacts to give the cation **252**, where the ion is much better described as an iminium ion.

Using the Hammett σ^+ parameter as a predictor of regiochemistry

One useful predictor of the ability of the substituent to stabilize the cationic center is the σ^+ constant of the substituent, defined on the basis of the rates of the solvolysis of substituted cumyl chlorides.[6–10] The alkenes that exhibit the greatest regioselectivity in electrophilic addition reactions are those that can give the most strongly stabilized carbocation intermediates. The more negative the value of σ^+, the more efficiently the substituent can stabilize a carbocation—these alkenes therefore give the most regioselective additions. They are also strongly activating and *o,p*-directing substituents in electrophilic aromatic substitution. Representative σ^+ values for *o,p*-directing, activating groups are gathered in Table 8.2.

Table 8.2 Substituent σ^+ constants for groups on an aromatic ring.

R	σ^+	R	σ^+	R	σ^+	R	σ^+
Alkyl Groups		Unsaturated Groups		3rd-Row and Heavier Pnictogens and Chalcogens		2nd-Row Groups	
Me	−0.31	$H_2C{=}CH$	−0.16	SMe	−0.60	O^-	−2.30
Et	−0.30	allyl	−0.22	SPh	−0.55	NMe_2	−1.70
n-Pr	−0.29	Ph	−0.18	SePh	−0.47	NH_2	−1.30
i-Pr	−0.28	HC≡C	+0.18	SH	−0.03	NHPh	−1.40
t-Bu	−0.26			PH_2	+0.06	OH	−0.92
c-C_3H_5	−0.41	Si- and Sn-				OEt	−0.81
c-C_4H_7	−0.29	CH_2SnMe_3	−0.92	Halogens		OMe	−0.78
c-C_5H_9	−0.30	CH_2SiMe_3	−0.62	Cl	+0.11	NHAc	−0.60
c-C_6H_{11}	−0.29	$SnMe_3$	−0.12	I	+0.14	OPh	−0.50
		$SiMe_3$	+0.02	Br	+0.15	OAc	−0.19
						F	−0.17

One particularly useful group of this type is the trimethylsilyloxy group, which is accessible by silylation of a lithium enolate. The values of the Hammett σ_p and the Taft σ_R constants for the trimethylsilyloxy group are practically identical to those of the methoxy group: $OSiMe_3$: σ_p −0.27 (σ_R−0.58); OMe: σ_p−0.27 (σ_R−0.56).[10] The trimethylsilyloxy group leads to strongly Markovnikov orientation in the addition to the trimethylsilyl enol ether. Representative alkylations of trimethylsilyl enol ethers using a *tert*-alkyl halide and a strong Lewis acid are shown in Fig. 8.2.[11–14]

The ketone silyl enol ethers **253**, **255**, **258**, and **261** all react regiospecifically to give the ketones **254**, **227**, **260**, and **262**, respectively. The *tert*-alkyl halides can range from acyclic *tert*-alkyl halides to cyclic halides such as 1-methylcyclohexyl chloride (**256**) and the 1-adamantyl bromide derivative **259**. The enol trimethylsilyl ethers of esters (the trimethylsilyl ketene acetals, **263**), also react with tertiary halides and a Lewis acid to provide a route to very hindered esters (**264**) (Scheme 8.1).

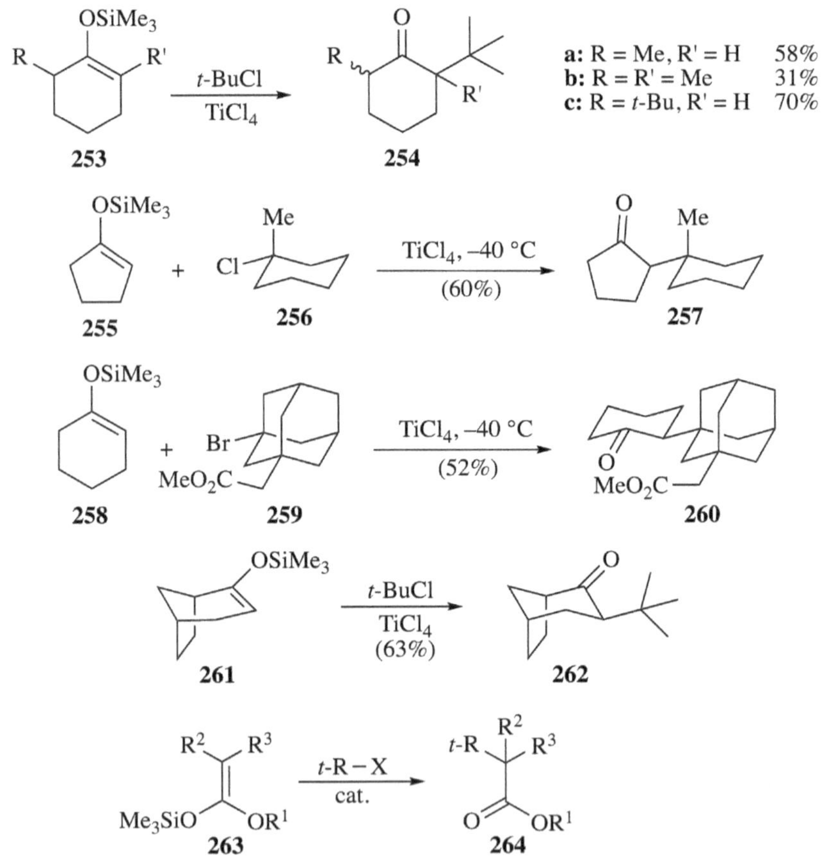

a: $R^1 = R^2 = Et$, $R^3 = i\text{-}Pr$; *t*-RX = *t*-BuCl, cat. = $ZnCl_2$ 12%
b: $R^1 = Me$, $R^2 = Et$, $R^3 = i\text{-}Pr$; *t*-RX = *t*-BuCl, cat. = $ZnCl_2$ 32%
c: $R^1 = Me$, $R^2 = H$, $R^3 = i\text{-}Pr$; *t*-RX = *t*-BuCl, cat. = $ZnCl_2$ 38%
d: $R^1 = Me$, $R^2 = H$, $R^3 = t\text{-}Bu$; *t*-RX = 1-AdBr, cat. = $ZnCl_2$ 52%
e: $R^1 = Me$, $R^2 = H$, $R^3 = t\text{-}Bu$; *t*-RX = Et_3CCl, cat. = $ZnCl_2$ 54%

Figure 8.2 Addition of *tert*-alkyl carbocations to enol trimethylsilyl ethers occurs with Markovnikov regiochemistry.

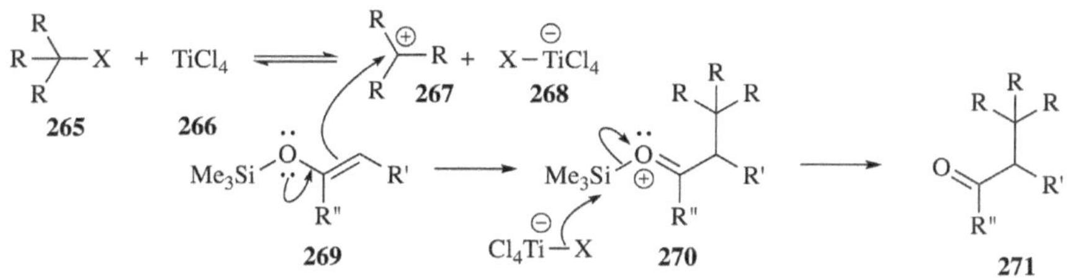

Scheme 8.1 The alkylation of an enol trimethylsilyl ether with a tertiary carbocation.

Conversely, substituents with large positive values of σ^+ (Table 8.3) tend to destabilize a positive charge on the adjacent atom; they are deactivating *m*-directors.

Table 8.3 Electron-withdrawing substituents.

R	σ^+	R	σ^+
NMe_3^+	+0.41	CN	+0.66
CO_2H	+0.42	CHO	+0.73
CO_2R	+0.48	COCl	+0.79
CF_3	+0.61	$OCOCF_3$	+0.85
NO_2	+0.65	SO_2NMe_2	+0.86

The effect of an electron-withdrawing substituent on the regiochemistry of addition is illustrated by the addition of an electrophile, E^+, to 1-nitropropene (**272**) (Fig. 8.3).[15] The addition of the electrophile to the α-carbon produces the carbocation **273**, where the positive charge is on the β-carbon. On the other hand, the addition of the electrophile to the β-carbon produces the carbocation **274**, where the positive charge is on the α-carbon—directly next to the positively charged nitrogen. The activation energy of this reaction is higher.

Silicon and tin as directing groups in electrophilic addition reactions of alkenes

Simple hyperconjugation is seldom the most efficient method for stabilization of a carbocation; nevertheless, it is responsible for the order of stability of simple carbocations when the atom β to the cationic carbon is carbon or

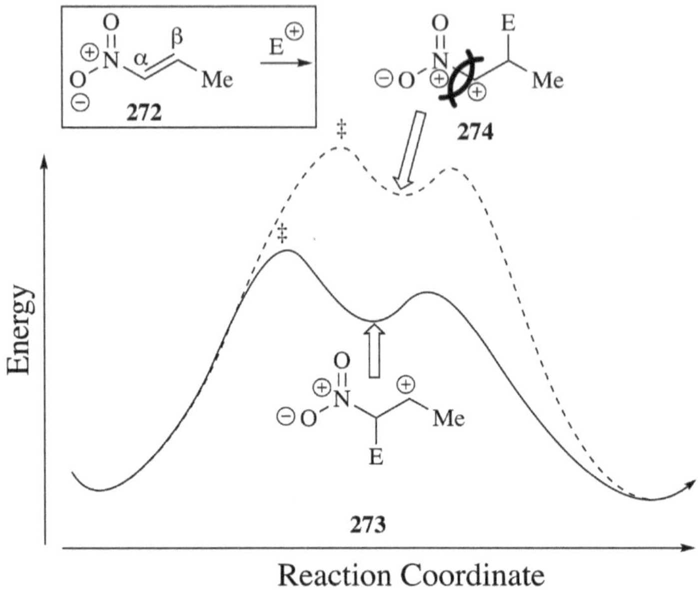

Figure 8.3 The reaction coordinate diagram for regioisomeric additions of E^+, an electrophile, to 1-nitropropene, which has an electron-withdrawing substituent on the double bond.

hydrogen: 3 ° > 2 ° > 1 °. However, if this atom is significantly less electronegative than carbon (E.N. 2.55), hyperconjugation can become a highly effective mechanism for stabilizing a carbocation. Two such elements are silicon (E.N. 1.90) and tin (E.N. 1.96) due to the accumulation of the partial negative charge on carbon.

The percent ionic character of C—Si and C—Sn bonds is predicted to be in the range of 8%—10%, which makes C—Si and C—Sn bonds excellent stabilizers of a positive charge at the β-position by hyperconjugation. This, in turn, makes them excellent directing groups for electrophilic additions to vinyl and allyl groups. In these reactions, the electrophile adds to the π bond to give a β-silyl or β-stannyl carbocation; this stabilization of a positive charge by these Group IV elements at the β-position has become known as the "β-effect."[16,17]

The Hosomi—Sakurai reaction

The use of silicon as a directing group became widespread in the 1990s and has continued into the 21st century.[18—23] The first exploitation of the β-effect in synthesis was the 1976 report by Akira Hosomi (1943–2018) and Hideki Sakurai (b. 1931) of the displacement of silicon from an allylsilane by an electrophile, with concomitant allylic rearrangement.[24] This was followed in the subsequent 12 years by several more papers,[25–35] exploring, among other things, the diversity electrophiles that could be used.

The initial step of both reactions (Scheme 8.2) is the addition of the electrophile (E^+) to the C–C π bond to give the β-silyl carbocation. In the cation (**275**) derived from the allyl silane **274**, the nucleophile attacks at silicon to give the desilylated product (**276**), with allylic migration of the double bond. In the cation **278** derived from the vinylsilane **277**, the attack of the nucleophile on silicon returns the alkene **279** with retention of configuration. When the electrophile is a halogen, the three-membered halonium ion (**280**) is formed. Desilylation of this onium ion gives the corresponding halide (**281**) with inversion of configuration.

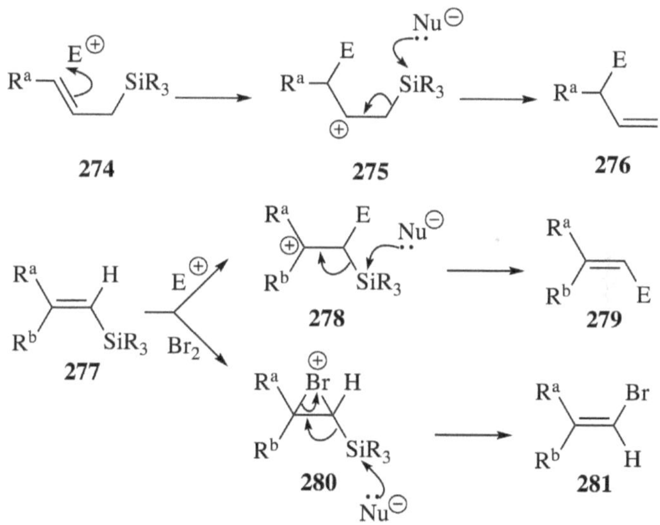

Scheme 8.2 The mechanisms of the reactions of allyl- and vinyl-trimethylsilanes with electrophiles.

The work of Kumada and his coworkers[36] showed that allylsilanes reacted with tertiary carbocations to give the products of an S_E2' reaction, with stereospecific antarafacial displacement of the silicon (Scheme 8.3).

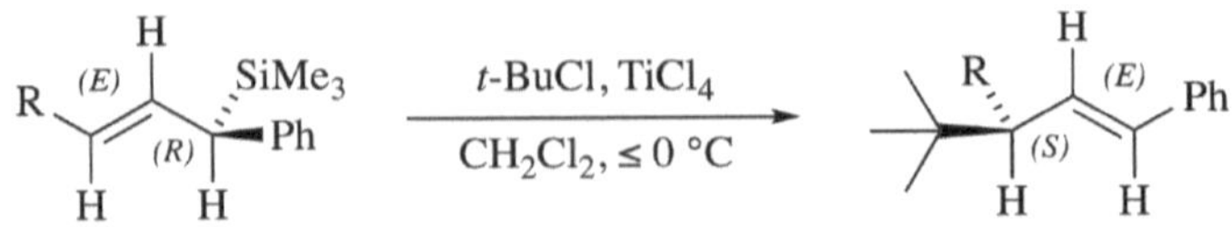

282a: R = Me
282b: R = Ph

283a: R = Me (61%)
283b: R = Ph (40%)

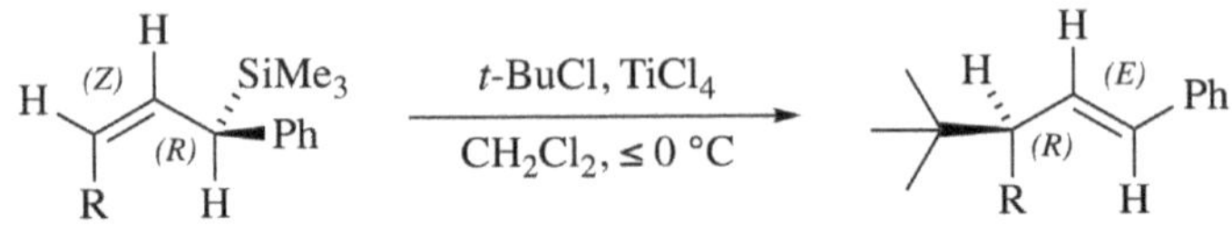

284a: R = Me
284b: R = Ph

285a: R = Me (75%)
285b: R = Ph (25%)

Scheme 8.3 The stereochemical outcome of S_E2' reactions of allyltrimethylsilanes with *tert*-butyl cation.

By far the most useful reactions of allyl- and vinyl-silanes involve the generation of oxygen-stabilized carbocations as intermediates. Carbocations stabilized by an adjacent heteroatom have been used as electrophiles since the early years of the 20th century. The efficiency of stabilization of a carbocation has been computed for several substituents at the B3LYP/6-31G* level by calculating the enthalpy of hydride transfer to a methyl cation in the gas phase (Fig. 8.4).[37]

These types of cations are most readily available by the reaction between a suitable oxygenated precursor and a strong Lewis acid (Scheme 8.4). The treatment of a simple aldehyde, ketone, ester, or amide (**288**) with a metal halide gives the corresponding Lewis acid complex **289**, which is strongly stabilized by resonance. Acid chlorides (**290**) form adduct ions (**291**), which lose the halogen to give acylium ions (**292**). Acetals and ketals (**293**) react by loss of an alkyloxy group as the metal alkoxide to give the resonance-stabilized cation **294**.

286a: X = OCHO
286b: X = NHCHO
286c: X = SMe
286d: X = OMe
286e: X = NHMe

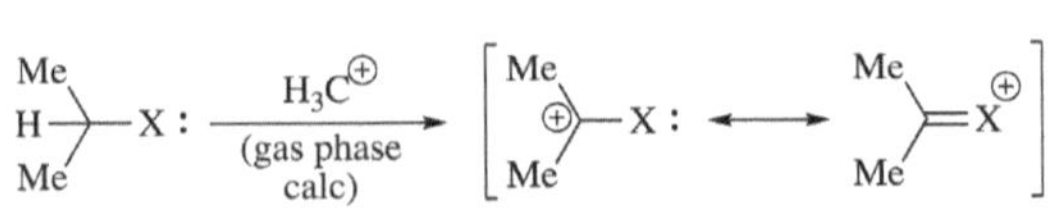

287a: X = OCHO
287b: X = NHCHO
287c: X = SMe
287d: X = OMe
287e: X = NHMe

X = OCHO, $\Delta H_{calc} = -236.2$ kcal mol^{-1}
X = NHCHO, $\Delta H_{calc} = -257.3$ kcal mol^{-1}
X = SMe, $\Delta H_{calc} = -258.7$ kcal mol^{-1}
X = OMe, $\Delta H_{calc} = -262.8$ kcal mol^{-1}
X = NHMe, $\Delta H_{calc} = -286.6$ kcal mol^{-1}

Figure 8.4 Computed enthalpies for hydride transfer to methyl cation.

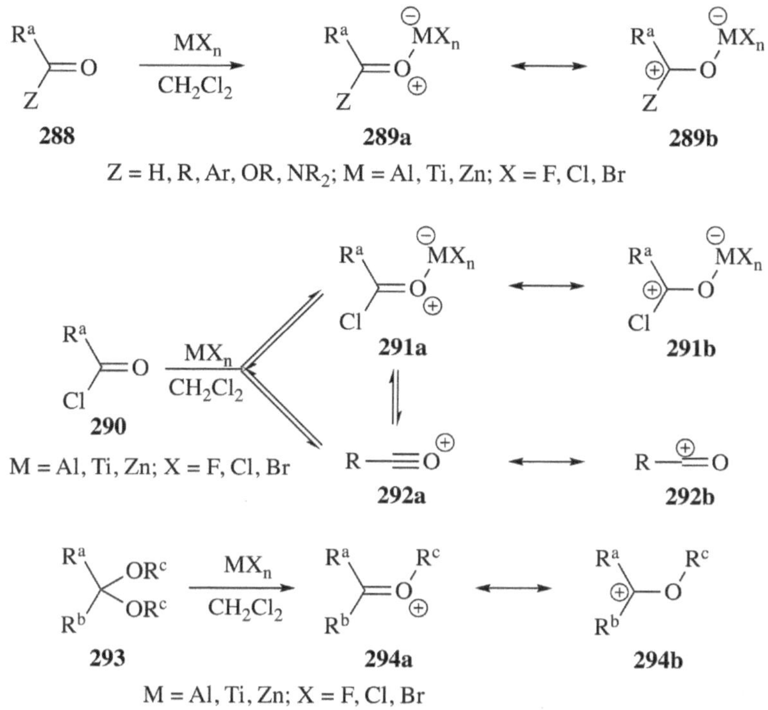

Scheme 8.4 The formation of oxygen-stabilized carbocations by means of Lewis acid complex formation.

Scheme 8.5 shows some representative Hosomi–Sakurai reactions: The first example shows the intramolecular reaction of 3-alkyl-3-(*Z*-4-trimethylsilyl-1-butenyl)ethanal (**295**) with a Lowry–Brønsted acid to give a cyclized product, the tetrahydropyran **296**, where the substituents at positions 2 and 5 of the tetrahydrpyran are *trans*-diequatorial, and the substituents at positions 4 and 5 are *cis* to each other, reflecting the stereochemistry of the double bond in the silane.[38] The effect of conjugation of the carbonyl precursor to the active electrophiles is illustrated in the second example. In such systems, the new σ bond is formed at the β-position of the original enone: the enones **297** give the allylated ketones **298**.[39] In this reaction, the active Lewis acid catalyst is formed by the reaction of indium metal with trimethylsilyl chloride. The acetals **299a** and **299b** and the **299c** in the third example react with allyltrimethylsilane in the presence of *o*-benzenedisulfonimide (**301**) as a Lowry–Brønsted acid catalyst to give the ethers **302**.[40]

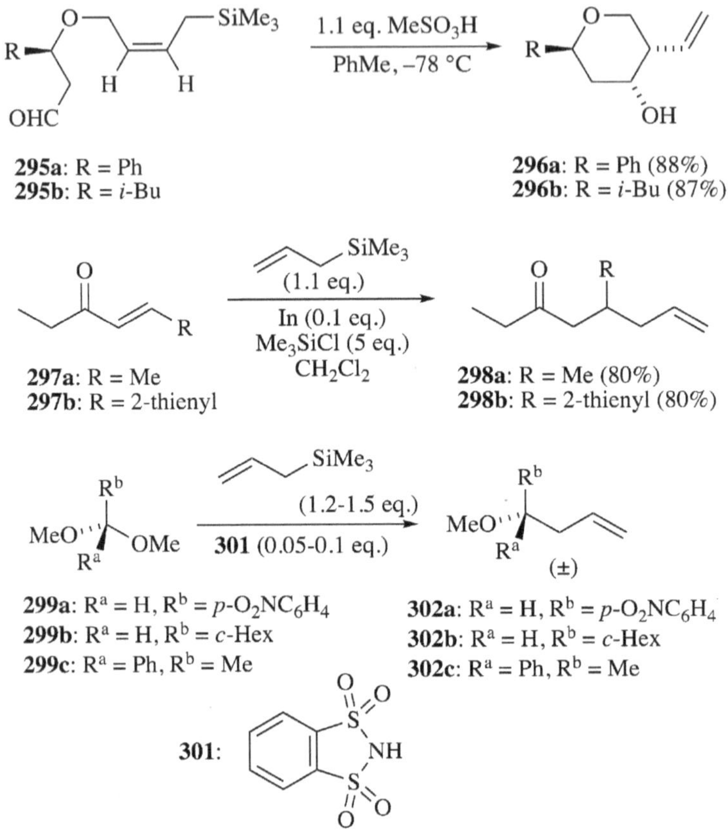

Scheme 8.5 Representative Hosomi–Sakurai reactions.

Heteroatom-stabilized electrophilic carbocations: The Prins, Mannich, and Mukaiyama reactions

The addition of heteroatom-stabilized carbocations (**287**) to alkenes that carry a substituent with a lone pair on a heteroatom bonded to the double bond (**303**) is extremely regioselective and generates a new stabilized carbocation (**304**) with Markovnikov regiochemistry. The products of such reactions often react with water during isolation, regenerating a carbonyl group in the product (**306**) (Scheme 8.6). Three widely used examples of this regiospecific addition are: (1) the Prins reaction; (2) the Mannich reaction; and (3) the Mukaiyama aldol reaction.

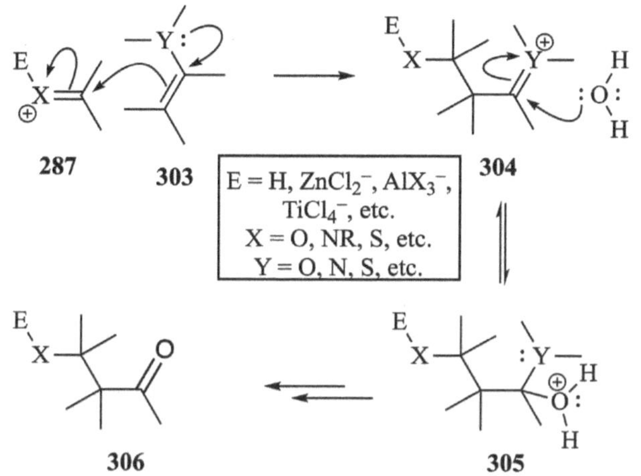

Scheme 8.6 The addition of a carbocation stabilized by a heteroatom to an alkene with a substituent carrying a lone pair is highly regioselective and gives the Markovnikov product.

The Prins reaction

In 1917, Dutch chemist Henrik Jacobus Prins (1889—1958) published the first report of the acid-catalyzed intramolecular reaction of citronellal and its isomers.[41,42] Prins (Fig. 8.5[43]) was educated at the University of Delft, in Holland, where he graduated in engineering in 1911 and where took his Dr. Ir. in 1912 for a dissertation on the addition reactions of alkenes and electron-deficient compounds in the presence of Lewis acids. Reactions of this type became a life-long pursuit of Prins, who joined the N. V. Polak & Schwartz Essencefabrieken, in Hilversum immediately after his graduation, working in an advisory capacity in animal rendering. In 1925, he left Polak & Schwartz to join Nederlandse Thermochemische Fabrieken, being appointed director in 1932. As director of the company, Prins helped improve the quality of the pensions for the employees. In 1951, he was honored by being made an Officer of the Order of Orange—Nassau. Prins reported the reaction between formaldehyde and alkenes in the presence of aqueous sulfuric acid—the reaction that now bears his name[44—47] —in *Chemisch Weekblad*, a major Dutch chemical journal, in 1919, and followed up his original report with several papers over the next three decades. He continued his research into the reactions of cations and alkenes through World War II, although publication of his work, which was always in Dutch journals, was interrupted by the war.

Figure 8.5 H. J. Prins. *From Van Genderen, H. (2001). Laboratorium aan der huis. In S. Rozendaal, H. van Bekkum & J. Reedijk (Eds.), Chemie achter de dijken: uitvindingen en uitvinders in de eeuw na van't Hoff (pp. 10–11). KNAW.*

In his 1917 papers, Prins reported the acid-catalyzed intramolecular addition of a formyl group to an alkene in citronellal (**307**), which gave a cyclohexanol derivative (**308**) and its terminal alkene isomer (**309**), which gave an isomeric cycloheptanol derivative (**310**) (Scheme 8.7). In 1919,[48,49] he reported that substituted styrenes (**311**) reacted with formaldehyde to give 1,3-diols (**312**), 1,3-dioxanes (**313**), or both.

Prins later reported the intermolecular reaction of formaldehyde with β-pinene (**314**), camphene (**316**), limonene (**318**), and cedrene (**320**) (Scheme 8.8).[50–52] The centennial of the use of the Prins reaction in the fragrance industry was celebrated in 2019.[53]

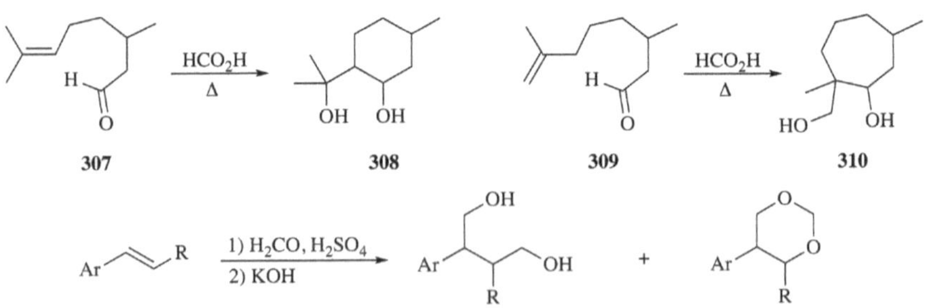

311a: Ar = Ph; R = H
311b: Ar = 4-MeOC_6H_4; R = Me
311c: Ar = 3,4-(CH_2O_2)C_6H_3; R = Me

312: Ar = Ph; R = H

313a: Ar = Ph, R = H
313b: Ar = 4-MeOC_6H_4; R = Me
313c: Ar = 3,4-(CH_2O_2)C_6H_3; R = Me

Scheme 8.7 The intramolecular Prins cyclization of isomers of citronellal.

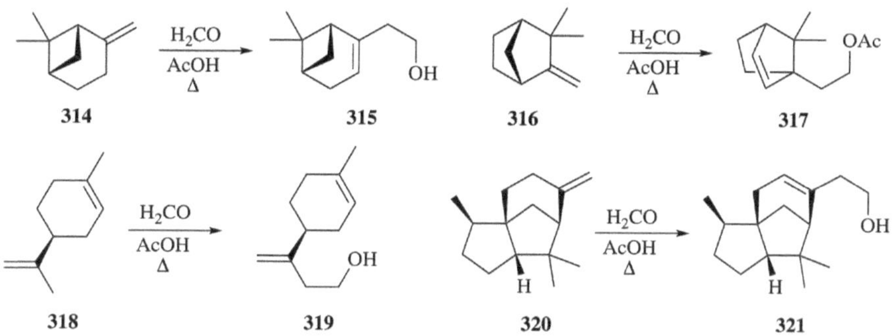

Scheme 8.8 The intramolecular Prins cyclization of isomers of citronellal.

The Prins reaction can be carried out with or without Lewis or Lowry–Brønsted acid catalysts; it generally does not involve a free carbocation intermediate. For example, the reaction of β-pinene (**314**) to give nopol (**315**) or its acetate (**322**)[54] proceeds without rearrangement (Scheme 8.9) with or without acid catalysis. However, the absence of carbocation intermediates in the reaction is not absolute, as the formation of the 10-hydroxymethyl-2-bornene (**317**) from camphene (**316**) shows.

The Mannich reaction

Five years before Prins first reported his eponymous reaction, Carl Ulrich Franz Mannich[55] (1877–1947) and his Ph.D. student, Walter Krösche (1882–1957),[56] reported the condensation of antipyrine (**323**) with ammonia and formaldehyde to give the methylene(*bis*-antipyrine) **324**.[57] In the same paper, they reported that an identical product could be obtained if the formaldehyde and ammonia were replaced by urotropine hydrochloride (**325**) (Scheme 8.10). The 1912 paper is taken verbatim from Krösche's

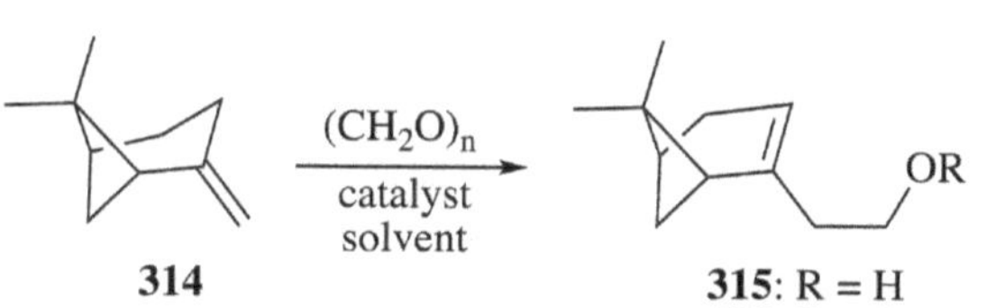

315: R = H
322: R = Ac

Reaction Conditions	Major Product: Yield
$(H_2CO)_n$, $ZnCl_2$, Δ	**315**: 57%
$(H_2CO)_n$, Δ	**315**: 96% ← KOH
$(H_2CO)_n$, AcOH, Δ	**322**: 38%

Scheme 8.9 The synthesis of nopol from β-pinene by the Prins reaction.

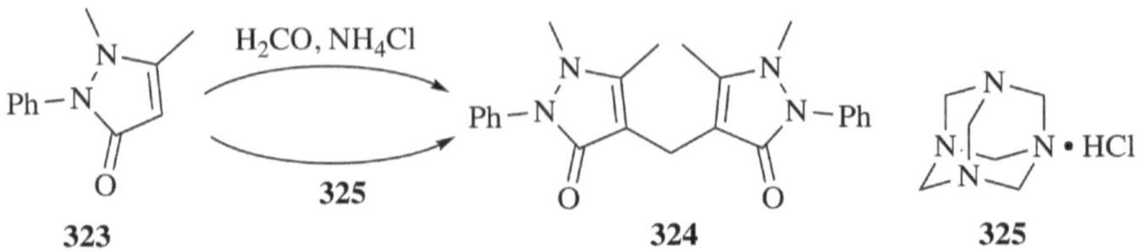

Scheme 8.10 Mannich's first report.

dissertation.[58] The reaction has since become an important synthetic method.[59–64]

Mannich was born in Breslau (now Wrocław in Poland) and began his education in Weimar and then Berlin. Before graduating from the Gymnasium, however, he left to take up an apprenticeship in pharmacy. From 1898 on, he studied in Marburg and Berlin. However, because he had not graduated from the Gymnasium, there was no possibility of him completing his doctorate in Berlin. In 1903, he received his doctoral degree from Faculty of Philosophy in Basel, Switzerland. His doctoral work, which resulted in three papers the same year,[65–67] was supervised by Hermann Thoms 1859–1931, the Docent in Pharmaceutical Chemistry at Berlin, whose assistant he became.

In 1905, he matriculated in Berlin and passed the State Examination for food chemists. In 1907, he received his *habilitation* and became Docent; he was promoted to Extraordinary Professor in 1910. In 1911, he moved to Göttingen as Extraordinary Professor of Pharmaceutical Chemistry, and he remained there until 1919, when he received the call as Professor of Pharmaceutical Chemistry to the newly founded University of Frankfurt. In 1927, he was called to the Chair of Pharmaceutical Chemistry at the University of Berlin. He held this post until his retirement in 1943. In 1947, was he persuaded to come out of retirement to take up the Chair of Pharmaceutical Chemistry at the Technische Hochschule Karlsruhe. From 1932 to 1934, he served a biennial term as President of the German Pharmaceutical Society (DPhG).

In 1917, Mannich published a systematic study of the reaction of formaldehyde and ammonium chloride and the hydrochlorides of primary and secondary amines[68] and the reaction of these compounds with ketones to give β-aminoketones or Mannich bases (Scheme 8.11). Mannich showed that dimethylamine and formaldehyde react with ketones (**326**) to give either β-dimethylaminoketones (**327**) or β,β′-bis-dimethylaminoketones (**328**). With alkyl methyl ketones (e.g., 2-butanone, **322b**), the reaction

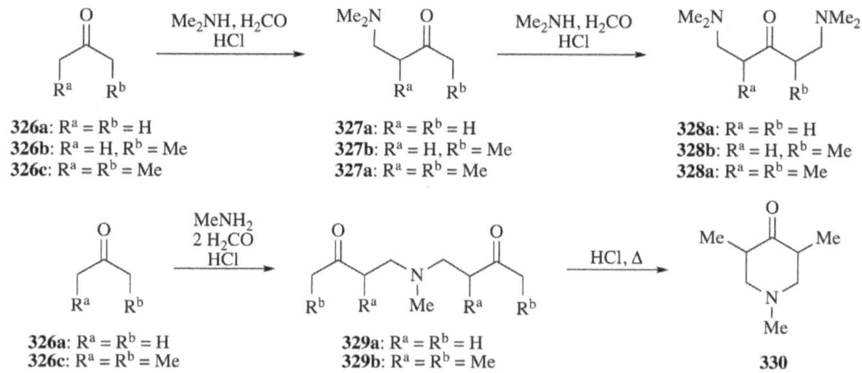

Scheme 8.11 Representative Mannich reactions with primary and secondary amines.

occurred preferentially at the methyl group. With primary amines, the tertiary amine formed by the reaction with two equivalents of both formaldehyde and the ketone (**329**) is the major product of the reaction. On heating with the ammonium chloride, this product cyclizes to the piperidin-4-one **330**.

In 2017, a special issue of *Archiv der Pharmazie* was published celebrating the centennial of Mannich's 1917 paper.[69] The currently accepted mechanism of the Mannich reaction involves the key addition of the iminium ion **332** to the enol **334** (Scheme 8.12).

The Mannich reaction has been used for decades in the synthesis of complex natural products, as illustrated by the examples in Scheme 8.13. In the Woodward synthesis of strychnine,[70] the heterocyclic ring of the indole **336** functions as an enol equivalent that then reacts with the iminium ion **332** (Scheme 8.12, $R^1 = R^2 =$ Me) to give the Mannich base **337**. In Evans'

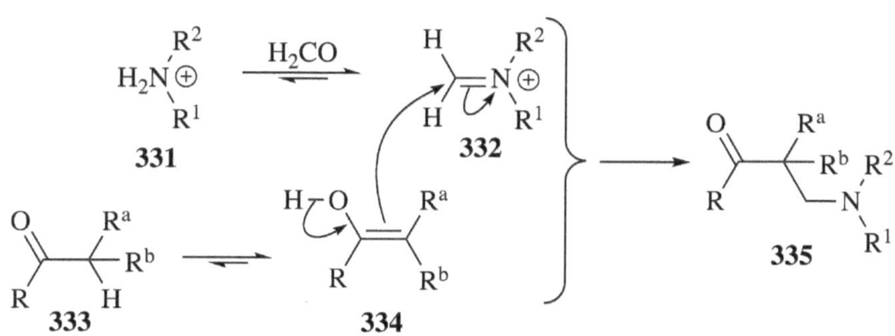

Scheme 8.12 The accepted mechanism of the Mannich reaction.

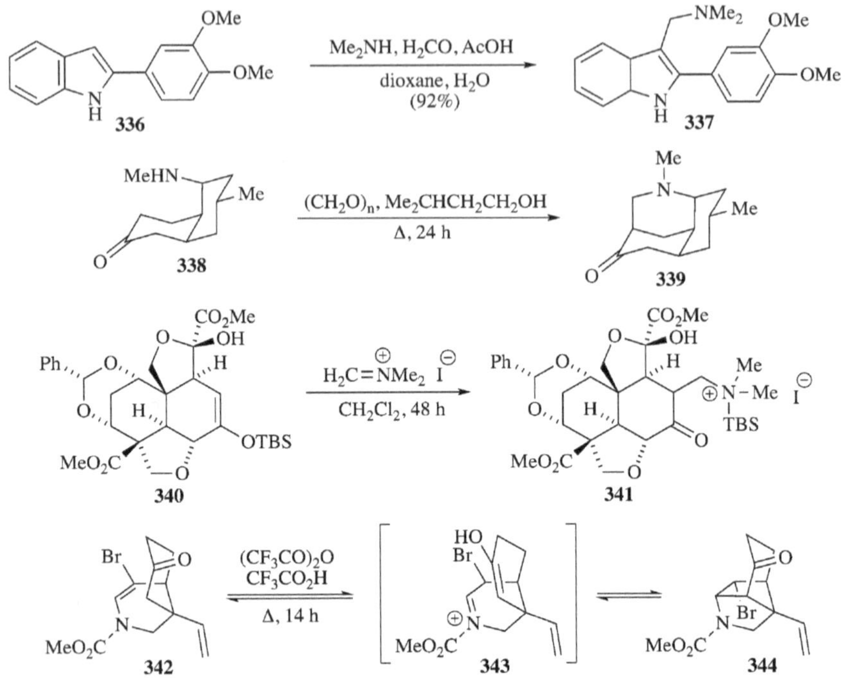

Scheme 8.13 Representative Mannich reactions in synthesis.

synthesis of luciduline (**339**),[71] the final ring is closed by means of the intramolecular Mannich reaction of the secondary amine **338**. Ley's synthesis of azadirachtin involved the Mannich reaction of the enol silyl ether **340** using a preformed iminium ion to give the Mannich base **341**.[72] The final example of the Mannich reaction comes from Overman's synthesis of gelsemine,[73] where the enamine **342** undergoes protonation to give the iminium ion with concomitant enolization of the ketone to give the cationic intermediate **343** that then undergoes cyclization to the tricyclic Mannich base **344**.

The Mukaiyama addition reactions

In Fig. 8.2, several representative additions of simple carbocations to enol trimethylsilyl ethers were summarized. In 1973, Teruaki Mukaiyama (1927–2018, Fig. 8.6) reported that the oxonium ion electrophile (**347**) formed by complexation of the oxygen atom of the carbonyl compound (**346**) by titanium tetrachloride[74] will react with an enol trimethylsilyl ether (**345**) to give the β-hydroxycarbonyl compound (**348**, an aldol) as the

Figure 8.6 Mukaiyama. *From https://commons.wikimedia.org/wiki/File:Mukaiyama,_Teruaki.jpg licensed under the Creative Commons Attribution-Share Alike 4.0 International licence.*

product (Scheme 8.14). A subsequent paper[75] described the scope and limitations of the reaction with respect to substrates, catalysts, and solvents.

Mukaiyama was born in the Nagano Prefecture of Japan and began his university education at the Tokyo Institute of Technology where he graduated with his bachelor's degree in 1948. In 1952, he joined the faculty of Gakushuin University as Assistant Professor of Chemistry. He continued research toward his graduate degree, and in 1956, he was awarded the

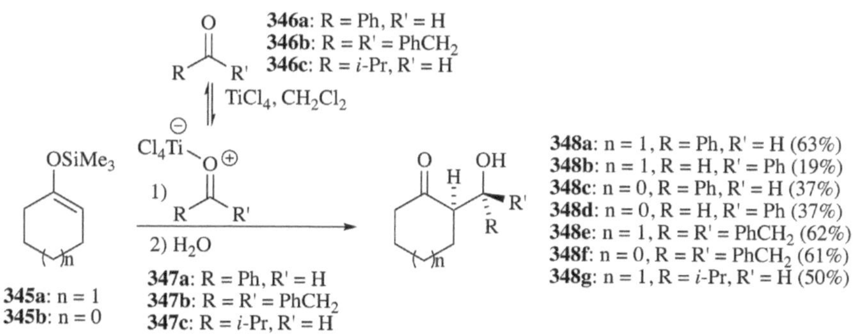

Scheme 8.14 Mukaiyama aldol reactions.

D.Sc. degree from the University of Tokyo. In 1962, he moved to Tokyo Institute of Technology as Professor of Chemistry, and in 1974, he moved to the University of Tokyo in the same capacity. Mukaiyama was one of the leaders of the new generation of Japanese organic chemists who have been responsible for the rise of Japanese organic chemistry as a creative force in the science. His work on the Lewis-acid-catalyzed reactions of covalent enolates of silicon and boron led to the development of several important synthetic methods, including the Mukaiyama reaction, a Lewis-acid catalyzed variant of the aldol addition reaction, and several reactions used for asymmetric synthesis.

As with the Mannich reaction, the final Markovnikov adduct is a carbonyl compound. Unlike the traditional aldol addition reaction, which gives poor yields of the adduct, the Mukaiyama aldol addition gives moderate to good yields of the aldol. In this, it is very much like the Reformatsky reaction,[76–79] which provides an entry into very hindered β-hydroxyesters.

An earlier related addition of a carbocation to an enol acetate was provided in Corey's synthesis of (±)-cedrol.[80] The treatment of the enol acetate **349** with boron trifluoride etherate gave the tertiary carbocation **350**, which underwent intramolecular Markovnikov addition to the electron-rich π bond of the enol acetate (Scheme 8.15). The cleavage of the acetyl group from the oxonium ion **351** then provided (±)-norcedranone (**352**).

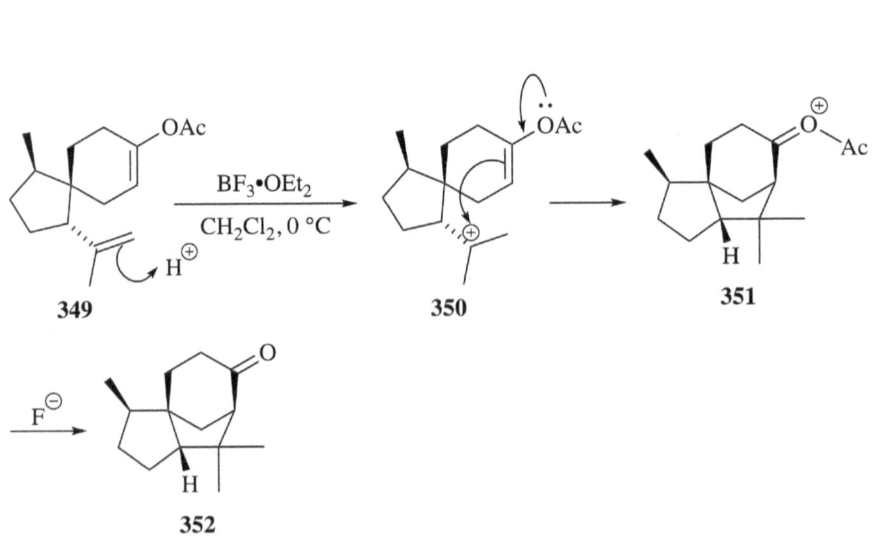

Scheme 8.15 Norcedranone synthesis.

Carbocation rearrangements in synthesis

Rearrangements in addition reactions are not always an unwelcome by-product of Markovnikov addition. One particularly informative example that shows how incorporating a strained ring into a carbocation increases its propensity to rearrange is provided by Pirrung's synthesis of isocomene[81,82] (**356**). Protonation of the alkene **353** with Markovnikov regiochemistry gives the tertiary cyclobutylcarbinyl carbocation **354**. This cation also has a *trans*-fused bicyclo[4.2.0]octane ring system, which is more energetic than the *cis*-fused isomer; the *trans* isomer rearranges rapidly to the isomeric tertiary carbocation **355**, with the much less strained triquinane ring system. The loss of a proton from **355** occurs with Zaitsev regiochemistry to give isocomene (Scheme 8.16).

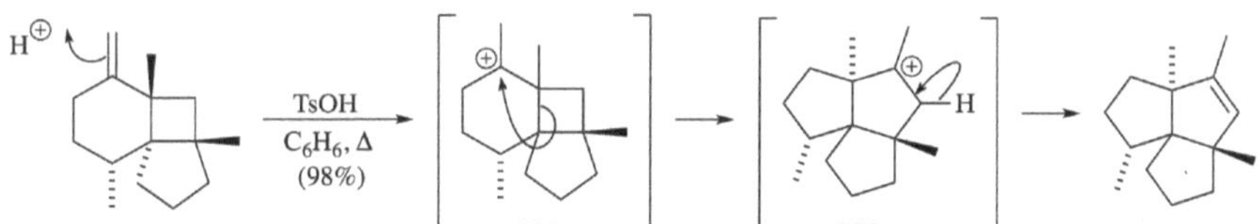

Scheme 8.16 Markovnikov protonation of an alkene as the key step of a synthesis of isocomene.

The pentaspiro alcohol **357** provides a very interesting application of carbocation rearrangements in synthesis. The protonation of this alcohol or the related alkene **358** by *p*-toluenesulfonic acid in solution in an alcohol initiates a cascade of cyclopropylcarbinyl-to-cyclobutyl cation rearrangements (Dem'yanov [Демьянов] rearrangements[83–85]), **359** ⇢ **360** ⇢ **361** ⇢ **362** ⇢ **363** ⇢ **364** ⇢ **365** (Scheme 8.17) The trapping of the final carbocation, **365**, by the solvent leads to the ether **366**.[86]

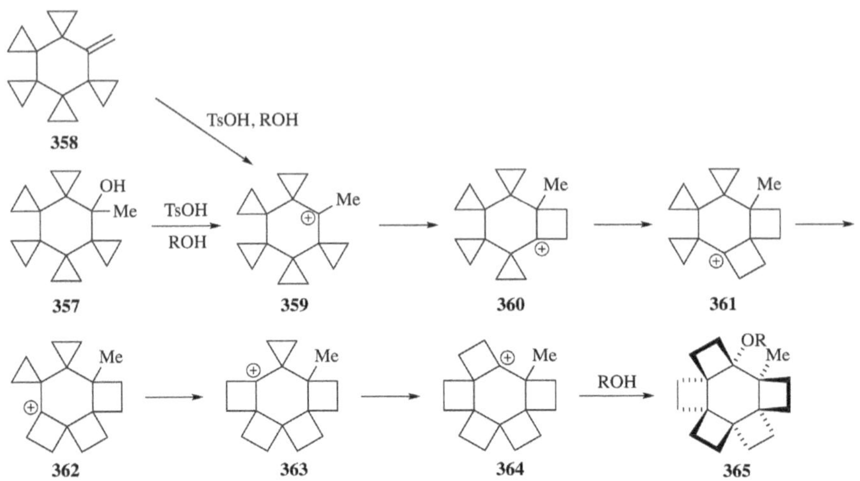

Scheme 8.17 Cascade rearrangements initiated by formation of the polyspiro carbocation **359**.

Anti-Markovnikov additions to alkenes

Free radicals differ from carbocations in one fundamental way: absent certain special structural features, free radicals do not rearrange. As discussed in Chapter 3, Kharasch showed that the additions of HBr to alkenes under the influence of peroxides proceed by a radical mechanism in which the bromine atom adds first. A 2008 report[87] details a method for the anti-Markovnikov addition of a dialkylphosphine oxide or H-phosphinate (**367**) to an alkene (**366**) in the presence of air to give the anti-Markovnikov adduct of the P—H bond across the alkene (**368**) (Scheme 8.18).

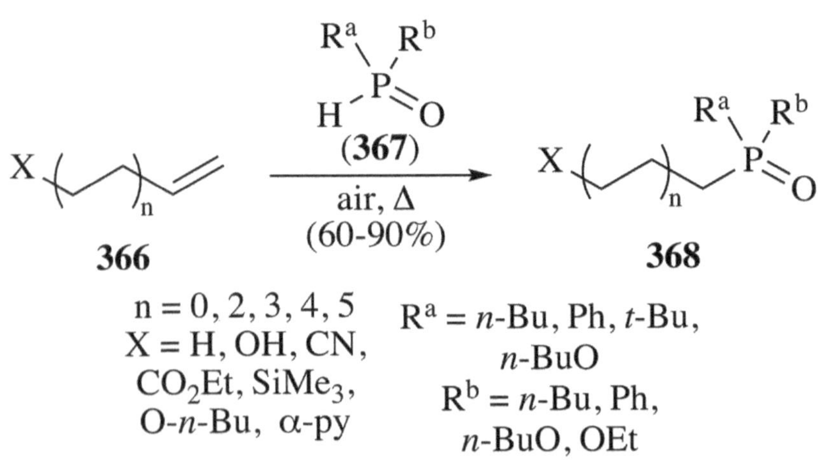

Scheme 8.18 The radical addition of the P—H bond of phosphines and phosphinates across an alkene π bond in the presence of air to give an anti-Markovnikov adduct.

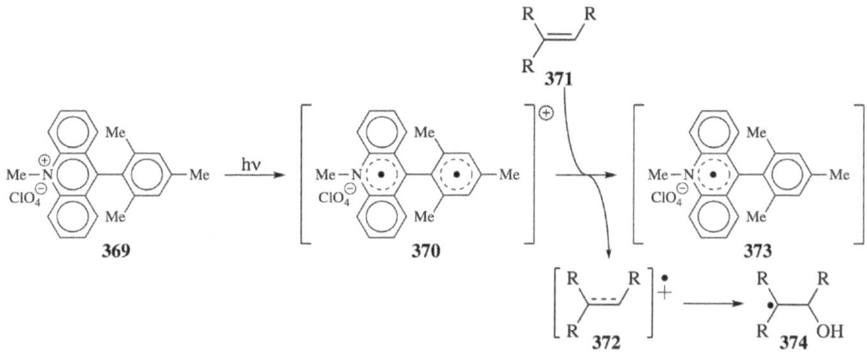

Scheme 8.19 Photocatalytic anti-Markovnikov hydration of an alkene.

Recently, a photocatalytic anti-Markovnikov hydration has also been reported (Scheme 8.19), with the acridinium salt **369** as the photocatalyst.[88]

In this reaction, the photocatalyst **369** undergoes excitation to an excited state **370**, followed by photoelectron transfer from the alkene (**371**) to give the radical cation **372** and the reduced photocatalyst (**373**). The radical cation **372** is then intercepted by water to give the tertiary radical **374** that goes on to give the anti-Markovnikov alcohol. Results from the hydration of the substituted styrenes **375** and **377** are gathered in Table 8.4.

Hydroboration of alkenes

Borane, BH_3, was the first of this type of reagent to achieve widespread use; its use in ethereal solution was developed by H. C. Brown.[89] The rise of hydroboration – oxidation, a formal anti-Markovnikov hydration, as an important synthetic reaction was aided by the fact that it also occurs with

Table 8.4 Photocatalytic anti-Markovnikov hydration of styrene derivatives.

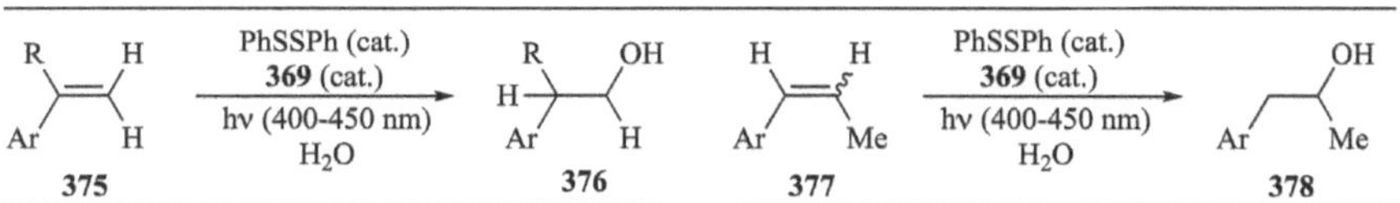

Substituents	Yield
a: Ar = R = Ph	86%
b: Ar = *p*-MeC_6H_4; R = Ph	62%
c: Ar = *p*-ClC_6H_4; R = Ph	50%
d: Ar = *p*-BrC_6H_4; R = Ph	74%
e: Ar = *p*-$F_3CC_6H_4$; R = Ph	75%
f: Ar = R = *p*-MeC_6H_4	43%
g: Ar = R = *p*-FC_6H_4	85%
h: Ar = R = *p*-ClC_6H_4	66%
i: Ar = *p*-$F_3CC_6H_4$; R = *p*-MeC_6H_4	62%

Substituents	Yield
a: Ar = Ph	87%
b: Ar = *m*-MeC_6H_4	43%
c: Ar = *p*-FC_6H_4	83%
d: Ar = *p*-ClC_6H_4	83%
e: Ar = *o*-BrC_6H_4	60%
f: Ar = *m*-BrC_6H_4	80%
g: Ar = *p*-BrC_6H_4	93%
h: Ar = *p*-IC_6H_4	76%
i: Ar = *p*-$F_3CC_6H_4$	80%
j: Ar = *p*-$MeO_2CC_6H_4$	84%

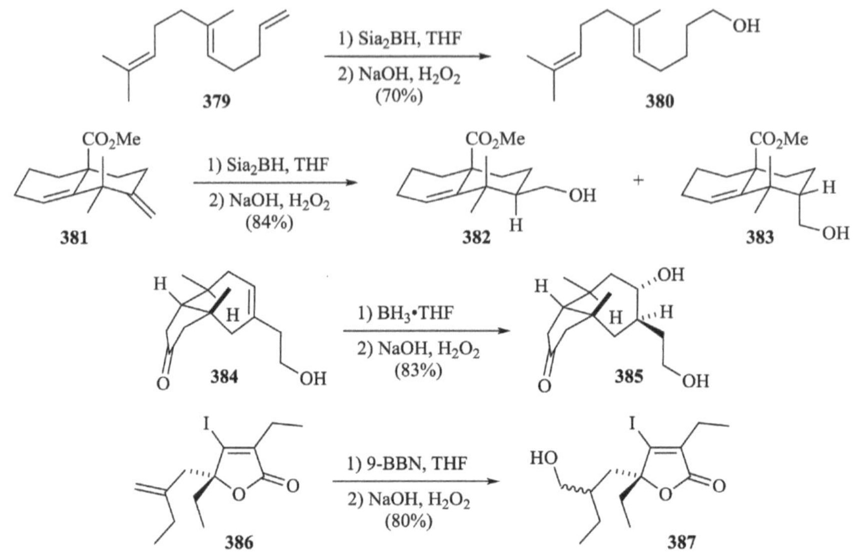

Scheme 8.20 Representative hydroboration-oxidation reactions.

stereospecific *syn* stereochemistry. Representative examples from the past half-century are shown in Scheme 8.20.

Grieco's synthesis of *dl*-sirenin[90] illustrates the strong steric effects on the hydroboration oxidation: the least-substituted double bond of the triene **379** reacted with the sterically hindered disiamylborane to give the anti-Markovnikov, primary alcohol **380** after oxidation of the initial borane product. The hydroboration-oxidation of the disubstituted double bond of the diene **381** gives approximately equal quantities of the two anti-Markovnikov alcohols **382** and **383** in Piers' synthesis of (±)-isolongifolene,[91] Majetich's 1993 synthesis of (±)-graveolide[92] had a key hydroboration oxidation; the conformation of the alkene **384** led to borane attacking the bottom face of the π bond to give the anti-Markovnikov alcohol **385**. The 2017 synthesis of gracilioether F by Wong and coworkers[93] again demonstrates the strong preference for attack of boron at the less substituted end of the least substituted, most electron-rich alkene π bond in the molecule. In this case, the methylene group of the side chain of the lactone **386** produces the primary alcohol **387** as a mixture of epimers.

When coupled with modern cross-coupling reactions, the formally anti-Markovnikov regiochemistry of hydroboration allows the regio-selective synthesis of boranes that can then be used in metal-catalyzed cross-coupling reactions such as the Suzuki–Miyaura coupling.[94] Their lack of air sensitivity has now made the boronic acids the preferred organoboron derivative for use in these coupling reactions.

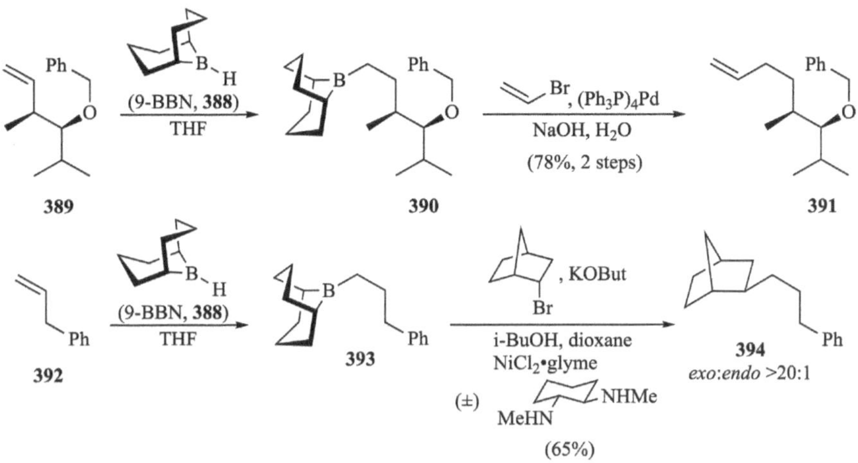

Scheme 8.21 Palladium-catalyzed coupling of a 9-alkyl-BBN with a vinyl halide.

In Rychnovsky's synthesis of roflamycoin, the hydroboration of the terminal alkene **389** gave the primary 9-alkyl-BBN **390**, which was then coupled with vinyl bromide in the presence of a Pd (0) catalyst to give the homologated alkene **391** (Scheme 8.21).[95] More recently, research has moved in the direction of using base metals (Fe, Co, Ni, etc.) as catalysts for the reaction, and this has allowed the cross-coupling of unactivated secondary halides with trialkylboranes in the presence of base.[96] These reactions appear to give the more stable stereoisomer of the product as the major product (e.g., the *endo*-2-bromobicyclo[2.2.1]-heptane) gives the *exo* product (**394**) from the borane (**393**).

Hydrometallations of alkynes

Building trisubstituted alkenes with defined stereochemistry has been a major task in synthesis. The use of boronic acids as partners in cross-coupling reactions has led to a resurgence of the use of catecholborane (**392a**) and the analogous pinacolborane (**392b**) as hydroborating agents. The hydroboration of terminal alkynes with these reagents gives the terminal *E*-vinylboranes, which can then be used in Suzuki-type couplings with halides to give *E*-alkenes. This is illustrated by the stereospecific synthesis of the *E,E*-diene **395** in early key steps of a synthesis of antillatoxin.[97] In this sequence, hydroboration of the terminal alkyne **393** with catecholborane gives the *E*-vinylboronate **394**, which is cross-coupled in situ with the iodide **395** to give the *E,E*-diene **396** as a single stereoisomer (Scheme 8.22).

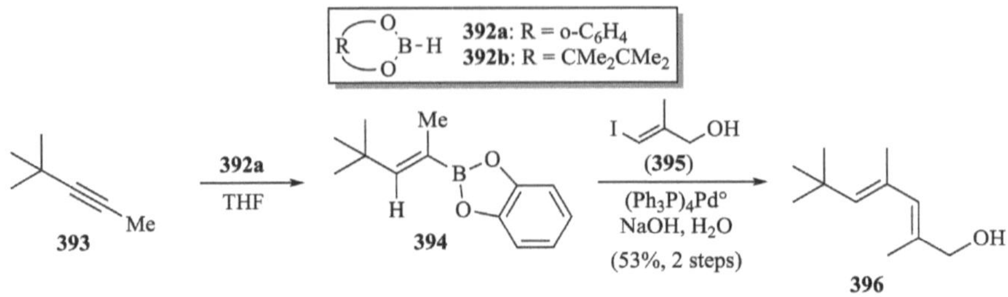

Scheme 8.22 Hydroboration and cross-coupling reactions.

However, despite the conventional wisdom that the vinylboronate ester reacts with retention of configuration, Lipshutz and coworkers found that the stereochemistry of the coupling depends on the ligand on the metal.[98]

The modern definition of Markovnikov's Rule is based on mechanisms where the initiating electrophile is a proton, so the hydrogen atom ends up on the less substituted carbon atom. The polarity of a hydrogen–heteroatom σ bond involving an electropositive element (B, Si, Sn, Cu, Zr, etc.) is the opposite of that of a hydrogen–heteroatom σ bond in Lowry-Brønsted acids: the partial charge on hydrogen is negative. These compounds add regioselectively to alkenes (**397**) and alkynes (**400**) to give the alkane (**398**) or alkene (**401**) with the heteroatom at the less substituted terminus of the original π bond (Scheme 8.23).[99] Subsequent replacement of the metal by a more electronegative atom (e.g., O, N, halogens) gives the (formally) anti-Markovnikov adduct (**399** and **402**, respectively).

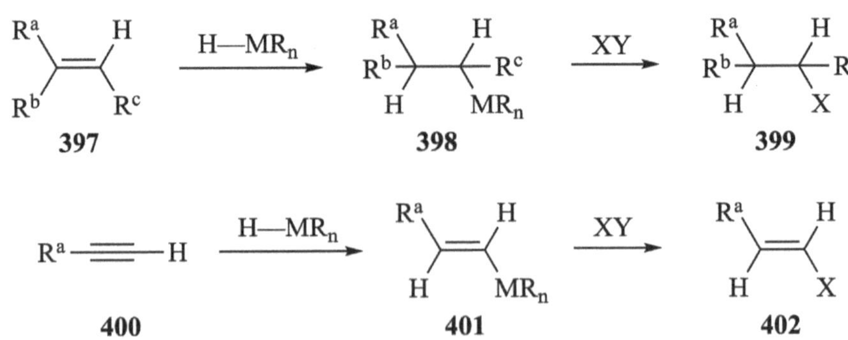

X = Br, Cl, I, OH, OR, SR, etc.
Y = Br, Cl, I, OH, OR, SR, etc.

Scheme 8.23 Hydrometallation of alkenes and alkynes.

Hydroalumination

The regioselective addition of the aluminum-hydrogen bond in alanes and hydridoaluminates to terminal alkynes gives derivatives of the corresponding terminal alane. In the 1981 synthesis of dendrobatid toxin by Overman,[100] the addition of diisobutylaluminum hydride to the alkyne group of trimethylsilylalkyne **403** gave the vinylalane **404** with stereospecific *syn* addition and the aluminum at the silylated carbon of the alkene (Scheme 8.24). The difficulties sometimes associated with hydroaluminations were addressed by Powers and his coworkers in 2014. In their study, the highly hindered, dimeric arylalane **409** permitted the conversion of the alkynes **405** and the alkenes **407** into the *syn*, (formally) anti-Markovnikov adducts **406** and **408**.[101]

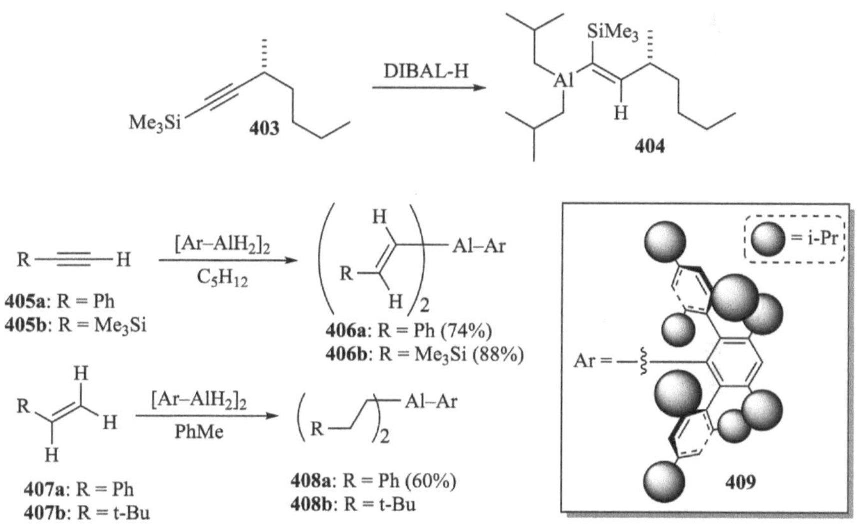

Scheme 8.24 Representative hydroaluminations of aliphatic hydrocarbons.

Hydrozirconation of alkynes

Schwartz reagent (Cp_2TiClH, **410**) has become a highly useful reagent for the stereocontrolled synthesis of *E*-vinyl iodides from alkynes, with (formal) anti-Markovnikov regiochemistry. Two representative examples of the hydrozirconation-iodinolysis sequence are given in Scheme 8.25. In a key step of their 2004 synthesis of 6,7-dehydrostipiamide,[102] Negishi and coworkers used a hydrozirconation-iodinolysis sequence to convert the terminal alkyne **411** into the *E*-vinyl iodide **413** through the intermediate

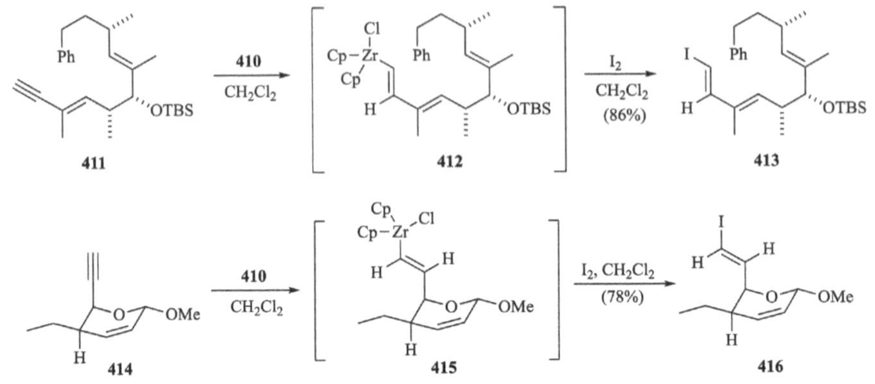

Scheme 8.25 Representative hydrozirconation-iodinolysis of alkynes.

terminal vinylzirconium reagent **412**. In this sequence, the hydrozirconation occurs as a strictly *syn*, anti-Markovnikov addition, and the iodinolysis occurs with retention of configuration of the double bond. A recent example using the same hydrozirconation-iodinolysis procedure to give is provided by the conversion of the terminal alkyne **414** into the *E*-vinyl iodide **416** as part of the Trost synthesis of leustroducsin B, published in 2015.[103] Again, the two steps are both stereospecific and essentially regiospecific. Zhao and Snieckus have also published a one-pot procedure for this transformation, where the Schwartz reagent is generated in situ from zirconocene dichloride.[104]

References

1. de la Mare, P. B. D.; Bolton, R., 2nd ed.; *Electrophilic Additions to Unsaturated Systems*, Vol. 9; Elsevier Science: New York, 1982.
2. Whitmore, F. C. The Common Basis of Intramolecular Rearrangements. *J. Am. Chem. Soc.* **1932,** *54* (8), 3274–3283. https://doi.org/10.1021/ja01347a037.
3. Ecke, G. G.; Cook, N. C.; Whitmore, F. C. Some Reactions of T-Butylethylene. *J. Am. Chem. Soc.* **1950,** *72* (4), 1511–1513. https://doi.org/10.1021/ja01160a023.
4. Fahey, R. C.; McPherson, C. A. Mechanism of the Hydrochlorination of Tert-Butylethylene and Styrene in Acetic Acid. *J. Am. Chem. Soc.* **1969,** *91* (14), 3865–3869. https://doi.org/10.1021/ja01042a030.
5. Whitmore, F. C.; Johnston, F. Secondary Isoamyl Chloride, 3-Chloro-2-methylbutane. *J. Am. Chem. Soc.* **1933,** *55* (12), 5020–5022. https://doi.org/10.1021/ja01339a053.
6. Brown, H. C.; Okamoto, Y. Substituent Constants for Aromatic Substitution. *J. Am. Chem. Soc.* **1957,** *79* (8), 1913–1917. https://doi.org/10.1021/ja01565a039.
7. Okamoto, Y.; Brown, H. C. A Quantitative Treatment for Electrophilic Reactions of Aromatic Derivatives. *J. Org. Chem.* **1957,** *22* (8), 485–494. https://doi.org/10.1021/jo01356a003.

8. Brown, H. C.; Okamoto, Y. Electrophilic Substituent Constants. *J. Am. Chem. Soc.* **1958,** *80* (18), 4979–4987. https://doi.org/10.1021/ja01551a055.
9. Yukawa, Y.; Tsono, Y. Resonance Effect in Hammett Relationship. II. Sigma Constants in Electrophilic Reactions and Their Intercorrelation. *Bull. Chem. Soc. Jpn.* **1959,** *32* (9), 965–971. https://doi.org/10.1246/bcsj.32.965.
10. Hansch, C.; Leo, R.; Taft, R. W. A Survey of Hammett Substituent Constants and Resonance and Field Parameters. *Chem. Rev.* **1991,** *91* (2), 165–195. https://doi.org/10.1021/cr00002a004.
11. Lion, C.; Dubois, J.-E. Le Tert-Butylation de Quelques Cétones Alicycliques et Aliphatiques via Leurs Éthers Triméthylsilylénoliques. Effet de l'environnement Carbonylé Sur l'évolution de La Réaction. *Bull. Soc. Chim. France* **1982,** *1982* (11), 375–379.
12. Reetz, M. T.; Maier, W. F.; Chatziiosifidis, I.; et al. Lewis-Säure-Bedingte α-Alkylierung von Carbonyl-Verbindungen, VII. Regio- und Positionsspezifische α-*tert*-Alkylierung von Ketonen. *Chem. Ber.* **1980,** *113* (12), 3741–3757. https://doi.org/10.1002/cber.19801131208.
13. Clements, M. T. M.; Klinck, R. E.; Peiris, S.; Ragauskas, A. J.; Stothers, J. B. Carbon-13 Magnetic Resonance Studies. 133. An Examination of the Effect of Conformation on β- Enolization in Some Bicyclo[3.2.1] Octan- 2- Ones. *Can. J. Chem.* **1988,** *66* (3), 454–460.
14. Lion, C.; Dubois, J.-E. Alkylation de Quelques Composés Carbonylés Par Des Groupes Tertiaires. Utilisation de La Réaction de Friedel-Crafts Dans La Synthèse d'ésters. *Tetrahedron* **1981,** *37* (2), 319–323. https://doi.org/10.1016/S0040-4020(01)92017-4.
15. Beller, M.; Seayad, J.; Tillack, A.; Jiao, H. Catalytic Markovnikov and Anti-Markovnikov Functionalization of Alkenes and Alkynes: Recent Developments and Trends. *Angew. Chem. Int. Ed.* **2004,** *43* (6), 3368–3398. https://doi.org/10.1002/anie.200300616.
16. Colvin, E. *Silicon in Organic Synthesis;* Butterworths: London, 1981.
17. Colvin, E. W. *Silicon Reagents in Organic Synthesis;* Academic Press: London, 1988.
18. Chan, T. H.; Fleming, I. Electrophilic Substitution of Organosilicon Compounds-Applications to Organic Synthesis. *Synthesis* **1979,** *1979* (10), 761–786. https://doi.org/10.1055/s-1979-28829.
19. Fleming, I.; Dunogués, J.; Smithers, R. The electrophilic substitution of allylsilanes and vinylsilanes. In *Organic Reactions,* Vol. 37, Kende, A. S., Ed.; Wiley: New York, 1989; pp 57–575. https://doi.org/10.1002/0471264180.or037.02.
20. Fleming, I. Allylsilanes, Allylstannanes and Related Systems. In *Comprehensive Organic Synthesis,* 2.2, Trost, B. M., Fleming, I., Eds.; Pergamon: Oxford, 1991; pp 563–593. https://doi.org/10.1016/b978-0-08-052349-1.00041-x.
21. Fleming, I.; Barbero, A.; Walter, D. Stereochemical Control in Organic Synthesis Using Silicon-Containing Compounds. *Chem. Rev.* **1997,** *97* (6), 2063–2192. https://doi.org/10.1021/cr941074u.
22. Gung, B. W. Additions of Allyl, Allenyl, and Propargylstannanes to Aldehydes and Imines. In *Organic Reactions,* Vol. 64, Overman, L. E., Ed.; Wiley: New York, 2004; pp 1–413. https://doi.org/10.1002/0471264180.or064.01.
23. Jacques, T.; Markó, I. E.; Pospísil, J. The Modified Sakurai and Related Reactions. In *Multicomponent Reactions;* Zhu, J., Bienaymé, H., Eds.; Wiley-VCH: Weinheim, 2005; pp 398–452. https://doi.org/10.1002/3527605118.ch13.
24. Hosomi, A.; Sakurai, H. Syntheses of γ,δ-Unsaturated Alcohols from Allylsilanes and Carbonyl Compounds in the Presence of Titanium Tetrachloride. *Tetrahedron Lett.* **1976,** *17* (16), 1295–1298. https://doi.org/10.1016/S0040-4039(00)78044-0.
25. Hosomi, A.; Endo, M.; Sakurai, H. Allylsilanes as Synthetic Intermediates. II. Syntheses of Homoallyl Ethers from Allylsilanes and Acetals Promoted by Titanium

Tetrachloride. *Chem. Lett.* **1976,** *5* (9), 941–942. https://doi.org/10.1246/cl.1976.941.
26. Hosomi, A.; Sakurai, H. Chemistry of Organosilicon Compounds. 99. Conjugate Addition of Allylsilanes to α,β-Enones. A New Method of Stereoselective Introduction of the Angular Allyl Group in Fused Cyclic α,β-Enones. *J. Am. Chem. Soc.* **1977,** *99* (5), 1673–1675. https://doi.org/10.1021/ja00447a080.
27. Hosomi, A.; Kobayashi, H.; Sakurai, H. Latent Acetonylation of α,β-Enones with Allyltrimethylsilane or 2-Methyl-2-Propenyltrimethylsilane: Synthesis of 1,5-Diketones and Annelation to Fused Cyclohexenones. *Tetrahedron Lett.* **1980,** *21* (10), 955–958. https://doi.org/10.1016/S0040-4039(00)77749-5.
28. Hosomi, A.; Iguchi, H.; Sasaki, J. -i.; Sakurai, H. Highly Regioselective Diels-Alder Reactions of 2-Trimethylsilylmethyl-1,3-Butadiene Catalyzed by a Lewis Acid and Applications to Syntheses of Terpenes. *Tetrahedron Lett.* **1982,** *23* (5), 551–554. https://doi.org/10.1016/S0040-4039(00)86886-0.
29. Sakurai, H. Reactions of Allylsilanes and Applications to Organic Synthesis. *Pure Appl. Chem.* **1982,** *54* (1), 1–22. https://doi.org/10.1351/pac198254010001.
30. Hosomi, A.; Sakurai, H.; Saito, M.; Sasaki, K.; Iguchi, H.; Sasaki, J.; Araki, Y. Chemistry of Organosilicon Compounds—165: 2-Trimethylsilyl-Methyl-1,3-Butadiene—A Versatile Building Block for Terpene Synthesis. *Tetrahedron* **1983,** *39* (6), 883–894. https://doi.org/10.1016/S0040-4020(01)88587-2.
31. Hosomi, A.; Sakata, Y.; Sakurai, H. Reactions of α-Heteroatom-Substituted Ethers and Sulfides with Silyl Enol Ethers. Chemoselectivity in the Cleavage of Heteroatom–Carbon Bonds by Iodotrimethylsilane and Trimethylsilyl Trifluoromethanesulfonate. *Chem. Lett.* **1983,** *12* (3), 405–408. https://doi.org/10.1246/cl.1983.405.
32. Sakurai, H.; Hosomi, A.; Hayashi, J. Conjugate Allylation of α,β-Unsaturated Ketones with Allylsilanes: 4-Phenyl-6-Hepten-2-One. *Org. Synth.* **1984,** *62,* 443. https://doi.org/10.15227/orgsyn.062.0086.
33. Hosomi, A.; Sakata, Y.; Sakurai, H. Highly Stereoselective C-Allylation of Glycopyranosides with Allylsilanes Catalyzed by Silyl Triflate or Iodosilane. *Tetrahedron Lett.* **1984,** *25* (22), 2383–2386. https://doi.org/10.1016/S0040-4039(01)80261-6.
34. Sakurai, H. Selective Carbon-Carbon Bond Formation Based on Organosilicon Reagents: Origin and Application. *Pure Appl. Chem.* **1985,** *57* (12), 1759–1770. https://doi.org/10.1351/pac198557121759.
35. Hosomi, A. Characteristics in the Reactions of Allylsilanes and Their Applications to Versatile Synthetic Equivalents. *Acc. Chem. Res.* **1988,** *21* (5), 200–206. https://doi.org/10.1021/ar00149a004.
36. Hayashi, T.; Konishi, M.; Ito, H.; Kumada, M. Optically Active Allylsilanes. 1. Preparation by Palladium-Catalyzed Asymmetric Grignard Cross-Coupling and Anti Stereochemistry in Electrophilic Substitution Reactions. *J. Am. Chem. Soc.* **1982,** *104* (18), 4962–4963. https://doi.org/10.1021/ja00382a045.
37. Lewis, D. E. *Advanced Organic Chemistry;* Oxford: New York, 2016; pp 322–323.
38. Jervis, P. J.; Kariuki, B. M.; Cox, L. R. Stereoselective Synthesis of 2,4,5-Trisubstituted Tetrahydropyrans Using an Intramolecular Allylation Strategy. *Org. Lett.* **2006,** *8* (20), 4637–4640. https://doi.org/10.1021/ol061957v.
39. Lee, P. H.; Seomoon, D.; Kim, S.; Nagaiah, K.; Damle, S. V.; Lee, K. Efficient Addition of Allylsilanes to α,β-Enones Using Catalytic Indium and Trimethylsilyl Chloride. *Synthesis* **2003,** *2003* (14), 2023–2026. https://doi.org/10.1055/s-2003-41017.
40. Barbero, M.; Bazzi, S.; Cadamuro, S.; Dughera, S.; Piccinini, C. O-Benzene-Disulfonimide as a Reusable Brønsted Acid Catalyst for Hosomi-Sakurai Reactions. *Synthesis* **2010,** *2010* (2), 315–319. https://doi.org/10.1055/s-0029-1217093.

41. Prins, H. J. Over Zwee Isomere Citronelalen. *Chem. Weekbl.* **1917,** *14,* 692–695.
42. Prins, H. J. Isomerisatie En Hydratie van Citronelal Door Zuren. *Chem. Weekbl.* **1917,** *14,* 627–630.
43. Van Genderen, H. Laboratorium Aan Der Huis. In *Chemie achter de dijken: uitvindingen en uitvinders in de eeuw na van't Hoff;* Rozendaal, S., van Bekkum, H., Reedijk, J., Eds.; *Koninklijke Akad. Wetenschap. Amsterdam: Amsterdam.* 2001; pp 10–11.
44. Arundale, E.; Mikeska, L. A. The Olefin-Aldehyde Condensation. The Prins Reaction. *Chem. Rev.* **1952,** *51* (3), 505–555. https://doi.org/10.1021/cr60160a004.
45. Pastor, I. M.; Yus, M. The Prins Reaction: Advances and Applications. *Curr. Org. Chem.* **2007,** *11* (10), 925–957. https://doi.org/10.2174/138527207781024067.
46. Pastor, I. M.; Yus, M. Focused Update on the Prins Reaction and the Prins Cyclization. *Curr. Org. Chem.* **2012,** *16* (10), 1277–1312. https://doi.org/10.2174/138527212800564196.
47. Adams, D. R.; Bhatnagar, S. P. The Prins Reaction. *Synthesis* **1977,** *1977* (10), 661–672. https://doi.org/10.1055/s-1977-24523.
48. Prins, H. J. Over de Condensatie van Formaldehyd met eenige unverzadigde Verbindingen. *Chem. Weekbl.* **1919,** *16,* 1072–1073.
49. Prins, H. J. The Reciprocal Condensation of Unsaturated Organic Compounds. *Chem. Weekbl.* **1919,** *16,* 1510–1526.
50. Prins, H. J. On the Condensation of Formaldehyde with Some Unsaturated Compounds. *Proc. Sect. Sci. Koninklike Akad. Wetenschap. Amsterdam* **1920,** *22,* 51–56.
51. Prins, H. J. Condensation of Formaldehyde with Some Unsaturated Compounds. *J. Chem. Soc. Abstr.* **1920,** *118,* i42–i43. https://doi.org/10.1039/CA9201800001.
52. Prins, H. J. Die Anlagerung von Formaldehyd an Di-, Tri- und Tetrachlor-Aethylen. *Rec. Trav. Chim. Pays-Bas* **1932,** *51,* 469–474. https://doi.org/10.1002/recl.19320510511.
53. Doro, F.; Ackeroyd, N.; Schiet, F.; Narula, A. Prins Reaction in the Fragrance Industry: 100th Anniversary (1919–2019). *Angew. Chem. Int. Ed.* **2019,** *58* (22), 7174–7179. https://doi.org/10.1002/anie.201814470.
54. Bain, J. P.; Nopol, I. The Reaction of β-Pinene with Formaldehyde. *J. Am. Chem. Soc.* **1946,** *68* (4), 638–641. https://doi.org/10.1021/ja01208a032.
55. Friedrich, C. Die Berliner Schule Der Pharmazie (1895–1943) von Hermann Thoms und Carl Mannich. *Beiträge Geschichte Pharm.* **1988,** *40* (4), 33–36.
56. Scholz, F. Walter Krösche (1882–1957) and the Mannich Reaction. *Pharmazie* **2008,** *63* (12), 916–918. https://doi.org/10.1691/ph.2008.8209.
57. Mannich, C.; Krösche, W. Ueber ein Kondensationsprodukt aus Formaldehyd, Ammoniak und Antipyrin. *Arch. Pharm.* **1912,** *250* (1), 647–667. https://doi.org/10.1002/ardp.19122500151.
58. Krösche, W. *Ueber ein Kondensationsprodukt aus Formaldehyd, Ammoniak und Antipyrin. Inaugural-Dissertation;* Friedrich-Wilhelms-Universität: Berlin, 1912.
59. Blicke, F. F. The Mannich Reaction. In *Organic Reactions;* Organic Reactions, Adams, R., Ed.; John Wiley & Sons: New York, 1942; pp 303e341. https://doi.org/10.1002/0471264180.or001.10.
60. Nie, J.; H, -C.; Cahard, D.; Ma, J.-A. Asymmetric Construction of Stereogenic Carbon Centers Featuring a Trifluoromethyl Group from Prochiral Trifluoromethylated Substrates. *Chem. Rev.* **2011,** *111* (2), 455–529. https://doi.org/10.1021/cr100166a.
61. Candeias, N. R.; Montalbano, F.; Cal, P. M. S. D.; Gois, P. M. P. Boronic Acids and Esters in the Petasis-Borono Mannich Multicomponent Reaction. *Chem. Rev.* **2010,** *110* (8), 6169–6193. https://doi.org/10.1021/cr100108k.
62. Kobayashi, S.; Mori, Y.; Fossey, J. S.; Salter, M. M. Catalytic Enantioselective Formation of C–C Bonds by Addition to Imines and Hydrazones: A Ten-Year Update. *Chem. Rev.* **2011,** *111* (4), 2626–2704. https://doi.org/10.1021/cr100204f.

63. Shibasaki, M.; Kanai, M. Asymmetric Synthesis of Tertiary Alcohols and α-Tertiary Amines via Cu-Catalyzed C−C Bond Formation to Ketones and Ketimines. *Chem. Rev.* **2008,** *108* (8), 2853−2873. https://doi.org/10.1021/cr078340r.
64. Subramaniapilla, S. G. Mannich Reaction: A Versatile and Convenient Approach to Bioactive Skeletons. *J. Chem. Sci.* **2013,** *125* (3), 467−482. https://doi.org/10.1007/s12039-013-0405-y.
65. Mannich, C. Ueberführung des Nonylmethylketons in das isomere Octyläthylketon. *Ber. Dtsch. Chem. Ges.* **1903,** *36* (2), 2551−2553. https://doi.org/10.1002/cber.190303602202.
66. Thoms, H.; Mannich, C. Ueber 2-Aminoundekan und 2-Aminononan. *Ber. Dtsch. Chem. Ges.* **1903,** *36* (2), 2554−2555. https://doi.org/10.1002/cber.190303602203.
67. Thoms, H.; Mannich, C. Ueber die Condensation hochmolekularer aliphatischer Ketone zu Verbindungen vom Typus des Mesityloxyds. *Ber. Dtsch. Chem. Ges.* **1903,** *36* (2), 2555−2558. https://doi.org/10.1002/cber.190303602204.
68. Mannich, C. Eine Synthese von β-Ketonbasen. *Arch. Pharm* **1917,** *255* (2−4), 261−276. https://doi.org/10.1002/ardp.19172550217.
69. Link, A. Centennial Anniversary of Mannich's Report on the Formation of β-Amino-Ketones in the Archiv Der Pharmazie. *Arch. Pharm. Chem. Life Sci.* **2017,** *350,* e1700152. https://doi.org/10.1002/ardp.201700152.
70. Woodward, R. B.; Cava, M. P.; Ollis, W. D.; Hunger, A.; Daeniker, H. U.; Schenker, K. The Total Synthesis of Strychnine. *J. Am. Chem. Soc.* **1954,** *76* (18), 4749−4751. https://doi.org/10.1021/ja01647a088.
71. Scott, W. L.; Evans, D. A. The Total Synthesis of (±)-Luciduline. *J. Am. Chem. Soc.* **1972,** *94* (13), 4779−4780. https://doi.org/10.1021/ja00768a083.
72. Veitch, G. E.; Beckmann, E.; Burke, B. J.; Boyer, A.; Maslen, S. L.; Ley, S. V. Synthesis of Azadirachtin: A Long but Successful Journey. *Angew. Chem. Int. Ed.* **2007,** *46* (40), 7629−7632. https://doi.org/10.1002/anie.200703027.
73. Earley, W. G.; Jacobsen, J. E.; Madin, A.; Meier, G. P.; O'Donnell, C. P.; Oh, T.; Old, D. W.; Overman, L. E.; Sharp, M. J. Aza-Cope Rearrangement− Mannich Cyclizations for the Formation of Complex Tricyclic Amines: Stereocontrolled Total Synthesis of (±)-Gelsemine. *J. Am. Chem. Soc.* **2005,** *127* (51), 18046−18053. https://doi.org/10.1021/ja055710p.
74. Mukaiyama, T.; Narasaka, K.; Banno, K. New Aldol Type Reaction. *Chem. Lett.* **1973,** *2* (9), 1011−1014. https://doi.org/10.1246/cl.1973.1011.
75. Mukaiyama, T.; Banno, K.; Narasaka, K. New Cross-Aldol Reactions. Reactions of Silyl Enol Ethers with Carbonyl Compounds Activated by Titanium Tetrachloride. *J. Am. Chem. Soc.* **1974,** *96* (24), 7503−7509. https://doi.org/10.1021/ja00831a019.
76. Shriner, R. L. The Reformatsky Reaction. In *Organic Reactions;* Adams, R., Ed.; John Wiley & Sons: New York, 1942, pp 1−37. https://doi.org/10.1002/0471264180.or001.01.
77. Rathke, M. W. The Reformatsky Reaction. *Org. React.* **1975,** *22,* 423−460. https://doi.org/10.1002/0471264180.or022.04.
78. Diaper, D. G. M.; Kukis, A. Synthesis of Alkylated Alkanedioic Acids. *Chem. Rev.* **1959,** *59* (1), 89−178. https://doi.org/10.1021/cr50025a003.
79. Ocampo, R.; Dolbier, W. R., Jr. The Reformatsky Reaction in Organic Synthesis. Recent Advances. *Tetrahedron* **2004,** *60* (42), 9325−9374. https://doi.org/10.1016/j.tet.2004.07.018.
80. Corey, E. J.; Girotra, N. N.; Mathew, C. T. Total Synthesis of *dl*-Cedrene and *dl*-Cedrol. *J. Am. Chem. Soc.* **1969,** *91* (6), 1557−1559. https://doi.org/10.1021/ja01034a062.
81. Pirrung, M. C. Total Synthesis of ±-Isocomene. *J. Am. Chem. Soc.* **1979,** *101* (23), 7130−7131. https://doi.org/10.1021/ja00517a087.

82. Pirrung, M. C. Total Synthesis of (±)-Isocomene and Related Studies. *J. Am. Chem. Soc.* **1981,** *103* (1), 82–87. https://doi.org/10.1021/ja00391a016.
83. Dem'yanov, N. Y.; Lushnikov, M. O. Deistvii Azotistoi Kisloty Na Tetrametilenilamin [On the Action of Nitrous Acid on Tetramethylenylamine]. *Zh. Russ. Fiz.-Khim. O-va.* **1903,** *35,* 26–42.
84. Demjanow, N. J. Die Ringerweiterung bei Den Cyclischen Aminen mit der Seitenkette $CH_2.NH_2$. Über den Alkohol aus dem Amin $(CH_2)_2CH.CH_2.NH_2$. *Ber. Dtsch. Chem. Ges.* **1907,** *40* (4), 4393–4397. https://doi.org/10.1002/cber.19070400473.
85. Demjanow, N. J.; Dojarenko, M. Über einige Umwandlungen des Cyclobutanols. *Ber. Dtsch. Chem. Ges.* **1908,** *41* (1), 43–46. https://doi.org/10.1002/cber.19080410113.
86. Fitier, L.; Wehle, D. Multiple Cyclopropylmethyl-Cyclobutyl Rearrangements in a Pentaspirohexadecanol. *Angew Chem. Int. Ed. Engl.* **1979,** *18* (11), 868–869. https://doi.org/10.1002/anie.197908682.
87. Takayoshi, H. T.; Han, L.-B. Air-Induced Anti-Markovnikov Addition of Secondary Phosphine Oxides and H-Phosphinates to Alkenes. *Org. Lett.* **2007,** *9* (1), 53–55. https://doi.org/10.1021/ol062505l.
88. Hu, X.; Zhang, G.; Bu, F.; Lei, A. Visible-Light-Mediated Anti-Markovnikov Hydration of Olefins. *ACS Catal.* **2017,** 7 (2), 1432–1437. https://doi.org/10.1021/acscatal.6b03388.
89. Brown, H. C. *Hydroboration*; W. A. Benjamin: New York, 1962.
90. Grieco, P. A. Total Synthesis of Dl-Sirenin. *J. Am. Chem. Soc.* **1969,** *91* (20), 5660–5661. https://doi.org/10.1021/ja01048a051.
91. Piers, E.; Zbozny, M. Intramolecular Alkylation of α,β-Unsaturated Ketones: A Total Synthesis of (±)-Isolongifolene and an Approach to the Synthesis of Zizaane-type Sesquiterpenoids. *Can. J. Chem.* **1979,** *57* (9), 2249–2255. https://doi.org/10.1139/v79-360.
92. Majetich, G.; Song, J. S.; Leigh, A. J.; Condon, S. M.; Annulations, A.-B. Direct Stereoselective Syntheses of (±)-Graveolide and (±)-Aromaticin. *J. Org. Chem.* **1993,** *58* (5), 1030–1037. https://doi.org/10.1021/jo00057a012.
93. Shen, X.-Y.; Peng, X.-S.; Wong, H. N. C. Total Synthesis of (±)-Gracilioether. *F. Org. Lett.* **2016,** *18* (5), 1032–1035. https://doi.org/10.1021/acs.orglett.6b00161.
94. Miyaura, N.; Suzuki, A. Palladium-Catalyzed Cross-Coupling Reactions of Organoboron Compounds. *Chem. Rev.* **1995,** *95* (7), 2457–2483. https://doi.org/10.1021/cr00039a007.
95. Rychnovsky, S. G.; Khire, U. R.; Yang, G. Total Synthesis of the Polyene Macrolide Roflamycoin. *J. Am. Chem. Soc.* **1997,** *119* (8), 2058–2059. https://doi.org/10.1021/ja963800b.
96. Saito, B.; Fu, G. C. Alkyl-Alkyl Suzuki Cross-Couplings of Unactivated Secondary Alkyl Halides at Room Temperature. *J. Am. Chem. Soc.* **2007,** *129* (31), 9602–9603. https://doi.org/10.1021/ja074008l.
97. Yokokawa, F.; Shioiri, T. Total Synthesis of Antillatoxin, an Ichthyotoxic Cyclic Lipopeptide, Having the Proposed Structure. What Is the Real Structure of Antillatoxin? *J. Org. Chem.* **1998,** *63* (24), 8638–8639 https://doi.org/10.1021/jo981744m.
98. Lu, G. P.; Voigtritter, K. R.; Cai, C.; Lipshutz, B. H. Ligand Effects on the Stereochemical Outcome of Suzuki–Miyaura Couplings. *J. Org. Chem.* **2012,** 77 (8), 3700–3703. https://doi.org/10.1021/jo300437t.
99. Lewis, D. E. *Advanced Organic Chemistry;* Oxford: New York, 2016; pp 716 730.
100. Overman, L. E.; Bell, K. L. Enantiospecific Total Synthesis of dendrobatid Toxin 251D. A Short Chiral Entry to the Cardiac-Active Pumiliotoxin A Alkaloids via Stereospecific Iminium Ion-Vinylsilane Cyclizations. *J. Am. Chem. Soc.* **1981,** *103* (7), 1851–1853. https://doi.org/10.1021/ja00397a052.

101. Fettinger, J. C.; Gray, P. A.; Melton, C. E.; Power, P. P. Hydroalumination of Alkenes and Alkynes by Primary Aluminum Hydrides under Mild Conditions. *Organometallics* **2014,** *33* (21), 6232–6240. https://doi.org/10.1021/om500911f.
102. Zeng, X.; Zeng, F.; Negishi, E.-I. Efficient and Selective Synthesis of 6,7-Dehydrostipiamide via Zr-Catalyzed Asymmetric Carboalumination and Pd-Catalyzed Cross-Coupling of Organozincs. *Org. Lett.* **2004,** *6* (19), 3245–3248. https://doi.org/10.1021/ol048905v.
103. Trost, B. M.; Biannic, B.; Brindle, C. S.; O'Keefe, B. M.; Hunter, T. J.; Ngai, M.-Y. A Highly Convergent Total Synthesis of Leustroducsin B. *J. Am. Chem. Soc.* **2015,** *137* (36), 11594–11597. https://doi.org/10.1021/jacs.5b07438.
104. Zhao, Y.; Snieckus, V. A Practical in Situ Generation of the Schwartz Reagent. Reduction of Tertiary Amides to Aldehydes and Hydrozirconation. *Org. Lett.* **2014,** *16* (2), 390–393. https://doi.org/10.1021/ol403183a.

CHAPTER NINE

Development of highly regiospecific alkene syntheses: elimination and its substitutes

Introduction

The Zaitsev and Hofmann eliminations were the first reproducibly regioselective methods for the formation of alkenes. The basic difference between the two reactions is defined by the differences in the base and the leaving group of the substrate. In general, leaving groups that form stronger bonds to carbon (e.g., F) and onium ions (e.g., Me_3N^+ and Me_2S^+) favor the Hofmann orientation by an E1cb or E1cb-like E2 pathway. This is also observed with the base strength: the stronger the base, the greater the proportion of the Hofmann alkene in the product.

Steric effects of alkoxide anion bases

The pK_a values of alcohols increase as the number of branches at the α-carbon increases, as shown in Table 9.1. The values show that each alkyl substituent at the α-carbon raises the pK_a by approximately one unit.

The steric bulk of the alkoxide anion also increases in the same order. The effects of base strength and size on the regiochemical course of *anti*-elimination reactions have been elucidated by Bartsch,[1] in particular (Scheme 9.1). These studies demonstrated that the most strongly associated base, potassium *tert*-butoxide anion in *tert*-butyl alcohol, gave the highest amount of the Hofmann regioisomer. The smallest (and weakest) base, potassium ethoxide in ethanol, gave the highest amount of the Zaitsev regioisomer. In dimethyl sulfoxide, the base is less strongly associated because of the stronger solvation of the potassium ion; in this solvent potassium *tert*-butoxide gave results between those of the potassium alkoxides in solution in the corresponding alcohols.

A similar effect is observed when the crown ether, dicyclohexyl-18-crown-6 (**421**), which is normally used as a mixture of diastereoisomers, is added to the reaction mixture.[2,3] This ether selectively chelates potassium

Addition, Elimination and Substitution: Markovnikov, Hofmann, Zaitsev and Walden
ISBN: 978-0-12-821027-7
https://doi.org/10.1016/B978-0-12-821027-7.00010-4

Table 9.1 Typical pKa values for alcohols.

ROH	pK_a	ROH	pK_a	ROH	pK_a	ROH	pK_a
MeOH	15.5	EtOH	15.9	*i*-PrOH	16.5	*t*-BuOH	18.5
		n-PrOH	15.9	2-BuOH	17.6	*t*-AmOH	18.5
		n-BuOH	16.1			Et_3COH	17.1

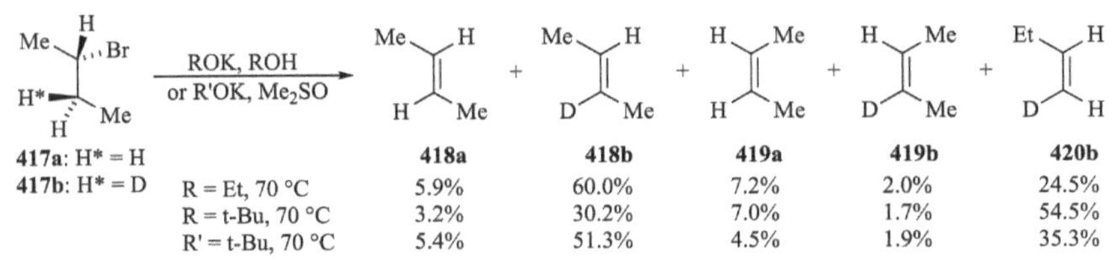

Scheme 9.1 The dehydrobromination of 2-bromobutane with alkoxide anion bases.

ion, leading to the much larger cation, **422**, and to decreased association of the more hindered bases, effectively reducing their size. In this case, the crown ether depressed the amount of 1-butene produced (Scheme 9.2), consistent with the results in Scheme 9.1.

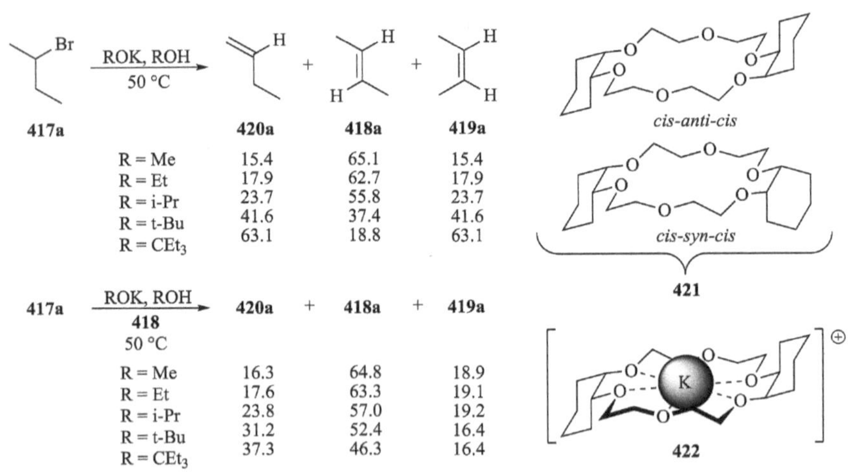

Scheme 9.2 The reaction of 2-bromobutane with alkoxide bases in the presence and absence of dicyclohexyl-18-crown 6 (**421**).

β-Eliminations not involving deprotonation: reductive elimination

Molecules (e.g., **423**) in which there are vicinal substituents consisting of a group (Y) capable of functioning as a leaving group (e.g., an alcohol, an ether, a halide, an ester, etc.) and a group (X) that can act as a potential precursor to a carbanion or its equivalent (e.g., a halogen, trialkylsilyl, etc.), react by either the E2 mechanism, where there is strict control over the stereochemistry (usually *anti*, to give a product such as **425** in Scheme 9.3) or by the E1cb mechanism, where more than one geometric isomer of the product (**425**, **427**) may be obtained. The general mechanistic overview of these reactions is shown in Scheme 9.3.

Note that the stereochemistry of the elimination from the carbanion (**424**) is determined by the rate of inversion of the anion carbon or the rate of rotation about the C—C σ bond to give the diastereomeric rotamer, **426**, compared to the rate of the elimination itself. If the loss of the leaving group is rapid, the *anti*-elimination product (**425**) is either the dominant or even the sole product. If the inversion or rotation to give **426** is more rapid, then a mixture of **425** and **427** is obtained, with the more stable alkene dominating.

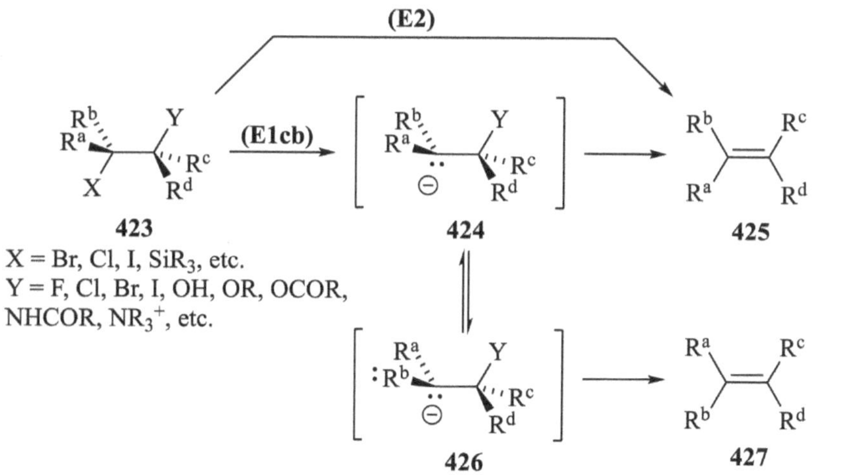

Scheme 9.3 The 1,2-elimination from vicinal compounds may occur by either the E2 or the E1cb mechanism. The E2 elimination is stereospecific and *anti*; the E1cb mechanism can give both geometric isomers of the product.

Elimination during Wolff–Kishner and related reactions[4]

The Wolff–Kishner reduction is known to proceed through a carbanion intermediate, and when this carbanion is generated adjacent to a leaving group, a regiospecific elimination occurs (Scheme 9.4). This reaction had been observed by Nikolai Matveevich Kizhner (Кижнер, Николай Матвеевич, 1867–1935) himself, in 1913,[5] and now bears his name.

In contrast to the E2 and E1cb reactions that have been discussed heretofore, this reaction works very well to eliminate α-hydroxy and α-amino substituents provided that the substituent can achieve a conformation where good overlap occurs with the HOMO of the hydrazone conjugate base (**426**). Stereoelectronic factors determine the ease of elimination: when the system can attain a fully extended zig-zag conformation, with the N–N–C–C–X system planar, the elimination is favored; if this conformation cannot be attained, the normal Wolff–Kishner deoxygenation occurs.

The elimination begins with the deprotonation of the hydrazone (**428**) to give the resonance-stabilized conjugate base **429**. This anion acts as the carbanion intermediate in an E1cb elimination, displacing the leaving group to give the vinyl diazene **430**. This diazene will react with base to generate the corresponding diazenide anion, **431**, which then loses molecular nitrogen to generate the vinyl anion **432**. In the final step, the vinyl anion is protonated by the solvent to give the alkene **433**.

The stereoelectronic effects in the Kishner reductive elimination of cyclohexanones are evidenced by the ready reductive elimination of the α-substituent when it is axial and its failure when the α-substituent is

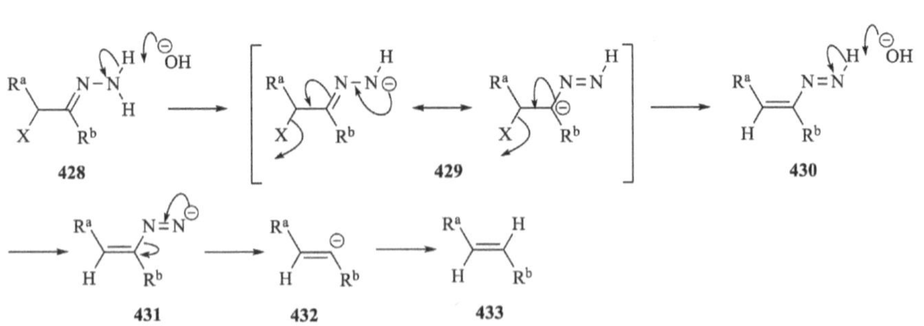

Scheme 9.4 A pathway for the formation of alkenes from α-substituted ketones in the Kishner reductive elimination.

equatorial.[6–8] This is shown by the Wolff–Kishner reduction of the ketones **434** and **436** in Scheme 9.5.

The ketone **434** has a rigid skeleton in which the oxygen atom in the pyranone is, of necessity, equatorial, and the nonbonding π orbital of the diazene anion moiety of **435** is close to coplanar with the σ* orbital of the axial α-C–O bond, which facilitates this elimination. The ketone **436**, on the other hand, is conformationally flexible and can exist as the equilibrium mixture of two chair conformations. In the presumptive minor chair conformation, the isopropoxy group is axial and can be eliminated.

There is, however, a limit to this reaction: when the leaving group can be deprotonated under the conditions of the reaction, so that it must be eliminated as a dianion, the reaction follows the normal course of the Wolff–Kishner reduction and does not give the alkene. This is illustrated by the conversion of **440** to **442** in the first synthesis of morphine by Gates and Tschudi[9,10] shown in Scheme 9.6. In this case, the initial step converts the lactam (**440**) into its conjugate base (**441**), which no longer contains a useable leaving group.

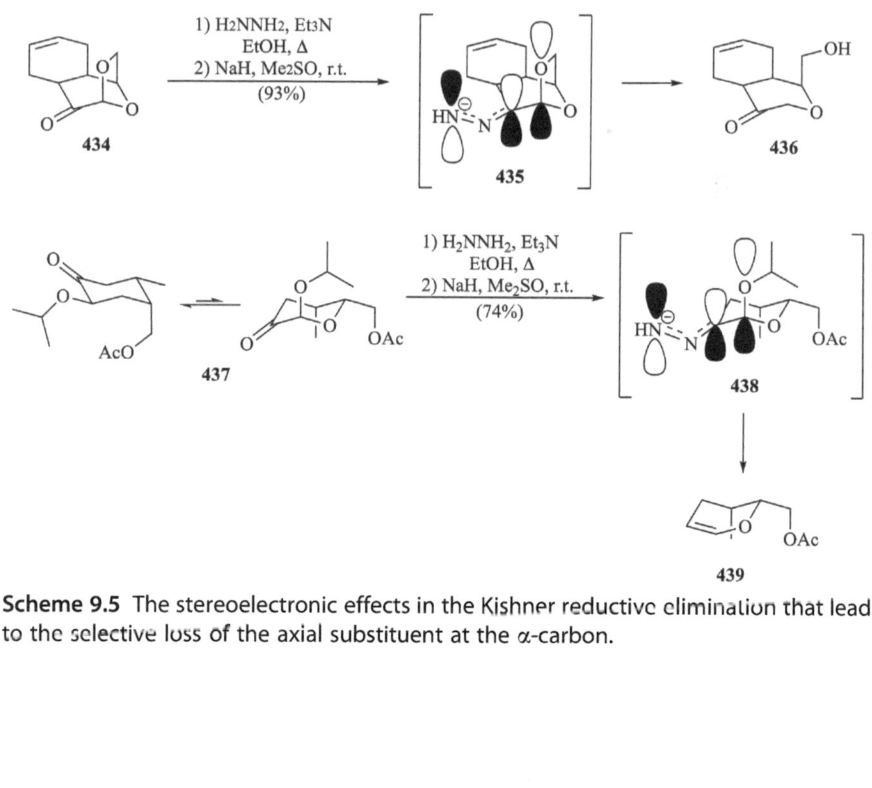

Scheme 9.5 The stereoelectronic effects in the Kishner reductive elimination that lead to the selective loss of the axial substituent at the α-carbon.

The Bamford–Stevens and Shapiro reactions

Two new incarnations of the Kishner reductive elimination emerged during the middle years of the 1950s. Both involved arylhydrazones (**443**), which carry the sulfonyl group as a masked sulfinate leaving group that is revealed by redox chemistry. The pathways of these two forms of the reaction are shown in Scheme 9.7. Both pathways begin with the same step: deprotonation of the sulfonylhydrazone to give the anion **444**. At this point, the fate of the anion is decided by the base.

Two new incarnations of the Kishner reductive elimination emerged during the middle years of the 1950s. Both involve arylhydrazones (**443**), which carry the sulfonyl group as a masked sulfinate leaving group that is revealed by redox chemistry, and begin with deprotonation of the sulfonylhydrazone to give the anion **444**. At this point, the fate of the anion is decided by the base. When the base is an alkoxide in alcohol solution, the reaction is known as the Bamford–Stevens reaction; it was reported by Scottish chemist Thomas Stevens Stevens (1900–2000)[11] and his student, William Randall Bamford, in 1952; it gives a diazoalkane as the initial product.[12] Replacing the alkoxide anion with an alkyllithium produces the Hofmann alkene,[13–15] as first reported by Shapiro in 1967.[16]

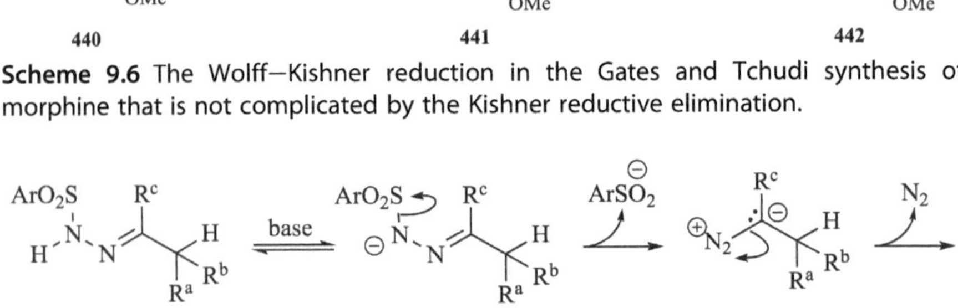

440 **441** **442**

Scheme 9.6 The Wolff–Kishner reduction in the Gates and Tchudi synthesis of morphine that is not complicated by the Kishner reductive elimination.

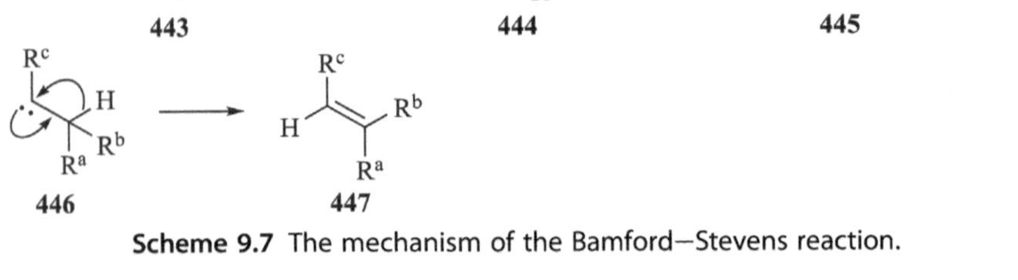

443 **444** **445**

446 **447**

Scheme 9.7 The mechanism of the Bamford–Stevens reaction.

The Bamford–Stevens reaction

When alkoxide anion bases are used in alcohol solution, the anion decomposes to a diazoalkane (**445**) that then decomposes to a carbene (**446**). The carbene then rearranges to the alkene (**447**). The regiochemistry of the alkene produced is decided by the migratory aptitude of the alkyl groups at the β position. In this form, the reaction is known as the Bamford–Stevens reaction (Scheme 9.7).[12]

The Shapiro reaction

When two equivalents of an alkyllithium or lithium diisopropylamide (LDA) are used as the base, the reaction is known as the Shapiro reaction. In this reaction, the initial deprotonation is followed by a second deprotonation to give a dianion that may exist as the chelated alkyllithium complex **448**. The loss of the lithium arenesulfinate from this complex results in the formation of the vinyllithium **449**, which can be trapped by a wide range of electrophiles to give alkenes (**450**) (Scheme 9.8).[13–16]

Robert H. Shapiro (1935–2004) was born in New Haven, Connecticut, and joined the U.S. Navy in 1952, at the age of 17. After his service in the Navy and the U.S. Marine Corps, he entered the University of Connecticut in 1956, graduating in 1960 with degrees in chemistry and pharmacy. He took his Ph.D. at Stanford in 1964. After a year in postdoctoral study in Copenhagen, he returned to the United States as Assistant Professor of Chemistry at the University of Colorado, Boulder, where he rose through the ranks to become Professor of Chemistry. In 1980, he moved to James Madison University in Harrisonburg, Virginia, as Head of Chemistry, then Dean, then Vice-President for Academic Affairs. In 1989, he moved to the Naval Academy as Academic Dean and then Provost. He retired in 1998 and moved back to Harrisonburg. Shapiro's research was largely focused on the reactions of strong bases with organic compounds.

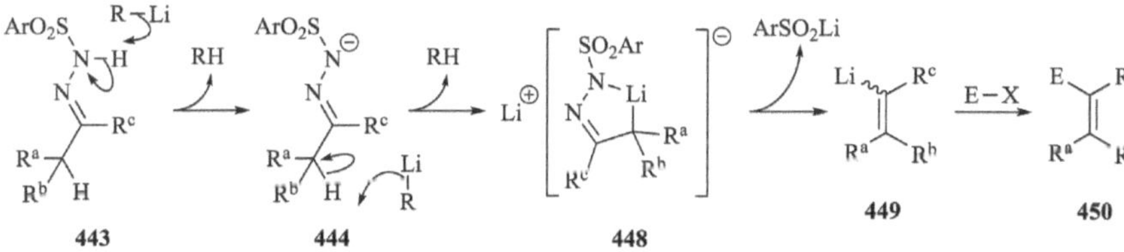

Scheme 9.8 The mechanism of the Shapiro reaction.

The reaction is highly regioselective for the less substituted α-carbon, corresponding to Hofmann regiochemistry in the elimination, as shown by the example in Scheme 9.9, taken from Evans' synthesis of the alkaloid luciduline.[17] In this reaction, a methylene proton is removed from the tosylhydrazone **451** in preference to a methine proton, leading to the disubstituted alkene **448** rather than the isomeric trisubstituted alkene **453**. In this reaction, epoxidation of the crude product with *m*-CPBA gave the epoxide from **452** in 80% yield from **451**, meaning that the ratio **452**:**453** cannot be less than 4:1. The next two examples are taken from syntheses of artemisinin by Avery[18] and Cook.[19] In both these reactions, the Shapiro reaction is used to effect the regiospecific generation of a vinyllithium reagent (**455** and **458**) from the corresponding ketone (**454** and **457**). The vinyllithium reagent is then trapped by DMF to give the aldehydes **456** and **459**. In these cases, also, the Shapiro reaction is essentially regiospecific for the less substituted alkene. The final example, which is a key step in Grieco's synthesis of a model lacking the A ring,[20] shows how the Shapiro reaction can proceed in preference to a Kishner reductive elimination reaction due to the effects of the stronger base.

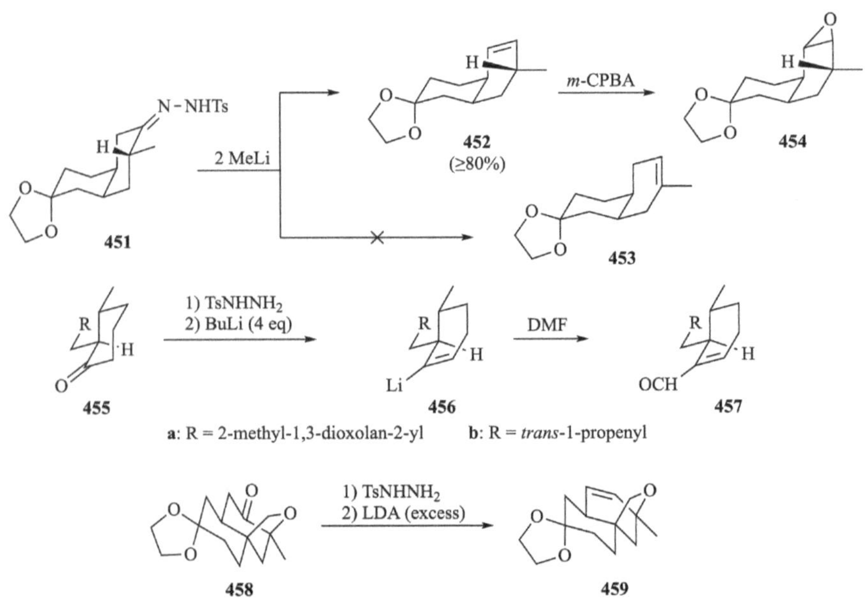

Scheme 9.9 Representative Shapiro reactions showing the preferred formation of the Hofmann alkene.

In general, the stereoselectivity of the Shapiro reaction is not high, but a recent modification using the ketone phenylaziridinylhydrazones (e.g., **461**) and a catalytic amount of LDA[21] leads to good yields of a *Z*-alkene. The example in Scheme 9.10 is illustrative. In this reaction, 3-nonanone (**460**) reacts with the *N*-aminoaziridine to give the corresponding hydrazone (**461**). When this hydrazone is treated with a catalytic amount of LDA in ether, the anion eliminates styrene and nitrogen, giving a mixture of *Z*-2-nonene (**462**) and *Z*-3-nonene (**463**) in the ratio 49:1, showing the high regioselectivity of this reaction. In addition, the *cis:trans* ratio was close to 250:1. This represents a major improvement in the regio- and stereoselectivity of the olefin-forming reaction.

The difficulties of obtaining trisubstituted alkenes by working with di-*sec*-alkyl ketones can be circumvented by using the hindered base, LDA, in place of the alkyllithium, and ensuring that one of the α-substituents is aryl. For example, under the standard conditions (2 equivalents of the strong base), the tosylhydrazone of the α,α′-diphenylcyclopentanone **464** reacts to give the trisubstituted alkene **465** in good yield (Scheme 9.11).[22]

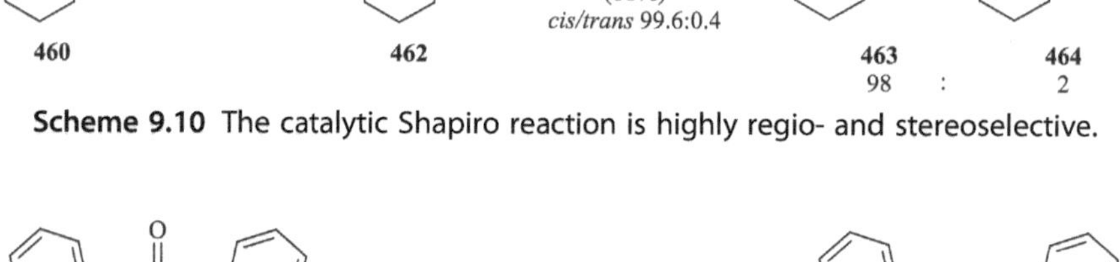

460 462 463 464

98 : 2

Scheme 9.10 The catalytic Shapiro reaction is highly regio- and stereoselective.

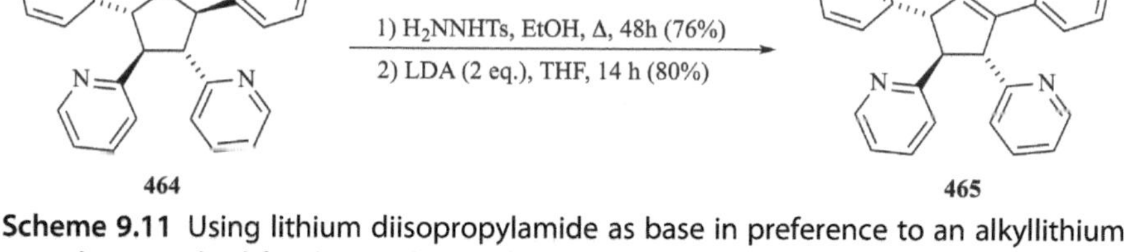

464 465

Scheme 9.11 Using lithium diisopropylamide as base in preference to an alkyllithium provides a method for the synthesis of trisubstituted olefin.

The Wharton reaction

The treatment of α,β-epoxyketones (**466**) with hydrazine results in a reductive elimination to give an allylic alcohol (**469**) with concomitant 1,3-transposition of the oxygen functionality to the β-position of the starting epoxyketone. The double bond finishes up between the carbonyl and α-carbon atoms of the starting epoxyketone (Scheme 9.12).

The discoverer of the reaction, Peter S. Wharton (1939–2014), was born and educated in England, earning his B.A. (1952) and M.A. (1956) at Queen's College of Cambridge University. Following his graduation, he moved to the United States, where he entered Yale University. He took his Ph.D. at Yale in 1959. In 1960, he was appointed as an Instructor at the University of Wisconsin–Madison and Assistant Professor a year later. In 1962, he left Wisconsin for a postdoctoral year with Gilbert Stork (1921–2017). He returned to Wisconsin in 1962, where he remained until 1968. Despite his groundbreaking reaction, he was denied tenure at Wisconsin, so he moved to Wesleyan University as Assistant Professor, eventually rising to the rank of Professor. He remained at Wesleyan for the rest of his career.

Thanks to its very mild reaction conditions, the reaction, which was first reported by Wharton in 1961,[23] has exhibited great longevity as the examples in Scheme 9.13—reported over a period of more than half a century—show. Wharton's original report was the conversion of 4β,5β-epoxycoprostan-3-one (**470**) to 5β-hydroxy-3-coprostene (**471**). Burke used the reaction to convert the epoxyketone **472** into the allylic alcohol **473** in his 1982 synthesis of quadrone,[24] and Tu used it to convert the morphinan-based epoxketone **474** to the allylic alcohol **475** in his approach to the enantioselective synthesis of (−)-morphine.[25]

Alkenes by 1,3-elimination and extrusion

The regiospecific synthesis of alkenes by elimination has come close to being achieved, in particular when it comes to forming Hofmann alkenes.

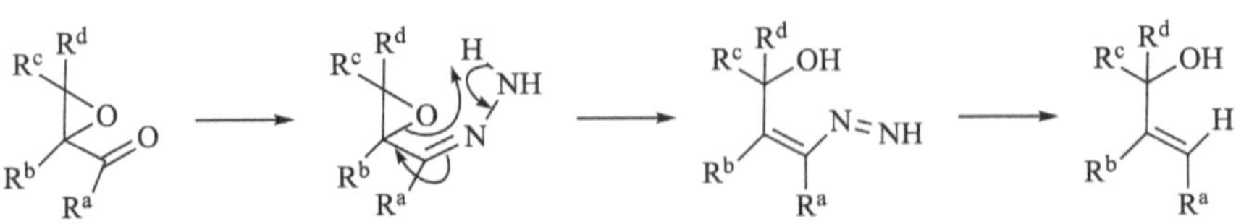

466 467 468 469

Scheme 9.12 The accepted mechanism of the Wharton reaction.

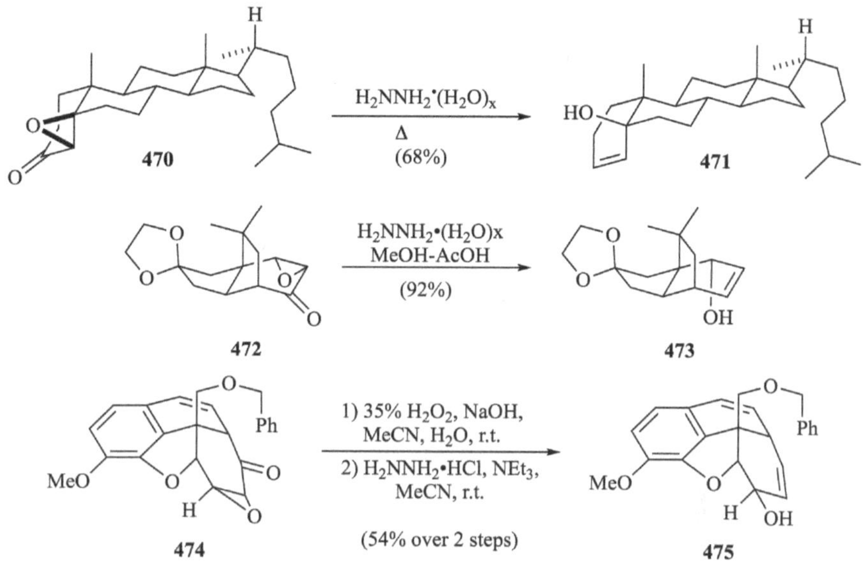

Scheme 9.13 Applications of the Wharton reaction in synthesis.

However, the middle of the 20th century saw a paradigm shift in the synthesis of alkenes from elimination to metathesis and extrusion strategies (Scheme 9.14), which were absolutely regiospecific in contrast to even the most regioselective elimination reactions. In the metathesis strategy, the first step is the formation of a four-membered cyclic intermediate (**478**) by formal cycloaddition of the starting compounds (**476** and **477**). The retro-cycloaddition of the four-membered ring generates the alkene **479**. In the extrusion mechanism, deprotonation of a precursor (**481**) gives a carbanion that leads to the formation of a three-membered heterocycle (**482**). The loss of group Z (usually SO_2) in **482** as a stable molecular species, the alkene (**479**) is formed. Because the metathesis approach does not formally involve an elimination, it will not be further discussed here.

The Ramberg–Bäcklund reaction

In 1940, Swedish chemist Ludwig Ramberg (1874–1940) and his student Birger Bäcklund (1908–97) reported first representative of a new strategy for the regiospecific formation of an alkene now known as the Ramberg–Bäcklund reaction.[26–28] The mechanism of the reaction is shown in Scheme 9.15. In the reaction, the halosulfone **483** reacts with strong base to give an anion (**484**) that effects an intramolecular S_N2 reaction to give the episulfide dioxide (thiirane-1,1-dioxide) **485**, Under the reaction

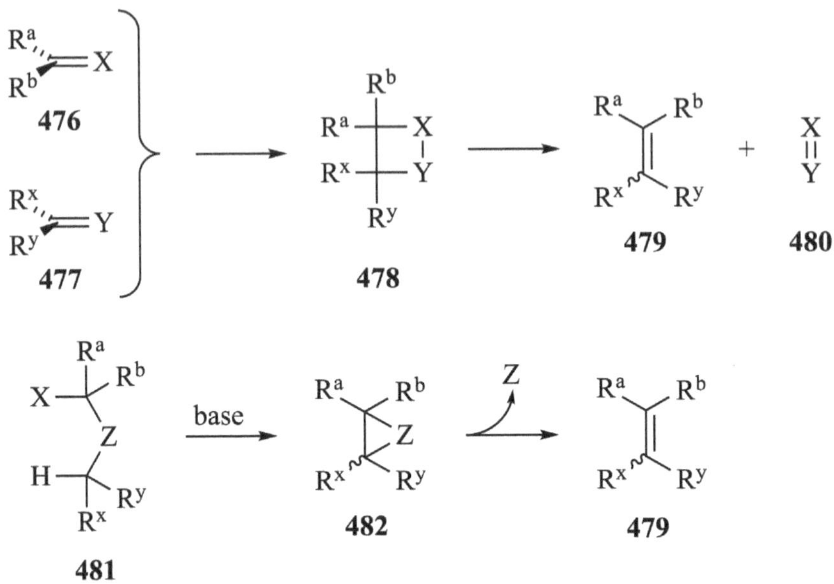

Scheme 9.14 Metathesis and extrusion strategies for regiospecific synthesis of alkenes.

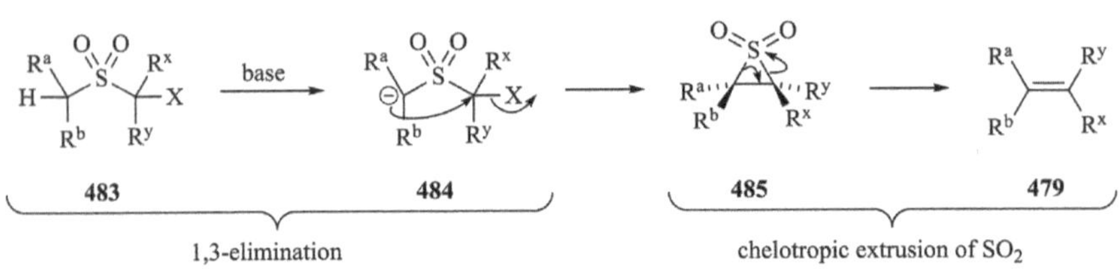

Scheme 9.15 The mechanism of the Ramberg–Bäcklund reaction.

conditions, this heterocycle undergoes suprafacial chelotropic extrusion of sulfur dioxide, thus giving the alkene (**486**) with retention of the thiirane dioxide stereochemistry.

The regiospecificity of the Ramberg–Bäcklund reaction has made it useful in the synthesis of widely divergent species (Scheme 9.16). These applications have ranged from cyclobutene derivatives: the propellane derivative **487**,[29] pterodactyladiene (**489**),[30] and a key intermediate (**492**) in the synthesis of the aspidosperma alkaloids.[31] The use of 1,2-dibromotetrachloroethane as an environmentally friendly halogenating agent in the reaction is illustrated by the synthesis of the resveratrol derivative **494**.[32] The synthesis of the eremantholide A (**497**)[33] illustrates the use of the reaction in the formation of medium-sized carbocyclic rings.

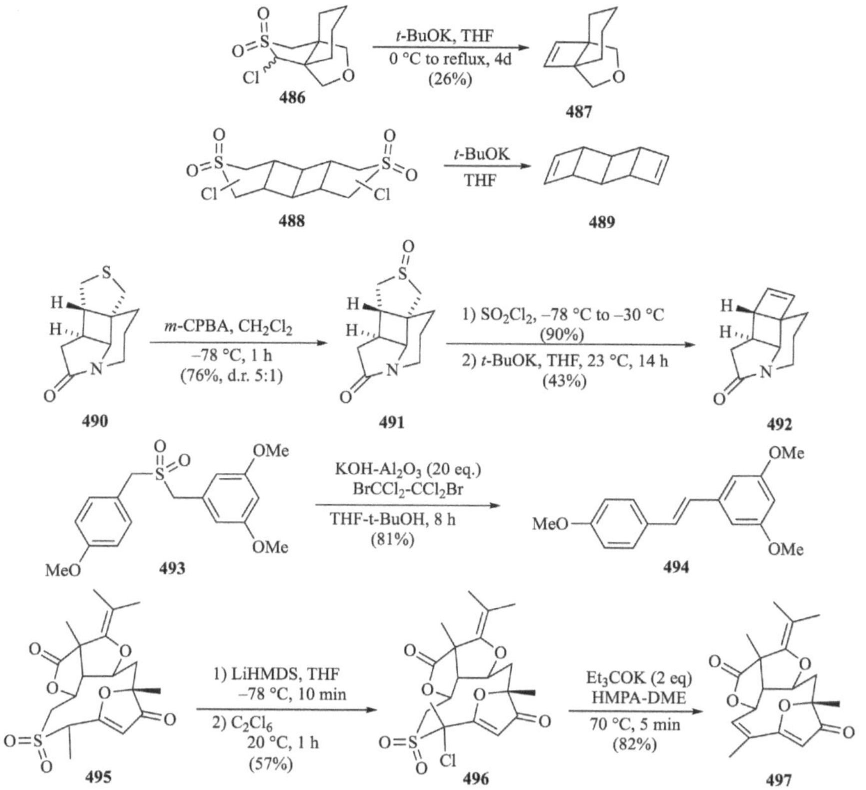

Scheme 9.16 The mechanism of the Ramberg–Bäcklund reaction.

Where do we stand now?

The advent of these more modern approaches to controlling the regioselectivity of alkene formation has largely superseded traditional elimination reactions except in particular circumstances because the reactions allow the regiospecific placement of the alkene double bond. The one problem that these newer methods face is that there is often less control over the stereochemistry of the new double bond. The control of stereochemistry in these reactions remains an active field of research.

References

1. Bartsch, R. A. Stereochemistry of Base-Catalyzed β Elimination from 2-Bromobutane. *J. Am. Chem. Soc.* **1971,** *93* (15), 3683–3686. https://doi.org/10.1021/ja00744a022.
2. Bartsch, R. A. Ionic Association in Base-Promoted β-Elimination Reactions. *Acc. Chem. Res.* **1975,** *8* (7), 239–245. https://doi.org/10.1021/ar50091a004.

3. Bartsch, R. A.; Pruss, G. M.; Cook, D. M.; Buswell, R. L.; Bushaw, B. A.; Wiegers, K. E. Orientation in Base-Promoted β Eliminations from 2-Alkyl Halides and P-Toluenesulfonates. Role of Base Association. *J. Am. Chem. Soc.* **1973,** *95* (20), 6745–6749. https://doi.org/10.1021/ja00801a035.
4. Lewis, D. E. The Wolff-Kishner Reduction and Related Reactions. *Elsevier: Air Qual Monogr. (Amst)* **2019**.
5. Kizhner, N. O razlohenie alkilidengidrazinov [On the Decomposition of Alkylidenehydrazines]. *Zh. Russ. Fiz.-Khim. O-va.* **1913,** *45,* 973–986.
6. Leonard, N. J.; Sentz, R. C. Rearrangement of α-Aminoketones during Clemmensen Reduction. VIII. The Fate of an Eight-Membered Ring. *J. Am. Chem. Soc.* **1952,** *74,* 1704–1709. https://doi.org/10.1021/ja01127a027.
7. Leonard, N. J.; Gelfand, S. The Kishner Reduction-Elimination. I. Cyclic and Open Chain α-Aminoketones. *J. Am. Chem. Soc.* **1955,** 77 (12), 3269–3271. https://doi.org/10.1021/ja01617a035.
8. Leonard, N. J.; Gelfand, S. The Kishner Reduction-Elimination. II. α-Substituted Pinacolones. *J. Am. Chem. Soc.* **1955,** 77 (12), 3272–3278. https://doi.org/10.1021/ja01617a036.
9. Gates, M.; Tschudi, G. The Synthesis of Morphine. *J. Am. Chem. Soc.* **1952,** *74* (4), 1109–1110. https://doi.org/10.1021/ja01124a538.
10. Gates, M.; Tschudi, G. The Synthesis of Morphine. *J. Am. Chem. Soc.* **1956,** *78* (7), 1380–1393. https://doi.org/10.1021/ja01588a033.
11. Haslam, E.; Morris, D. G. Thomas Stevens Stevens. 8 October 1900 – 13 November 2000. *Biogr. Mem. Fellows R. Soc.* **2003,** *49,* 521–535. https://doi.org/10.1098/rsbm.2003.0031.
12. Bamford, W. R.; Stevens, T. S. The Decomposition of Toluene-*p*-sulphonylhydrazones by Alkali. *J. Chem. Soc.* **1952,** 4735–4740. https://doi.org/10.1039/JR9520004735.
13. Adlington, R. M.; Barrett, A. G. M. Recent Applications of the Shapiro Reaction. *Acc. Chem. Res.* **1983,** *16* (2), 55–59. https://doi.org/10.1021/ar00086a004.
14. Shapiro, R. H. Alkenes from Tosylhydrazones. *Org. React.* **1976,** *23,* 405–507. https://doi.org/10.1002/0471264180.or023.03.
15. Shapiro, R. H.; Lipton, M. F.; Kolonko, K. J.; Buswell, R. L.; Capuano, L. A. Tosylhydrazones and Alkyllithium Reagents: More on the Regiospecificity of the Reaction and the Trapping of Three Intermediates. *Tetrahedron Lett.* **1976,** *16* (22–23), 1811–1814. https://doi.org/10.1016/S0040-4039(00)75263-4.
16. Shapiro, R. H.; Heath, M. J. Tosylhydrazones. V. Reaction of Tosylhydrazones with Alkyllithium Reagents. A New Olefin Synthesis. *J. Am. Chem. Soc.* **1967,** *89* (22), 5734–5735. https://doi.org/10.1021/ja00998a601.
17. Scott, W. L.; Evans, D. A. The Total Synthesis of (±)-Luciduline. *J. Am. Chem. Soc.* **1972,** *94* (13), 4779–4780. https://doi.org/10.1021/ja00768a083.
18. Avery, M. A.; Jennings-White, C.; Chong, W. K. M. The Total Synthesis of (+)-Artemisinin and (+)-9-Desmethyltemesinin. *Tetrahedron Lett.* **1987,** *28* (40), 4629–4632. https://doi.org/10.1016/S0040-4039(00)96582-1.
19. Zhu, C.; Cook, S. P. A Concise Synthesis of (+)-Artemisinin. *J. Am. Chem. Soc.* **2012,** *134* (33), 13577–13579. https://doi.org/10.1021/ja3061479.
20. Kawabata, T.; Grieco, P. A.; Sham, H. L.; Kim, H.; Jaw, J. Y.; Tu, S. Model Studies in the Quassimarin Series: Total Synthesis of De-A-quassimarin. *J. Org. Chem.* **1987,** *52* (15), 3346–3354. https://doi.org/10.1021/jo00391a033.
21. Maruoka, K.; Oishi, M.; Yamamoto, H. The Catalytic Shapiro Reaction. *J. Am. Chem. Soc.* **1996,** *118* (9), 2289–2290. https://doi.org/10.1021/ja951422p.

22. Siemeling, U.; Neumann, B.; Stammler, H.-G. First Example of a High-Yield Shapiro Reaction with a Substrate Containing Only Tertiary α-Carbon Atoms. *J. Org. Chem.* **1997,** *62* (10), 3407−3408. https://doi.org/10.1021/jo9617373.
23. Wharton, P. S.; Bohlen, D. H. Hydrazine Reduction of α,β-Epoxy Ketones to Allylic Alcohols. *J. Org. Chem.* **1961,** *26* (9), 3615−3616. https://doi.org/10.1021/jo01067a117.
24. Burke, S. D.; Murtiashaw, C. W.; Saunders, J. O.; Dike, M. S. A Novel and Efficient Entry to (±)-Quadrone. *J. Am. Chem. Soc.* **1982,** *104* (3), 872−874. https://doi.org/10.1021/ja00367a049.
25. Zhang, Q.; Zhang, F. M.; Zhang, C. S.; et al. Enantioselective synthesis of *cis*-hydrobenzofurans bearing all-carbon quaternary stereocenters and application to total synthesis of (−)-morphine. *Nat. Commun.* **2019,** *10* (1). https://doi.org/10.1038/s41467-019-10398-4.
26. Paquette, L. A. The Ramberg−Bäcklund Rearrangement. *Org. React.* **1977,** *25,* 1−71. https://doi.org/10.1002/0471264180.or025.01.
27. Ramberg, L.; Bäcklund, B. Några Monohalogenderivat Av Dietylsulfon Och Deras Reaktioner. [Some Monohalogen Derivatives of Diethyl Sulfone and Their Reactions]. *Ark. kemi, mineralog. geolog* **1940,** 1−50. No. 13A.
28. Taylor, R. J. K.; Casey, G. The Ramberg−Bäcklund Rearrangement. *Org. React.* **2003,** *62,* 357−475. https://doi.org/10.1002/0471264180.or062.02.
29. Weinges, K.; Pill, J.; Klessing, K.; et al. Ringsysteme. Über den Einfluß des Carbocyclus auf den Reaktionsverlauf der Ramberg-Bäcklund-Reaktion von Oxathiapropellanen. *Chem. Ber.* **1977,** *110* (9), 2969−2977.
30. Martin, H.-D.; Mayer, B.; Pütter, M.; et al. Pterodactyladiene (Tetracyclo [4.4.0.02,5.07.10]deca-3,8-diene). *Angew. Chem. Int. Ed.* **1981,** *20* (8), 677−678. https://doi.org/10.1002/anie.198106771.
31. Reuß, F.; Heretsch, P. Synthesis of Aspidodispermine via Pericyclic Framework Reconstruction. *Org. Lett.* **2020,** *22* (10), 3956−3959. https://doi.org/10.1021/acs.orglett.0c01242.
32. Söderman, S. C.; Schwan, A. L. 2-Dibromotetrachloroethane: An Ozone-Friendly Reagent for the in Situ Ramberg−Bäcklund Rearrangement and its Use in the Formal Synthesis of E-Resveratrol. *J. Org. Chem.* **2012,** 77 (23), 10978−10984. https://doi.org/10.1021/jo3021852.
33. Boeckman, R. K.; Yoon, S. K.; Heckendorn, D. K. Synthetic Studies Directed toward the Eremantholides. 2. A Novel Application of the Ramberg-Bäcklund Rearrangement to a Highly Stereoselective Synthesis of (+)-Eremantholide A. *J. Am. Chem. Soc.* **1991,** *113* (25), 9682−9684. https://doi.org/10.1021/ja00025a049.

CHAPTER TEN

Inversion of configuration in modern synthesis

Introduction

The control of stereochemistry in nucleophilic substitution reactions is effectively limited to S_N2 reactions, and considerable effort has gone into improving the level of stereo control afforded by the Walden inversion. In Chapter 6, the discovery of the Walden inversion was discussed in depth, and in Chapter 7, the work of Hughes and Ingold, in particular, in elucidating the mechanism of the reaction was discussed. Now it is time to discuss the ensuing developments in controlling the stereochemistry of nucleophilic substitution reactions.

Effects of solvent on the mechanism and rates of nucleophilic substitutions

Aprotic dipolar solvents

By the early 1960s, it had become well recognized that solvent played a key role in nucleophilic substitution reactions.[1–6] Although considerable effort had been expended on the interactions between ions and solvent in aqueous solutions, there was much less focus on the interactions of ions with aprotic solvents (e.g., issue No. 24 of the *Discussions of the Faraday Society*, *Interaction in Ionic Solutions*[7] contains no paper in which the solvation of ions in aprotic media is explicitly addressed). It took another 4 years before the solvation of ions and their sizes in aprotic media was actually addressed. In a 1961 paper by Prue and Sherrington, a sharp dichotomy between the Stokes law radii[8] of monovalent anions in protic (methanol) and aprotic dipolar (dimethylacetamide, Me_2NAc, and dimethyl sulfoxide, Me_2SO) solvents was noted.[9] The data for sodium and potassium cations, and bromide and iodide anions, collected in Table 10.1, reveal a dichotomy in the ion solvation. In methanol, the Stokes radii (r_s) of all the ions are larger than the crystallographic radii (r_c)—by at least 1.6 Å for the cations and a lesser amount for anions—consistent with them carrying a solvation shell. The behavior in the aprotic dipolar solvents, however, is different: whereas the

Addition, Elimination and Substitution: Markovnikov, Hofmann, Zaitsev and Walden
ISBN: 978-0-12-821027-7
https://doi.org/10.1016/B978-0-12-821027-7.00008-6

Table 10.1 Effects of solvent on the Stokes radii of ions.

Ion	$r_{cryst.}$	Me_2NAc	Me_2SO	MeOH
Na^+	0.95	3.5	3.4	3.3
K^+	1.33	3.5	3.3	2.9
Br^-	1.95	2.1	1.7	2.7
I^-	2.16	2.1	1.7	2.4

Stokes radii of the cations are still larger than the crystallographic radii by 1.9 Å or more, the Stokes radii for the anions in dimethyl acetamide are comparable to their crystallographic radii, and in dimethyl sulfoxide, both are at least 0.2 Å smaller than the crystallographic radii. These data must be interpreted as revealing that cations in these aprotic solvents are still strongly solvated, anions are more weakly solvated, making them effectively smaller.

Phase transfer catalysis

An alternative technique for denuding a nucleophilic anion of solvent first appeared in the early 1960s. In this approach, known as phase transfer catalysis,[10–18] the anion is carried across the interface between two immiscible liquids by a lipophilic quaternary ammonium ion (Fig. 10.1A) in the

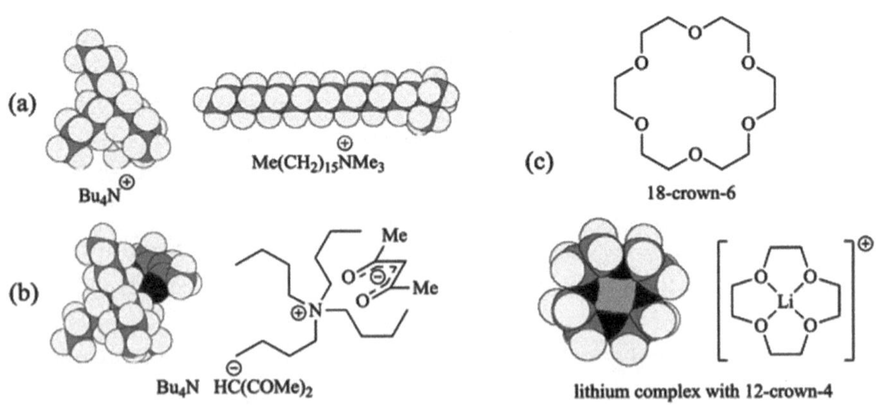

Figure 10.1 Representative phase transfer catalysts: (A) highly lipophilic cations: the tetra-*n*-butylammonium ion (left) and the cetyltrimethylammonium ion (right); (B) tetra-*n*-butylammonium acetylacetonate; (C) the potassium-selective crown ether, 18-crown-6 (upper) and the lithium complex of 12-crown-4 (lower).

form of a loose ion pair (Fig. 10.1B) (liquid—liquid phase transfer catalysis). Cations can be extracted from a solid matrix as a complex with a lipophilic crown ether (solid—liquid phase transfer) (Fig. 10.1C). The crown ethers form highly selective complexes with alkali metal ions: 12-crown-4 (12-C-4) with Li^+, 15-C-5 with Na^+, and 18-C-6 with K^+.

In both embodiments, the role of the phase transfer catalyst is to enhance the solubility of the ionic nucleophile in a low-polarity solvent, thus increasing its concentration, enhancing its nucleophilicity, and increasing the rate of the bimolecular substitution reaction. However, at the same time, the basicity of the nucleophile is also raised, so phase transfer catalysis has not been widely used in S_N2 reactions because elimination strongly competes with substitution in secondary halides and sulfonates.

S_N2 displacement driven by the formation of a phosphoryl group

The bond energy of a P—O bond is 335 kJ mol^{-1} and of the phosphoryl (P=O) group is 544 kJ mol^{-1}. These numbers mean that the formation of a phosphoryl group from tricoordinate or pentacoordinate phosphorus compounds is a highly exothermic process that can be used to drive a nucleophilic substitution reaction. Walden himself used the reaction between an alcohol and phosphorus pentachloride to produce the halide with inversion. The reaction, $ROH + PCl_5 \rightarrow RCl + HCl + POCl_3$, has an enthalpy of approximately −340 kJ mol^{-1}, depending on the alkyl group of the alcohol.

Phosphine-based halogenating reagents

The addition of an oxidant, X—Y, to a trialkylphosphine (**498**) leads to the formation of a phosphonium salt containing a P—X bond (**499**). These types of phosphonium ions react with alcohols (**500**) to give a pentacoordinate phosphorus compound (**501**) and H—Y. These pentacoordinate species enter into fragmentation reactions promoted by the nucleophile, Y:−, to give the inverted substitution product (**502**) and the trialkylphosphine oxide (**503**) (Scheme 10.1).

There have been several forms of this reaction since the 1960s. The reagents for the more important versions of the reactions that give halides are gathered in Table 10.2.

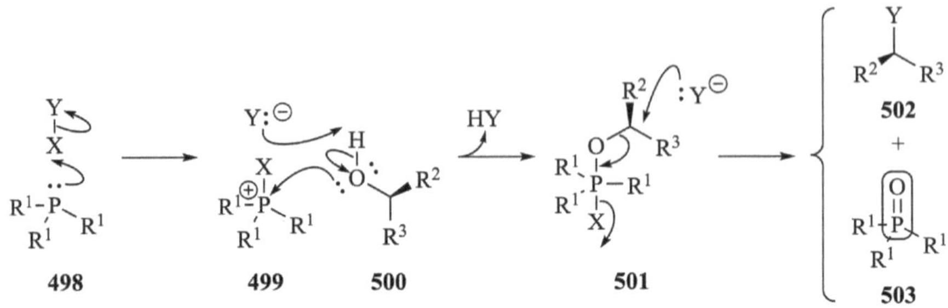

Scheme 10.1 The mechanism of the S_N2 displacement of an alcohol OH group by a reagent formed from triphenylphosphine and a nucleophile source.

Table 10.2 Phosphine-based halogenating agents for alcohols.

X	Y	Typical reagent	Products
Cl	CCl_3	Ph_3P, CCl_4, Δ	R–Cl
Br	CBr_3	$(n\text{-}C_8H_{17})_3P$, CBr_4, Et_2O	R–Br
Br	Br	Ph_3P, Br_2, DMF	R–Br
I	I	Ph_3P, I_2, DMF	R–I

In 1966, Downie, Holmes, and Lee reported the conversion of alcohols to alkyl chlorides under mild conditions by treating the alcohol with triphenylphosphine in carbon tetrachloride.[19] Two years later, Hooz and Gilani[20] reported the reaction between alcohols, carbon tetrahalides, and an alcohol to give alkyl halides with inversion of configuration, as shown by the conversion of *S*-(+)-2-octanol (**504**) to *R*-(−)-2-chlorooctane (**505**). The reaction now carries the name of Rolf Appel, whose review of the reaction appeared in almost a decade later.[21] The reaction with the carbon tetrahalides was used in the synthesis of histrionicotoxin by Stork and Zhao,[22] where the conversion of the alcohol **506** to the inverted bromide **507** was a critical step in assembling the piperidine ring of the alkaloid and the synthesis of the chlorinated marine natural product, (+)-brasilenyne, by Denmark and Yang.[23] The reaction with bromine as the source of the halogen was used in the synthesis of cyclosativene by Baldwin and Tomesch to convert the hindered alcohol **510** into the corresponding bromide, **511**.[24] These examples are gathered in Scheme 10.2.

Although this reaction does not usually lead to rearrangements, under appropriate circumstances it can lead to products that could be viewed as arising from solvolysis. An example of this is provided by the final ring closure step in the synthesis of iceane by Hamon and Taylor, where the

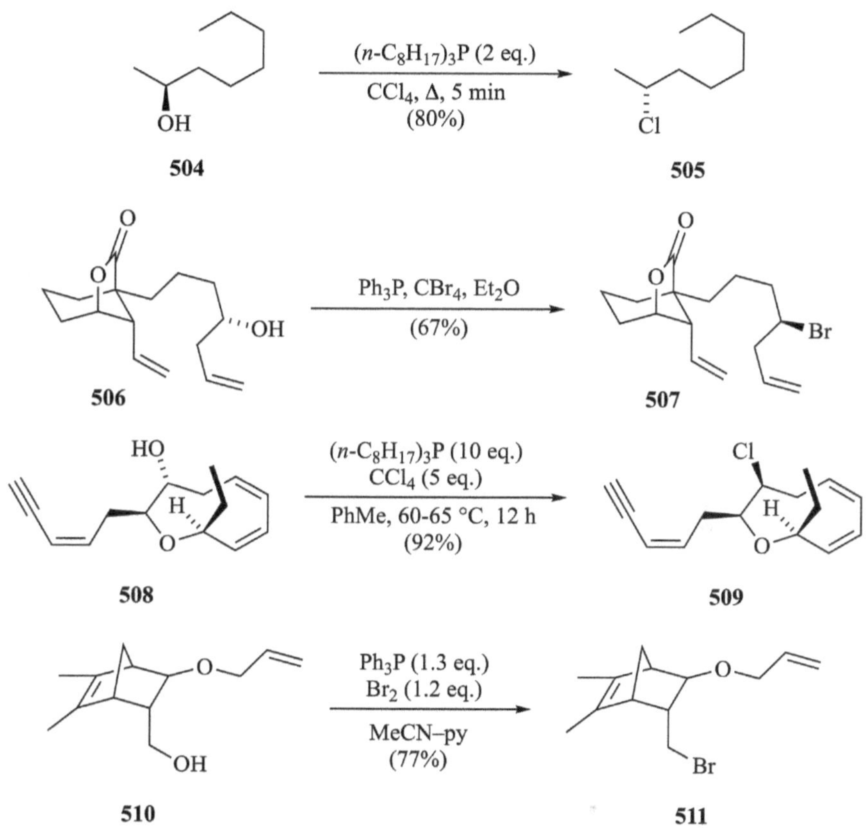

Scheme 10.2 Conversion of alcohols to halides with inversion using trialkylphosphine-based halogenating agents.

reaction of the alcohol **512** with triphenylphosphine in carbon tetrachloride gives a mixture of the epimeric chlorides **514**, the formal products of a transannular solvolysis (Scheme 10.3).[25]

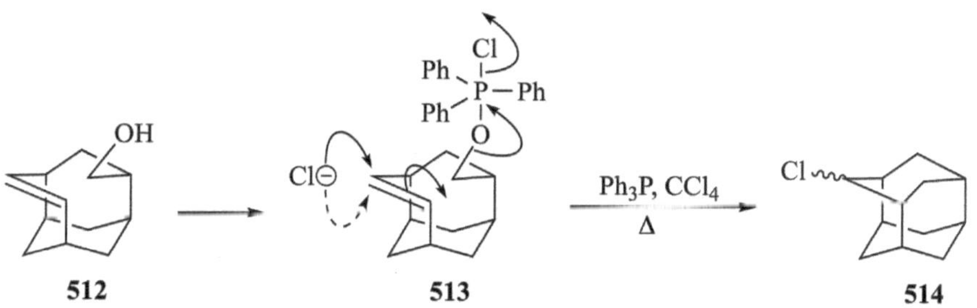

Scheme 10.3 Transannular cyclization in the reaction of the unsaturated alcohol 503 with triphenylphosphine in carbon tetrachloride.

The Mitsunobu reaction

The next major improvement in the use of phosphine came from the work of Japanese chemist, Ōyō Mitsunobu (1934–2003, Fig. 10.2). In 1967, Mitsunobu and Yamada reported that when alcohols react with the reagent formed from a dialkyl azodicarboxylate (**515**) and triphenylphosphine, the product reacts with a wide range of pronucleophiles (conjugate acids of anionic nucleophiles) to give the inverted product.[26–33]

Ōyō Mitsunobu (1934–2003) was born in Such, Tokyo, and educated at Gakushūin University. He graduated with his bachelor's degree in 1959. Following his graduation, he began study under Mukaiyama at Tokyo Institute of Technology. Immediately following his graduation with his D.Sc. in 1964, he was appointed to the faculty of the Chemistry Department of Aoyama Gakuin University in Tokyo. He remained there for his entire career. During his years as a graduate student, he began what became a lifelong study of the reactions of trivalent phosphorus compounds and their use in synthesis. He is best recognized for his development of his eponymous reaction for activating alcohols toward nucleophilic substitution using a tertiary phosphine and an azodicarboxylic acid derivative.

Figure 10.2 Ōyō Mitsunobu (1934–2003). *Modified from J. M. McBride, Chem 125. License: Creative Commons BY-NC-SA 3.0.*

The Mitsunobu reaction proceeds by the mechanism shown in Scheme 10.4. The initial step involves the addition of the phosphine to the electron-deficient π bond of the azodicarboxylate ester, which may be diethyl azodicarboxylate (DEAD) or diisopropyl azodicarboxylate (DIAD), to give a betaine (**516**), which has an approximate pK_a of 13. The betaine deprotonates the pronucleophile to give the phosphonium ion **517**, which reacts with the alcohol (**500**) to give the pentacoordinate phosphorane **518**. The fragmentation of this phosphorane by the nucleophile formed by the deprotonation of the pronucleophile to give the nucleophilic substitution product **502**, triphenylphosphine oxide, and the hydrazodicarboxylate ester **519**. The driving force for the reaction comes from the redox dehydration reaction: $RO_2CN = NCO_2R + PPh_3 + H_2O \rightarrow RO_2CNH{-}NHCO_2R + O = PPh_3$.

In contrast to most nucleophilic substitution reactions, where basic nucleophiles are more strong, the Mitsunobu reaction performs best with the conjugate bases of pronucleophiles with pK_a values below about 11. These are relatively weakly basic nucleophiles, so in some ways the Mitsunobu reaction is in some ways complementary to other S_N2 reactions. The range of pronucleophiles that are useful in the reaction is shown in Table 10.3.

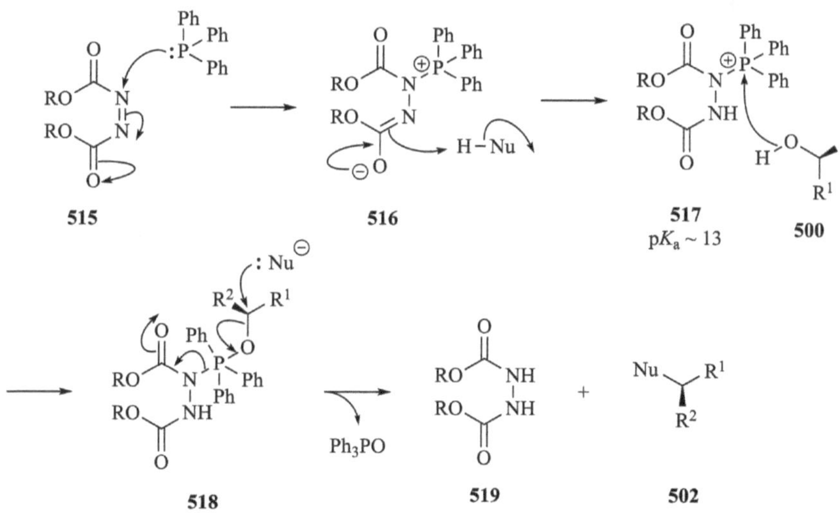

Scheme 10.4 The accepted mechanism of the Mitsunobu reaction.

Table 10.3 Useful pronucleophiles for the Mitsunobu reaction.

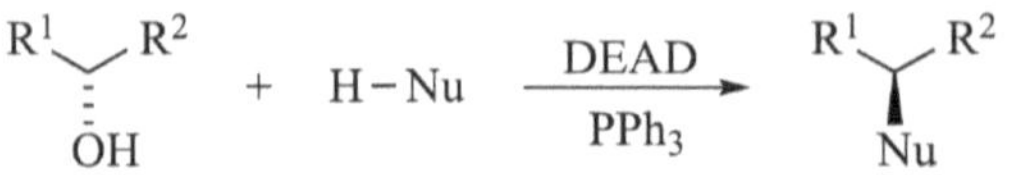

H–Nu	Product	H–Nu	Product
RCO_2H	$RCO_2{-}R$	$ArCO_2H$	$ArCO_2{-}R$
ArOH	ArO–R	ArSH	ArS–R
R′ C(O)SH	R′ C(O)S–R	R′ SO_2NHR''	R′ $SO_2NR''{-}R$
$(R'\ CO)_2NH$	$(R'\ CO)_2N{-}R$	NH_4SCN	R–SCN + R–NCS
$(R'\ CO)_2CH_2$	$(R'\ CO)_2CH{-}R$	$Me_2C(OH)CN$	R–CN
HN_3 (Me_3SiN_3)	$R{-}N_3$		

One of the more important uses of the Mitsunobu reaction in synthesis has been the inversion of configuration of alcohols, which is accomplished by formation of the inverted carboxylate ester, followed by base hydrolysis. In an interesting variant of this reaction, *p*-diphenylphosphinobenzoic acid (**521**) serves as both the phosphine and the pronucleophile.[34] In this example (Scheme 10.5), the *R* alcohol **520** (91% e.e.) is converted to the *S* ester **523**, which is then hydrolyzed to the *S* alcohol **524** (91% e.e.), a net inversion of configuration of not less than 98%. This reagent also allows the facile removal of the phosphine oxide, which is frequently a difficult and tedious task, by simple acid–base extraction.

The use of the reaction in the formation of other functional groups from alcohols is illustrated by the examples in Scheme 10.6. The intramolecular Mitsunobu reaction of the (*S*)-*p*-nitrosulfonamide **525** proceeds with inversion to give the (*R*)-*p*-nitrosulfonylpyrrolidine derivative **526**.[35] Although both DEAD and DIAD failed to give acceptable yields of product, the reaction of the *R*-α-hydroxyester **527** with hydrazoic acid and trimethylphosphine using ADDP gives the *S*-α-azidoester **528**

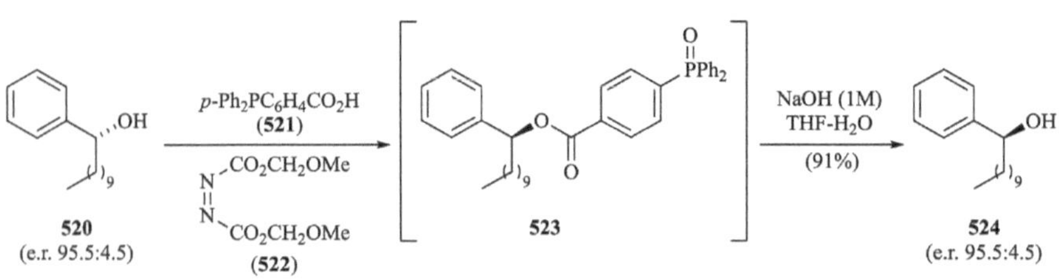

Scheme 10.5 The inversion of configuration of an alcohol using a bifunctional phosphine that also serves as the pronucleophile.

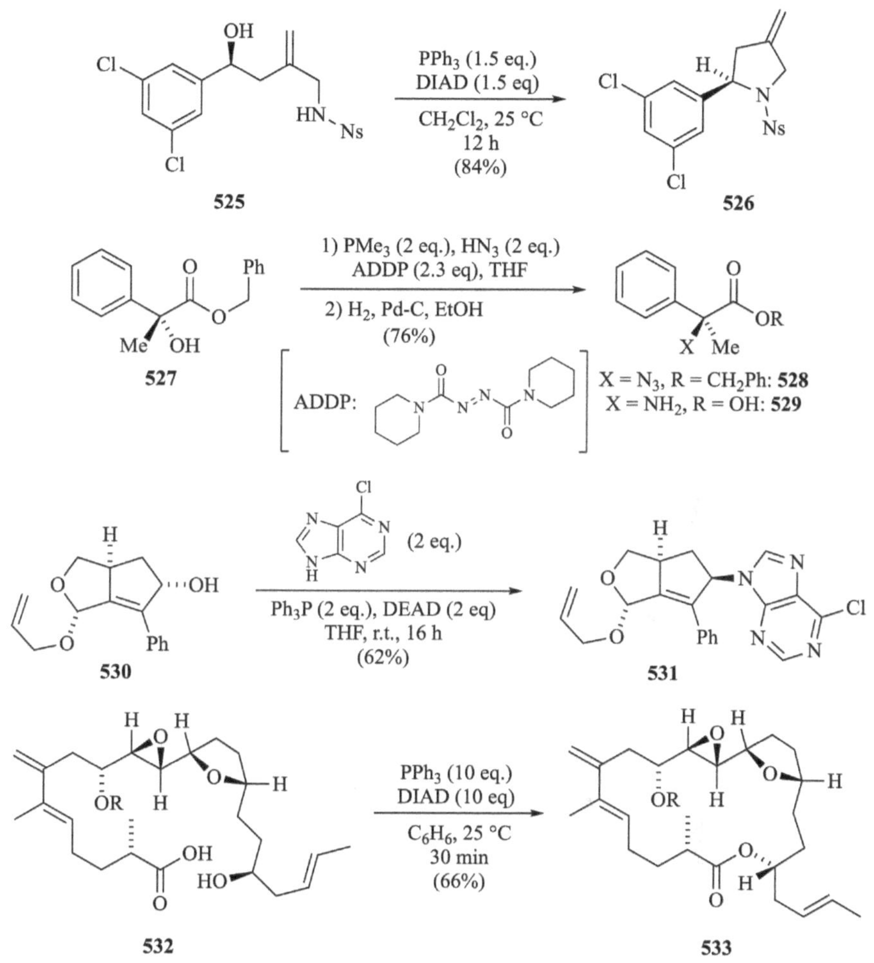

Scheme 10.6 Representative Mitsunobu reactions in synthesis.

in >99% e.e.; hydrogenolysis of the azide and benzyl groups gives the *S*-α-amino acid **529** in 76% overall yield and >99% e.e.[36] The Mitsunobu reaction has been successfully applied to the synthesis of nucleoside analogues,[37] as illustrated by the synthesis of the precursor for a carbocyclic nucleoside analogue **531** from the bicyclic alcohol **530**.[38] As expected, there is complete inversion of configuration at the carbinol carbon in this reaction. The Mitsunobu reaction is relatively tolerant of functional groups that do not function as pronucleophiles. This is illustrated nicely by the macrolactonization of the hydroxyacid **532** to the macrolactone

533 in the Williams and Meyer synthesis of (+)-amphidinolide K.[39] In this reaction, the ester formation follows the "normal" course of the reaction, where inversion of configuration at the carbinol carbon occurs. However, in lactone formation, this is not always the case, as observed in De Brabender's synthesis of peloruside A.[40]

Immobilized Mitsunobu reagents[41,42]

One approach that has been applied successfully to problems associated with the purification of the products of the reaction has been to employ immobilized reagents, typically polymer-bound reagents (Scheme 10.7). Modified Merrifield-type cross-linked polystyrenes have typically been used for the immobilization,[43–45] but a 2005 paper reported that both polystyrene-supported-azodicarboxylic esters and -triphenylphosphines can fail to produce isopropyl *p*-nitrobenzoate when used with the unsupported complementary reagent. In those cases, soluble oligomeric species obtained by ring-opening metathesis of norbornenes were more successful.[46] Soluble oligomer-supported reagents and catalysts have also been useful.[47–50]

The substoichiometric Mitsunobu reaction

The status of the Mitsunobu reaction as the premier method for carrying out transformations of alcohols with inversion stands essentially unsurpassed. The term "Mitsunobu reaction" returned 43,600 hits on Google Scholar and 11,000 hits on Scifinder Scholar on January 31, 2022, and the authors of a 100-page, 2009 review with 1855 references itself,[32] make the point that the field has become so large that modern reviewers are tending toward specialized reviews.

Still, despite its high value as a synthetic method, the reaction still presents problems. First, the reaction is stoichiometric in two reagents: the phosphine and the azodicarboxylate. Second, the reaction presents major hazards that limit its use to the small scale: The azodicarboxylates are toxic and also present the hazard associated with shock and thermal sensitivity—they are explosive[51] (in fact, the transportation of DEAD by air or as freight in passenger transports is prohibited; the pure compound is no longer sold and it is now only commercially available as a solution). Third, the phosphine oxide and hydrazodicarboxylate products can be extremely difficult to remove from the product; chromatography is routinely required to obtain the pure product. For these reasons, a range of alternative reagents have been sought to reduce these problems.

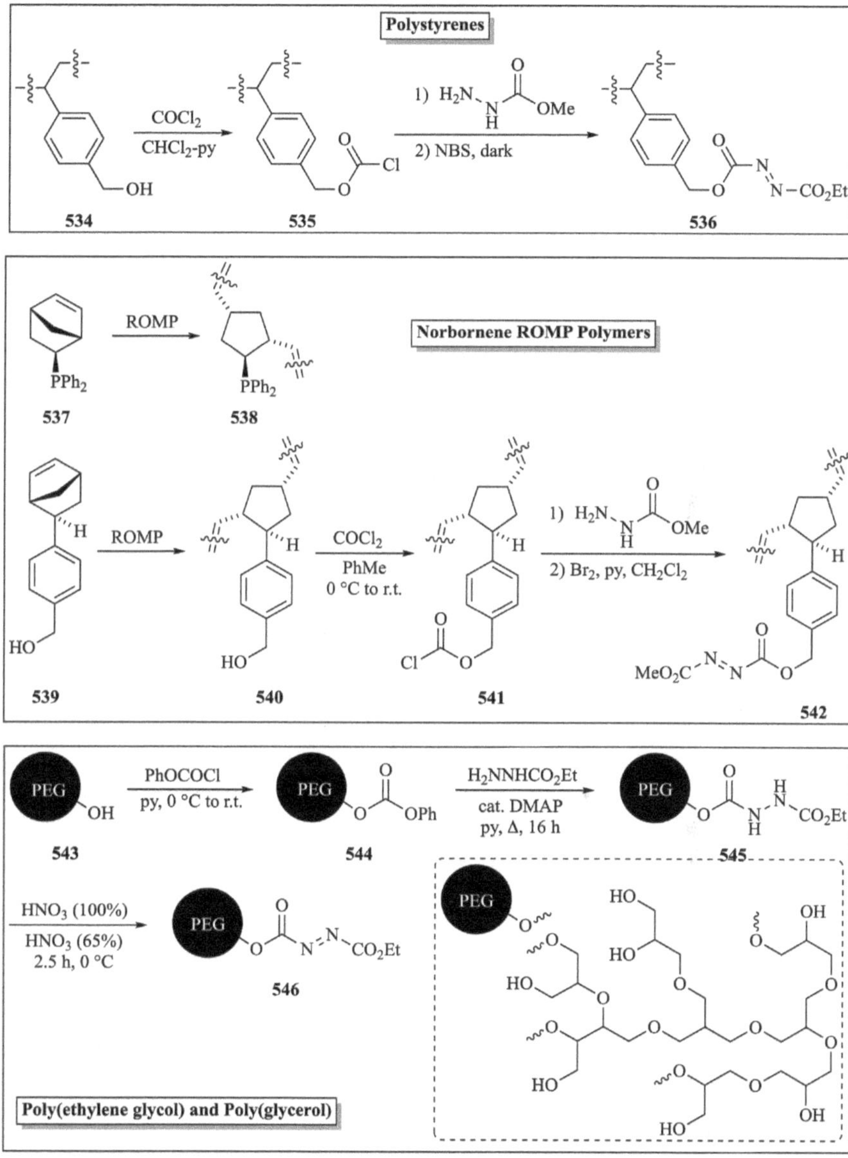

Scheme 10.7 Representative polymer-immobilized Mitsunobu reagents.

The approach to reducing the problems with the reaction first focused their efforts on methods to render the reaction substoichiometric in either the oxidant (azodicarboxylate) or the reductant (phosphine). The attempts to develop catalytic versions of the reaction have met with some success, as indicated in a recent review by Denton and coworkers.[52]

The Toy and Taniguchi variants

In two papers in 2006 and 2010, Toy and coworkers revealed a version of the reaction[53,54] in which the stoichiometric oxidant was reduced to sub-stoichiometric amounts by the addition of a sacrificial stoichiometric oxidant, The stoichiometric oxidant, (diacetoxyiodo)benzene, regenerated the azodicarboxylate, as shown in Fig. 10.3. The importance of the Toy variant comes from its status as the first major advance in the Mitsunobu reaction in four decades.

One limitation of the Toy procedure is the formation of acetic acid as a potentially competing pronucleophile. This was resolved by replacing the iodine-based oxidant with molecular oxygen. In 2010, Taniguchi first reported the use of molecular oxygen in the presence of an iron (II) porphyrin catalyst to accomplish the reoxidation of a hydrazo group to an azo group (Fig. 10.4). His group pioneered the use of an arylazocarboxylate ester as an oxidant in the reaction, as an alternative to the azodicarboxylic esters. His approach offers the major advantage that the by-product of the

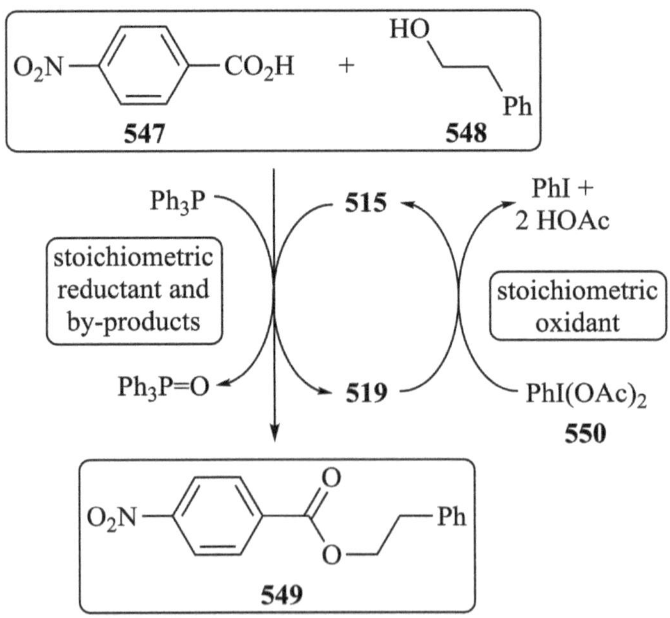

Figure 10.3 The Toy variant of the Mitsunobu reaction with substoichiometric quantities of the azodicarboxylate (**515**) and di(acetoxyiodo)benzene as a stoichiometric oxidant to recycle the hydrazodicarboxylate (**519**).

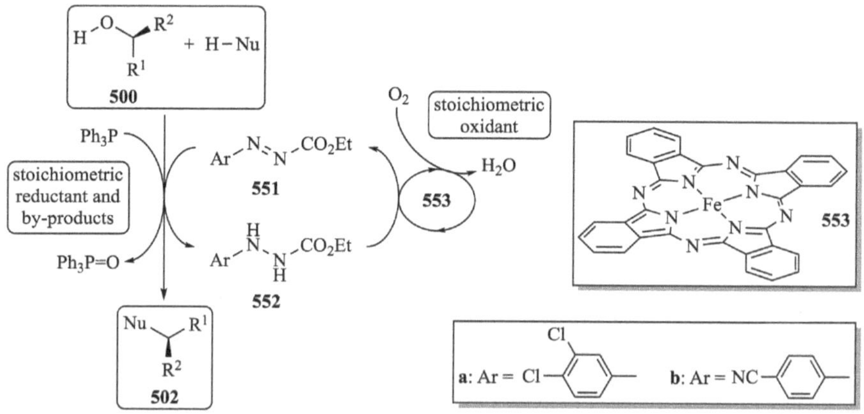

Figure 10.4 The model studies of the Taniguchi variant of the Mitsunobu reaction with substoichiometric quantities of an arylazocarboxylate (**551**) and molecular oxygen with an iron phthalocyanine catalyst (**553**) as a stoichiometric oxidant to recycle the hydrazodicarboxylate (**552**).

oxidation is water, rendering the reaction somewhat more green, and the arylazocarboxylate esters are less hazardous than the azodicarboxylate esters.[55–58]

Using this method, ethyl *S*-lactate (**554**) was converted to the 3,5-dinitrobenzoate ester of ethyl *R*-lactate (**556**).[58] The reaction gave a high optical yield, but was not stereospecific like the stoichiometric reaction (Scheme 10.8). In this example, the selectivity for inversion was at least 49:1.

A bona fide totally catalytic variant: the Aldrich variant

Both the Toy and the Taniguchi variants of the reaction still require stoichiometric amounts of the phosphine reductant, generating stoichiometric amounts of the phosphine oxide as waste. The next advance was disclosed by Aldrich in 2015.[59] The Aldrich variant was based on combining two

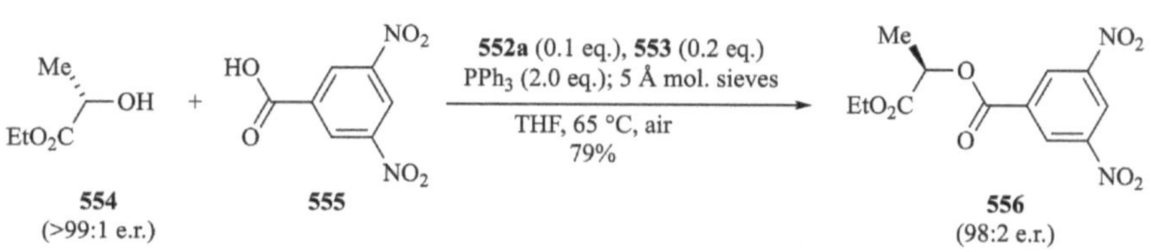

Scheme 10.8 The use of the Taniguchi variant of the Mitsunobu reaction resulted in formation of the dinitrobenzoate **556** with at least 98:2 selectivity for inversion.

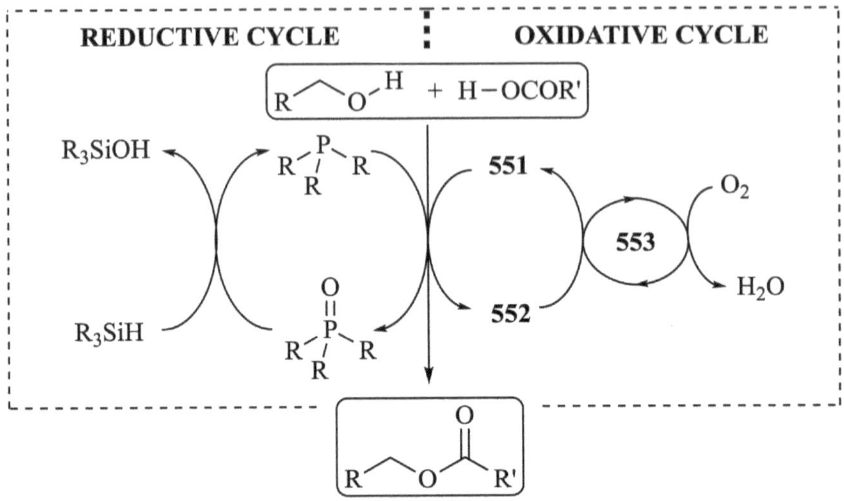

Figure 10.5 The Aldrich variant of the Mitsunobu reaction.

complementary catalytic cycles: an oxidation cycle, in which the hydrazo product is recycled to the azo compound, and a reduction cycle, in which the phosphine oxide is recycled to the phosphine (Fig. 10.5).

The oxidative cycle of the Aldrich variant was based on the Taniguchi variant, and the reductive cycle was based on the silane reduction of phosphine oxides. The two cycles were combined successfully, albeit with low yields. However, by using phospholane oxide **559** instead of tiphenylphosphine as the precursor of the phosphine reductant, phenylsilane to reduce **559** to the phosphine **560**, and carrying out the reaction under an atmosphere of oxygen, satisfactory yields of the esters are obtained. Under the optimized conditions, the *p*-substituted benzyl *p*-nitrobenzoates **558** were obtained in moderate yield from the corresponding benzyl alcohols **557** (Scheme 10.9). The effectiveness of the catalytic reaction with secondary alcohols has not yet been reported.

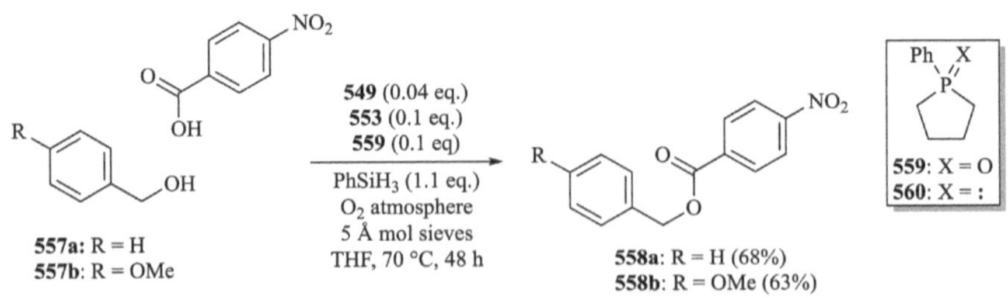

Scheme 10.9 The optimized reaction conditions for the Aldrich catalytic variant of the Mitsunobu reaction.

Changing the paradigm: the Denton variant

The "Holy Grail" of research on the Mitsunobu reaction, as the preceding discussion has made clear, has been to convert the stoichiometric reaction, which generates 2 mole equivalents of waste per mole of coupled product, into a catalytic reaction. In 2019, the first reagent that really fulfilled these requirements was disclosed by the research group of Ross M. Denton at Nottingham.[52,60] This advance was a counterintuitive modification of the reaction because the oxidation state of the phosphorus reagent remains constant at +5 throughout the reaction—the reaction is redox-neutral. Instead of attempting to modify the phosphine and azodicarboxylate reagents, the method is based on using an *o*-hydroxybenzyldiphenylphosphine oxide (**561**) as the catalyst. The catalytic cycle of the reaction is shown in Fig. 10.6. The mechanism proposed by Denton and coworkers has been confirmed computationally.[61]

The first step of the catalytic cycle is the reaction between the pronucleophile and the phosphine oxide **561**. Optimization of the reaction conditions revealed 2,4-dinitrobenzoic acid (pK_a 1.4) as the best choice for the pronucleophile; benzoic acid (pK_a 4.2) was totally ineffective. Protonation of the phosphoryl oxygen of the catalyst by the pronucleophile is accompanied by dehydration to give the cyclic phosphonium ion **562**, analogous to

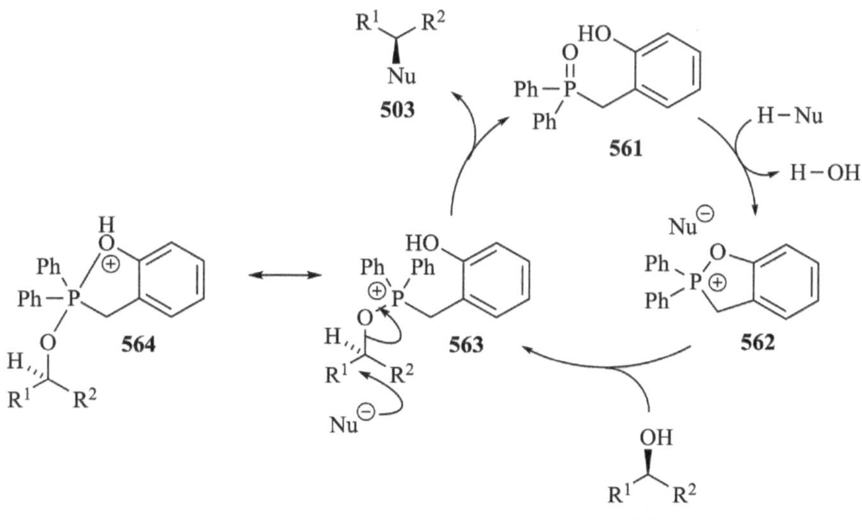

Figure 10.6 The catalytic cycle of the redox-neutral esterification of alcohols with inversion of configuration using the Denton organocatalyst.

the phosphonium ion **518** (Scheme 10.4). This ion then reacts with the alcohol to give the oxonium ion **563**, which may be in resonance with the pentacoordinate phosphorane **564**. The phosphonium ion **563** then reacts with the nucleophile to give the inverted product, **553**, and the regenerated catalyst, **561**.

The reaction requires that the water generated be removed because the phosphonium salt **560** is susceptible to hydrolysis to regenerate the starting catalyst. The dehydration, however, exhibits one fascinating characteristic: the addition of every external drying agent tested reduced the stereochemical control and even led to predominant retention of configuration. On the other hand, simple azeotropic removal of the water led to clean inversion of configuration. The higher-boiling solvents gave higher yields of the inverted product with shorter reaction times; xylene was the preferred solvent. Representative examples of the reaction are gathered in Scheme 10.10.

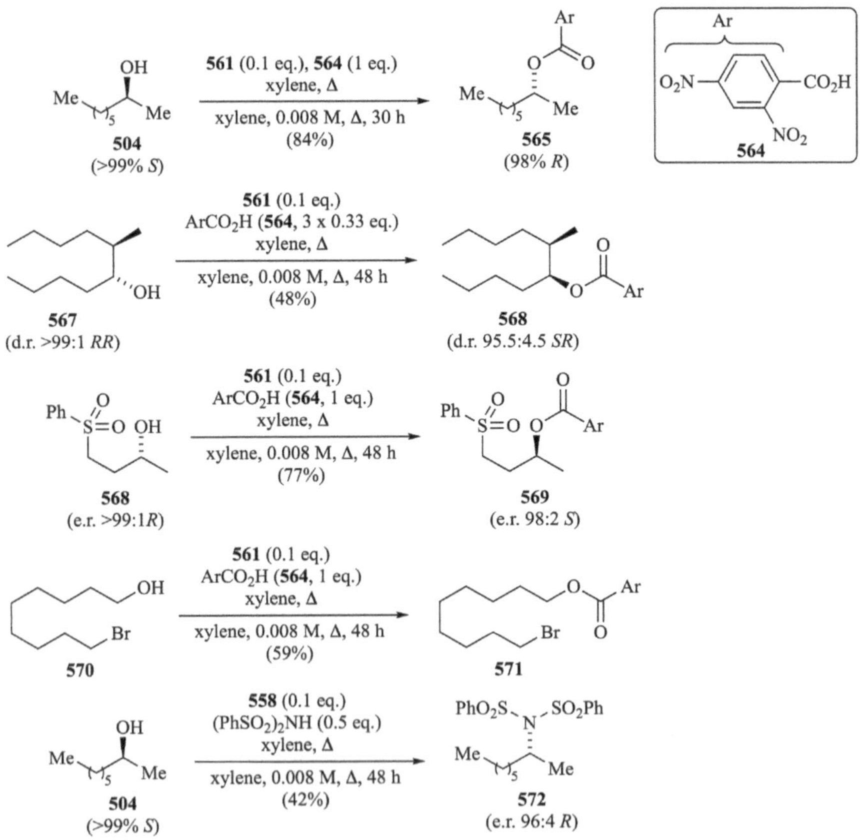

Scheme 10.10 Representative examples of the Denton variant of the Mitsunobu reaction.

The conversion of alcohol **570** to ester **571** in Scheme 10.10 is worth noting because under the standard Mitsunobu reaction conditions or any variant where a phosphine is used, it is unlikely that the primary bromide would survive. The last example in Scheme 10.10 illustrates the use of the Denton modification in the formation of a C—N bond with inversion.

Since the emergence of the Denton variant of the Mitsunobu reaction, it is hardly surprising that work to immobilize the catalyst to facilitate its recovery and removal from the reaction mixture would become a focus of research. This appears to be the case, with new papers appearing in 2021, one on the redox-neutral (Denton) Mitsunobu reaction[62] and one on the Appel reaction.[63] After so many decades as a "backwater" of organic chemistry, this area has become an active field for research.

Thioethers directly from alcohols

Recently, the search for direct nucleophilic substitution of alcohols with inversion of configuration has expanded to the use of *S*-alkylisothiouronium salts in the activation and substitution of alcohols. The reaction is proposed to proceed by the mechanism shown in Fig. 10.7. The first step is the reaction between the alcohol **501** and the isothiouronium salt **574** in the presence of *tert*-butyltetramethylguanidine (**573**) to give the adduct **575**. This adduct carries two possible leaving groups, of which the alkylthiolate anion (**577**) is the best. The ionization of **575** gives the *O*-alkylisoiuronium cation **576** that then reacts with the thiolate anion to give the thioether **578**, with inversion of configuration.[64]

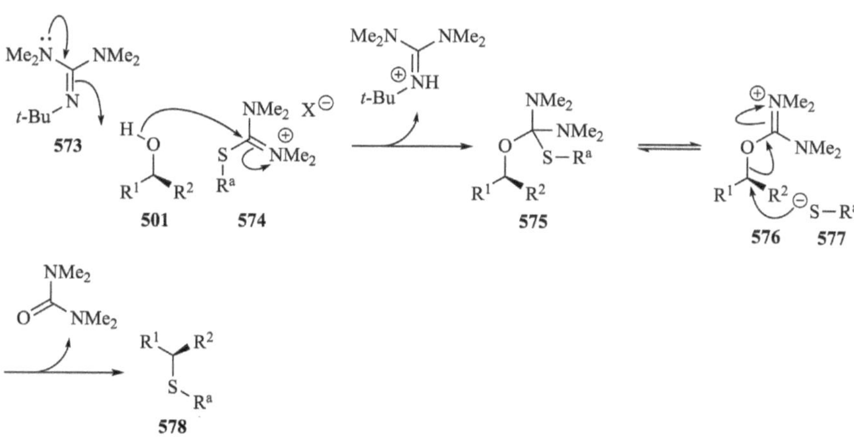

Figure 10.7 The direct conversion of an alcohol to a thioether with inversion of configuration.

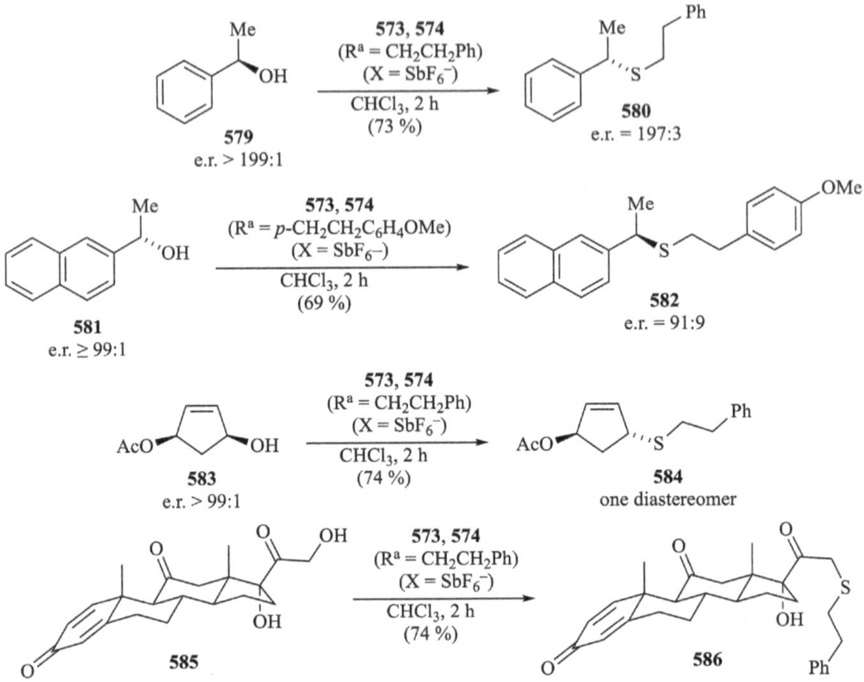

Scheme 10.11 Representative examples of the direct conversion of alcohols to thioethers using isothiouronium hexafluorophosphates.

Some representative examples of this reaction are given in Scheme 10.11. Several of the reactions were carried out with both the hexafluorophosphate salts and with the bromide salts. In all cases, the chemical and optical yields of the reactions with the hexafluorophosphate salts were superior to those with the bromide salts. The reaction of prednisone (**585**) with *S*-(2-phenylethyl) isothiouronium hexafluorophosphate gives only the product of displacement at the neopentyl hydroxyl group and has no effect on the cross-conjugated dienone.

References

1. Alexander, R.; Ko, E. C. F.; Parker, A. J.; Broxton, T. J. Solvation of Ions. XIV. Protic-Dipolar Aprotic Solvent Effects on Rates of Bimolecular Reactions. Solvent Activity Coefficients of Reactants and Transition States at 25°. *J. Am. Chem. Soc.* **1968,** *90* (19), 5049–5069 https://doi.org/10.1021/ja01021a002.
2. Parker, A. J. The Effects of Solvation on the Properties of Anions in Dipolar Aprotic Solvents. *Q. Rev. Chem. Soc.* **1962,** *16* (2), 163–187. https://doi.org/10.1039/qr9621600163.
3. Parker, A. J. Protic-Dipolar Aprotic Solvent Effects on Rates of Bimolecular Reactions. *Chem. Rev.* **1969,** *69* (1), 1–32. https://doi.org/10.1021/cr60257a001.

4. Reichardt, C.; Welton, T. *Solvents and Solvent Effects in Organic Chemistry;* Wiley-VCH, 2011.
5. Westaway, K. C. Solvent Effects on the Structure of SN2 Transition States. *Can. J. Chem.* **1978,** *56* (20), 2691–2699. https://doi.org/10.1139/v78-443.
6. Westaway, K. C.; Lai, Z.-G. Solvent Effects on SN2 Transition State Structure. II: The Effect of Ion Pairing on the Solvent Effect on Transition State Structure. *Can. J. Chem.* **1989,** *67* (2), 345–349. https://doi.org/10.1139/v89-056.
7. Contents Pages. *Discuss. Faraday Soc.* **1957,** *24,* 1–6. https://doi.org/10.1039/DF9572400001.
8. Atkins, P. W.; de Paola, J. *Atkins' Physical Chemistry;* W. H. Freeman, 2006.
9. Prue, J. E.; Sherrington, P. J. Test of the Fuoss-Onsager Conductance Equation and the Determination of Ion Sizes in Dimethyl-Formamide. *Trans. Faraday Soc.* **1961,** *57,* 1795–1808. https://doi.org/10.1039/TF9615701795.
10. Dockxx, J. Quaternary Ammonium Compounds in Organic Synthesis. *Synthesis* **2002,** *1973* (08), 441–456. https://doi.org/10.1055/s-1973-22233.
11. Dehmlow, E. V. Phase-Transfer Catalyzed Two-Phase Reactions in Preparative Organic Chemistry. *Angew Chem. Int. Ed. Engl.* **1974,** *13* (3), 170–179. https://doi.org/10.1002/anie.197401701.
12. Dehmlow, E. V. Advances in Phase-Transfer Catalysis [New Synthetic Methods (20)]. *Angew Chem. Int. Ed. Engl.* **1977,** *16* (8), 493–505. https://doi.org/10.1002/anie.197704933.
13. Dehmlow, E. V.; Dehmlow, S. S. *Phase Transfer Catalysts;* VCH, 1993.
14. Halpern, M. E., Ed. *Phase-Transfer Catalysis: Mechanisms and Syntheses;* American Chemical Society: Washington, DC, 1997.
15. Mąkosza, M.; Fedoryński, M. Catalysis in Two-phase Systems: Phase Transfer and Related Phenomena. *Adv. Catal.* **1987,** *35* (C), 375–422. https://doi.org/10.1016/S0360-0564(08)60097-8.
16. Nelson, A. Asymmetric Phase-Transfer Catalysis. *Angew. Chem. Int. Ed.* **1999,** *38* (11), 1583–1585. https://doi.org/10.1002/(SICI)1521-3773(19990601)38:11<1583::AID-ANIE1583>3.0.CO;2-E.
17. Sasson, Y.; Neumann, R. *Handbook of Phase Transfer Catalysis;* Springer: Netherlands, 1997. https://doi.org/10.1007/978-94-009-0023-3.
18. Starks, C. M.; Liotta, C. L.; Halpern, M. *Phase Transfer Catalysis: Fundamentals, Applications and Industrial Perspectives;* Chapman and Hall, 1994.
19. Downie, I. M.; Holmes, J. B.; Lee, J. B. Preparation of Alkyl Chlorides under Mild Conditions. *Chem. Ind.* **1966,** 900–901. No. 22.
20. Hooz, J.; Gilani, S. S. H. A rapid, mild procedure for the preparation of alkyl chlorides and bromides. *Can. J. Chem.* **1968,** *46* (1), 86–87. https://doi.org/10.1139/v68-017.
21. Appel, R. Tertiary Phosphane/Tetrachloromethane, a Versatile Reagent for Chlorination, Dehydration, and P–N Linkage. *Angew Chem. Int. Ed. Engl.* **1975,** *14* (12), 801–811. https://doi.org/10.1002/anie.197508011.
22. Stork, G.; Zhao, K. Total Syntheses of (-)-Histrionicotoxin and (-)-Histrionicotoxin 235A. *J. Am. Chem. Soc.* **1990,** *112* (15), 5875–5876. https://doi.org/10.1021/ja00171a035.
23. Denmark, S. E.; Yang, S. M. Total Synthesis of (+)-Brasilenyne. Application of an Intramolecular Silicon-Assisted Cross-Coupling Reaction. *J. Am. Chem. Soc.* **2004,** *126* (39), 12432–12440. https://doi.org/10.1021/ja0466863.
24. Baldwin, S. W.; Tomesch, J. C. Alkyne Capture of an Homoallylic Carbonium Ion. *Tetrahedron Lett.* **1975,** 1055–1058. https://doi.org/10.1016/S0040-4039(00)72643-8. No. 12.

25. Hamon, D. P. G.; Taylor, G. F. A Synthesis of Tetracyclo[5,3,1,1(2,6),0(4,9)]dodecane (Iceane). *Aust. J. Chem.* **1976,** *29* (8), 1721–1734. https://doi.org/10.1071/CH9761721.
26. Mitsunobu, O. The Use of Diethyl Azodicarboxylate and Triphenylphosphine in Synthesis and Transformation of Natural Products. *Synthesis* **1981,** *1981* (01), 1–28. https://doi.org/10.1055/s-1981-29317.
27. Mitsunobu, O.; Yamada, M. Preparation of Esters of Carboxylic and Phosphoric Acid via Quaternary Phosphonium Salts. *Bull. Chem. Soc. Jpn.* **1967,** *40* (10), 2380–2382. https://doi.org/10.1246/bcsj.40.2380.
28. Fletcher, S. The Mitsunobu Reaction in the 21st Century. *Org. Chem. Front.* **2015,** *2* (6), 739–752. https://doi.org/10.1039/c5qo00016e.
29. Guo, H.; Fan, Y. C.; Sun, Z.; Wu, Y.; Kwon, O. Phosphine Organocatalysis. *Chem. Rev.* **2018,** *118* (20), 10049–10293. https://doi.org/10.1021/acs.chemrev.8b00081.
30. Hughes, D. L. Progress in the Mitsunobu Reaction. A Review. *Org. Prep. Proced. Int.* **1996,** *28* (2), 127–164. https://doi.org/10.1080/00304949609356516.
31. Hughes, D. L. The Mitsunobu Reaction. *Org. React.* **1992,** *42,* 335–656. https://doi.org/10.1002/0471264180.or042.02.
32. Swamy, K. C. K.; Kumar, N. N. B.; Balaraman, E.; Kumar, K. V. P. P. Mitsunobu and Related Reactions: Advances and Applications. *Chem. Rev.* **2009,** *109* (6), 2551–2651. https://doi.org/10.1021/cr800278z.
33. Panday, S. K. Advances in the Mitsunobu Reaction: An Excellent Organic Protocol with Versatile Applications. *Minirev. Org. Chem* **2019,** *16* (2), 127–140. https://doi.org/10.2174/1570193X15666180612090313.
34. Muramoto, N.; Yoshino, K.; Misaki, T.; Sugimura, T. Mitsunobu Reaction with 4-(Diphenylphosphino)Benzoic Acid: A Separation-Friendly Bifunctional Reagent that Serves as Both a Reductant and a Pronucleophile. *Synthesis* **2013,** *45* (07), 931–935. https://doi.org/10.1055/s-0032-1318327.
35. Luo, G.; Xiang, M.; Krische, M. J. Successive Nucleophilic and Electrophilic Allylation for the Catalytic Enantioselective Synthesis of 2,4-Disubstituted Pyrrolidines. *Org. Lett.* **2019,** *21* (8), 2493–2497. https://doi.org/10.1021/acs.orglett.9b00508.
36. Green, J. E.; Bender, D. M.; Jackson, S.; O'donnell, M. J.; Mccarthy, J. R. Mitsunobu Approach to the Synthesis of Optically Active α,α-Disubstituted Amino Acids. *Org. Lett.* **2009,** *11* (4), 807–810. https://doi.org/10.1021/ol802325h.
37. de Sousa, E. C.; Rauter, A. P. Nucleobase Coupling by Mitsunobu Reaction towards Nucleoside Analogs. *Arkivoc* **2021,** *2021* (4), 241–267. https://doi.org/10.24820/ark.5550190.p011.377.
38. Lanver, A.; Schmalz, H. G. A Pauson-Khand Approach to New Carbocyclic Nucleoside Analogs. *Eur. J. Org Chem.* **2005,** 1444–1458. https://doi.org/10.1002/ejoc.200400886. No. 7.
39. Williams, D. R.; Meyer, K. G. Total Synthesis of (+)-Amphidinolide K. *J. Am. Chem. Soc.* **2001,** *123* (4), 765–766. https://doi.org/10.1021/ja005644l.
40. Liao, X.; Wu, Y.; De Brabander, J. K. Total Synthesis and Absolute Configuration of the Novel Microtubule-Stabilizing Agent Peloruside A. *Angew. Chem. Int. Ed.* **2003,** *42* (14), 1648–1652. https://doi.org/10.1002/anie.200351145.
41. Das, D.; Rajkumari, K.; Rokhum, L. Polymer-Bound Triphenylphosphine and 4,4′-Dinitroazobenzene as a Coupling Reagents for Chromatography-free Esterification Reaction. *Curr. Org. Synth.* **2019,** 7. https://doi.org/10.2174/1570179416666190919152424.
42. Reynolds, A. J.; Kassiou, M. Recent Advances in the Mitsunobu Reaction: Modifications and Applications to Biologically Active Molecules. *Curr. Org. Chem.* **2009,** *13* (16), 1610–1632. https://doi.org/10.2174/138527209789578144.

43. Arnold, L. D.; Vederas, J. C. Polymer-Supported Alkyl Azodicarboxylates and Their Use in Mitsunobu Reactions. *Eur. Pat. Appl. No. 88303818.4* **1988**.
44. Arnold, L. D.; Assil, H. I.; Vederas, J. C. Polymer-Supported Alkyl Azodicarboxylates for Mitsunobu Reactions. *J. Am. Chem. Soc.* **1989,** *111* (11), 3973–3976. https://doi.org/10.1021/ja00193a032.
45. Barrett, A. G. M.; Hopkins, B. T.; Köbberling, J. ROMPgel Reagents in Parallel Synthesis. *Chem. Rev.* **2002,** *102* (10), 3301–3324. https://doi.org/10.1021/cr0103423.
46. Harned, A. M.; He, H. S.; Toy, P. H.; Flynn, D. L.; Hanson, P. R. Multipolymer Solution-phase Reactions: Application to the Mitsunobu Reaction. *J. Am. Chem. Soc.* **2005,** *127* (1), 52–53. https://doi.org/10.1021/ja045188r.
47. Bergbreiter, D. E. Using Soluble Polymers to Recover Catalysts and Ligands. *Chem. Rev.* **2002,** *102* (10), 3345–3384. https://doi.org/10.1021/cr010343v.
48. Bergbreiter, D. E.; Tian, J.; Hongfa, C. Using Soluble Polymer Supports to Facilitate Homogeneous Catalysis. *Chem. Rev.* **2009,** *109* (2), 530–582. https://doi.org/10.1021/cr8004235.
49. Bergbreiter, D. E.; Yang, Y. C.; Hobbs, C. E. Polyisobutylene-Supported Phosphines as Recyclable and Regenerable Catalysts and Reagents. *J. Org. Chem.* **2011,** *76* (16), 6912–6917. https://doi.org/10.1021/jo201097x.
50. Roller, S.; Zhou, H.; Haag, R. High-Loading Polyglycerol Supported Reagents for Mitsunobu- and Acylation-Reactions and Other Useful Polyglycerol Derivatives. *Mol. Divers.* **2005,** *9* (4), 305–316. https://doi.org/10.1007/s11030-005-8117-y.
51. Berger, A.; Wehrstedt, K. D. Azodicarboxylates: Explosive Properties and DSC Measurements. *J. Loss Prev. Process. Ind.* **2010,** *23* (6), 734–739. https://doi.org/10.1016/j.jlp.2010.06.019.
52. Beddoe, R. H.; Sneddon, H. F.; Denton, R. M. The Catalytic Mitsunobu Reaction: A Critical Analysis of the Current State-of-the-Art. *Org. Biomol. Chem.* **2018,** *16* (42), 7774–7781. https://doi.org/10.1039/c8ob01929k.
53. But, T. Y. S.; Toy, P. H. Organocatalytic Mitsunobu Reactions. *J. Am. Chem. Soc.* **2006,** *128* (30), 9636–9637. https://doi.org/10.1021/ja063141v.
54. But, T. Y. S.; Lu, J.; Toy, P. H. Organocatalytic Mitsunobu Reactions with 3,5-Dinitrobenzoic Acid. *Synlett* **2010,** 1115–1117. https://doi.org/10.1055/s-0029-1219795. No. 7.
55. Hirose, D.; Gazvoda, M.; Košmrlj, J.; Taniguchi, T. Advances and Mechanistic Insight on the Catalytic Mitsunobu Reaction Using Recyclable Azo Reagents. *Chem. Sci.* **2016,** 7 (8), 5148–5159. https://doi.org/10.1039/c6sc00308g.
56. Hirose, D.; Gazvoda, M.; Košmrlj, J.; Taniguchi, T. Systematic Evaluation of 2-Arylazocarboxylates and 2-Arylazocarboxamides as Mitsunobu Reagents. *J. Org. Chem.* **2018,** *83* (8), 4712–4729. https://doi.org/10.1021/acs.joc.8b00486.
57. Hirose, D.; Taniguchi, T.; Ishibashi, H. Recyclable Mitsunobu Reagents: Catalytic Mitsunobu Reactions with an Iron Catalyst and Atmospheric Oxygen. *Angew. Chem. Int. Ed.* **2013,** *52* (17), 4613–4617. https://doi.org/10.1002/anie.201300153.
58. Taniguchi, T.; Sugiura, Y.; Zaimoku, H.; Ishibashi, H. Iron-Catalyzed Oxidative Addition of Alkoxycarbonyl Radicals to Alkenes with Carbazates and Air. *Angew. Chem. Int. Ed.* **2010,** *49* (52), 10154–10157. https://doi.org/10.1002/anie.201005574.
59. Buonomo, J. A.; Aldrich, C. C. Mitsunobu Reactions Catalytic in Phosphine and a Fully Catalytic System. *Angew. Chem. Int. Ed.* **2015,** *54* (44), 13041–13044. https://doi.org/10.1002/anie.201506263.

60. Beddoe, R. H.; Andrews, K. G.; Magné, V.; Cuthbertson, J. D.; Saska, J.; Shannon-Little, A. L.; Shanahan, S. E.; Sneddon, H. F.; Denton, R. M. Redox-Neutral Organocatalytic Mitsunobu Reactions. *Science* **2019,** *365* (6456), 910–914. https://doi.org/10.1126/science.aax3353.
61. Zou, Y.; Wong, J. J.; Houk, K. N. Computational Exploration of a Redox-Neutral Organocatalytic Mitsunobu Reaction. *J. Am. Chem. Soc.* **2020,** *142* (38), 16403–16408. https://doi.org/10.1021/jacs.0c07487.
62. Zhou, L.; Perulli, S.; Mastandrea, M. M.; Llanes, P.; Lai, J.; Pericàs, M. A. Development of a Robust Immobilized Organocatalyst for the Redox-Neutral Mitsunobu Reaction. *Green Chem.* **2021,** *23* (22), 8859–8864. https://doi.org/10.1039/d1gc02819g.
63. Xie, C.; Smaligo, A. J.; Song, X. R.; Kwon, O. Phosphorus-Based Catalysis. *ACS Cent. Sci.* **2021,** 7 (4), 536–558. https://doi.org/10.1021/acscentsci.0c01493.
64. Merad, J.; Matyašovský, J.; Stopka, T.; Brutiu, B. R.; Pinto, A.; Drescher, M.; Maulide, N. Stable and Easily Available Sulfide Surrogates Allow a Stereoselective Activation of Alcohols. *Chem. Sci.* **2021,** *12* (22), 7770–7774. https://doi.org/10.1039/d1sc01602d.

Index

Note: 'Page numbers followed by "f" indicate figures and "t" indicate tables.'

H

I

K

L

M

N

O

P

Kolibri 360 GmbH
Ferdinand-Jühlke-Straße 7
99095 Erfurt, Deutschland
produktsicherheit@kolibri360.de